LIBERAL ARTS PHYSICS

LIBERAL ARTS PHYSICS
Invariance and Change

John M. Bailey
Beloit College

W.H. FREEMAN AND COMPANY
San Francisco

Cover: *Sky and Water 1, 1938* by M. C. Escher.
Courtesy of the Escher Foundation,
Haags Gemeentemuseum, The Hague.

Library of Congress Cataloging in Publication Data

Bailey, John M.   1928–
    Liberal arts physics: invariance and change.

    1. Physics.   I. Title.
QC23.B105   530   73–21531
ISBN 0–7167–0343–2

Printed in the United States of America

9 8 7 6 5 4 3 2 1

Contents

12   Magnetism and Field Theory

13   Waves

14   The Interaction of Waves

THERMAL PHYSICS

15 Temperature and Thermal Energy

16 Using Thermal Energy; Disorder and the Direction of Time

QUANTUM PHENOMENA AND KNOWLEDGE

17 Physics, Knowledge, and Symbol

Appendixes

Preface

Without question, one of the biggest problems in college physics teaching is how to reach and draw in students not majoring in the sciences. The size of the opportunity and problem are attested to by the rash of "non-calculus" physics textbooks which has broken out since 1970. Some of these books have taken the approach of making physics *easy*. This has meant removing most of the numerical aspects. In many cases what is left is a discussion *about* physical phenomena, not physics itself. Crucial formulae whose omission would leave just about nothing to talk about are injected without justification. In my opinion this not only insults the student but fortifies the belief that science is foreign, inhuman, magic, something cooked up by and for geniuses and that the best the student can hope for is to memorize enough of it to pass the course.

Liberal Arts Physics, written for a semester or two-quarter course, tries a different approach. It is based on two assumptions:

(1) College students know when they're being talked down to. If they take a physics course, they expect it to be adult physics. This need not mean a lot of algebra, but it does mean that they like to see where the ideas and formulae come from. (Or, at least, have the possibility of seeing where they come from.)

(2) College students are beginning to know something about *some* field of knowledge: its style, presuppositions, evidence, and history. One good way to make physics interesting to them is to point out that it too has styles, presuppositions, evidence and history. Since it is dreamed up by people, it also shares common elements with other modes of thought.

It is invention as much as discovery. It is one of the basic humanities.

Putting it another way, this text tries to emphasize the historical, philosophical and aesthetic dimensions of physics. For several years I have used it for a course entitled "Physics as Natural Philosophy".

The subject matter herein is largely macroscopic physics. Special relativity is included to emphasize the invariant, search-for-pattern nature of physical theory. Microscopic or atomic ideas are included only to help make sense of fusion and fission, radiation of light, thermal phenomena, and, in the concluding chapter, the uncertainty principle. As you read into the book, you will see that much of it is concerned with particle dynamics and conservation principles. These topics provide clear examples of the evolution of physical thinking. Fields and waves come next, pointing out that physics developed in a less mechanistic direction during the nineteenth century. Thermal physics is included to provide a contrast to the solid Newtonian determinism on which the earlier parts of the book are based.

I try to emphasize themes such as simplicity, symmetry, determinism, probability, coherence, and order-disorder which occur and recur in physics. To keep things attached to the late twentieth century world, I discuss topics such as automobile crashes, thermal and atmospheric pollution, energy costs of transportation, lasers, astrology, and include a breather chapter on politics, society, and science.

The 17 chapters herein may be a bit too much for a one-semester course. I use the book for a 14-week course, and sometimes omit portions of Chapters 9, 12, 14, or 15. (Skipping the discussion of interference in Chapter 14 is regrettable because it provides such a good illustration of the fact that physics is concerned with wholes, not just analysis. However, there seems to be no satisfactory way to go over it lightly.) The subject of Chapter 16—entropy—usually generates a lot of student interest. Chapter 17 contains a review section, but goes on to discuss the epistemologically important uncertainty principle.

Please do not be put off by the occasional "sin" or "cos" which appears in the book. I feel that learning what they are is like learning whether the fork goes on the left or right side of the plate when you set the table: hard to remember, but not intellectually overwhelming. Even so, in most cases save Snell's Law (Chapter 13) and interference-diffraction (Chapter 14), the instructor can avoid the trig functions entirely by speaking in terms of components. To avoid speaking of vectors entirely would be depriving the student of a lot.

I have used a number of supplementary readings with this text, some of which you might also find helpful. They contain readings on the history and fabrication of physical theory (as well as other things):

Books or collections of articles:

Bork, Alfred M., *Science and Language*, D. C. Heath, 1966.

Bronowski, Jacob, *Insight*, MacDonald, London, 1964.

Gardner, Martin, *Great Essays in Science*, Washington Square Press, 1965.

Holton, G., et al., *The Project Physics Course* Readers (6 vol.), Holt, Rinehart and Winston, 1970.

Karplus, Robert, *Physics and Man*, W. A. Benjamin, 1970.

Olson, Richard, *Science as Metaphor*, Wadsworth, 1971.

Schroeer, Dietrich, *Physics and Its Fifth Dimension: Society*, Addison-Wesley, 1972.

Vavoulis, Alexander, and Colver, A. Wayne, *Science and Society*, Holden-Day, 1966.

Young, Louise B., *Exploring the Universe*, McGraw-Hill, 1963.

Articles:

Koestler, Arthur, "The Eureka Process", from *The Act of Creation*, Dell, 1964.

Mumford, Lewis, "The Crime of Galileo", from *The Pentagon of Power*, Harcourt Brace Jovanovich, 1970. Reprinted in *The New Yorker*, October 10, 1970.

You might also find articles referred to in *Further Readings* at the end of each chapter to be helpful in carrying certain ideas further.

One of the most successful elements of the course Physics as Natural Philosophy has been a research paper. Each student investigates a topic wherein physics relates to his or her prospective major field of study or main interest, whichever the case may be. To get the project moving I give each student a bibliography at the beginning of the course with pages headed "Physics and the Visual Arts", "Physics and Education", "Physics and Sociology", etc. A 1971 version of the bibliography was published in the *American Journal of Physics*, volume 39, page 1347 (November, 1971).

I would like to express my sincerest thanks to Bob Garrett, Jim Gerhart, Ron Palmer, Dan Schroeder, Joe Stoltzfus, Edwin Taylor, and many students in Physics 100 for criticizing and correcting portions of the manuscript during its long trip from outline to print.

I wish to pay homage to Jacob Bronowski and Gerald Holton for inspiring us all to look for the connections between ideas.

J. M. Bailey
October, 1973

1

INTRODUCTION

(to be read now and
then immediately after finishing
Chapter 17.)

On that August day in 1945 when the Enola Gay dropped a nuclear bomb on Hiroshima, a new age began. This act marked the completion of the transition of physics from a quiet laboratory discipline to an activity in full public view. In the decades since then, physics and physicists (along with the technology they helped spawn) have played an enlarging role in international relations, defense strategies, economic development, communication, the quality of life, and many other problems of contemporary society.

We are in an age when just about every person who can read has some idea of what physics is, what physicists do, and how they do it. But even though the last several decades have thrust physics into the arena of human affairs, the "man on the street" harbors some misconceptions (and misgivings) about it. This is especially true when issues connected with physics-based technology are being debated on a national scale. The burden on the layman to keep informed on and partially understand these scientific issues has grown heavier.[1] One commentator[2] questions whether the democratic form of government can keep going in a time when expert decisions may have to be made quickly or not at all. This book is dedicated to the propositions that people *can* still understand and cope with what goes on around them and in their societies; and that one function of education is to lead us away from specialized training and make us examine alternative ways of viewing the world.

(1) de Jouvenel, Bertrand, "The Political Consequence of the Rise of Science", *Bulletin of the Atomic Scientists 19*, pg. 2, December, 1963.
(2) Daddario, Emilio Q., "Technology and the Democratic Process", *Technology Review 73*, July/August, 1971, pg. 18.

One purpose of this book is to tell you something about physics and to discuss how physicists go about looking at and describing the world around them. Hopefully this will engender some trust, as well as understanding of physics. But another equally serious purpose is to expose some characteristics of science which bring it closer to, rather than set it apart from, other human intellectual or creative activities. They are all trying to give structure to some aspect of human experience.

What *is* physics? A traditional definition says that physics is the study of the laws of nature. But even this definition is misleading. It suggests that the "laws of nature" already exist somewhere around us, just waiting for the right Moses to dig them up. The laws and theories are far more man-made, thus, human, than this. They come closer to being invented than discovered even though practicing physicists often fall into thinking that the explanations they devise and use are a part of an unarguably "real" external world. This was especially true in the eighteenth and early nineteenth century when Newtonian mechanics, with its sharp predictions of the motions of planets and other celestial objects was at its peak. It was so successful that its mechanistic and deterministic thinking began to spread into other fields: the hope of some early psychologists was to understand and predict individual and collective human behavior on the basis of external pushes and pulls. Newtonian natural law affected the thinking of the men who wrote the United States Constitution.

However, the revolutionary changes in 20th century physics (relativity and quantum mechanics) have led physicists to be more humble in their claims regarding truth. "It works" has become a more active term in physics circles than "truth" and "reality". The physical view of the world now coexists with and has things in common with the view of the poet, the painter, the theologian, and the statesman. These views are all at least partially subjective, with the thinkers involved with (and participating in) the thing they are describing. The separate, objective observer is largely a myth.

The intermediate stages of science, including physics, involve careful controls, elimination of alternative possibilities one by one, and in general, the use of deductive logic. However, the initial and final stages of the process are more creative than logical. The hypotheses which physicists start with and the theory toward which they build are put together in much the same unpredictable manner as are poems and concertos. Physical explanations involve symmetries, rhythms, and thrusts, just as artistic creations do. The movement from pictorial to abstract art since 1880 is matched by a shift in physics from mechanistic, machine/model representation to more symbolic, mathematical and comprehensive descriptions. In other words, physics has become far more aesthetic and less materialistic than most non-scientists realize.

Two important elements in physics have been simplicity and

abstraction. Physics is the simplest of the sciences. This does not mean that its tools and methods are easy to learn, but that physicists are continually trying to make simple order out of complex fact. When doing physics, physicists try to see real systems, with all their friction, gusty winds, rust, sun going behind clouds, etc., as ideal systems with known disturbances added. Insofar as systems can be imagined as being ideally isolated from their surroundings and from these changing disturbances, their behavior can be understood better. This was one of Galileo's major realizations. More recently, high-speed computers have made possible the inclusion of more and more elements from the real world, but physics has grown largely by idealizing. The same criterion of simplicity is sometimes applied to physical theory itself. The more complex and patched up a theoretical explanation is, the more suspect it is. Physicists have no proof that their explanations will continue to be simple, but simplicity, or "elegance", as it is often called, remains one of the sought-after qualities in good scientific theory.

The technique of *abstracting*, or focusing in on one or two characteristics of a whole system has also been important in the growth of physics during the last three centuries. The motion of a ball is seen by physicists as the motion of a point coinciding with the center of the ball. The ball's color, history and ownership are ignored if they don't obviously affect the motion. Before the Renaissance, "physics" was more concerned with the whole picture. If a ball fell, it was because of a part it was destined to play in some universal scheme. It moved for innate reasons. This ancient-medieval view should not be scoffed at. It showed a coherence that our age lacks. It provided answers to all the questions which were generally considered important. But it was a view which discouraged further investigation. It was also a static view, changing little in the 2000 years between Aristotle and Galileo. But with Galileo in the 1600's, orderly, detailed experimentation began. He spent time measuring mundane things such as how a ball gathers speed as it rolls down an incline. Natural philosophers before him would have thought such activity to be child's play, but by concentrating on specific actualities Galileo was able to contribute much to the overthrow of Aristotelean physics, which made predictions by deduction from a world view rather than building up from detailed observation.

Modern physics is thus not concerned with the final "why" of things, even though physicists as humans certainly wonder about them. (There is probably more mystery in the world for the scientist than for the person who spends less time looking into the world.) Unlike the branch of philosophy called *metaphysics*, physics is not involved with the ultimate grounds of being, knowing, or substance. However, it certainly *exposes* fundamental questions in metaphysics which may be overlooked in other realms of research. The ideas of communicability and invariance arise repeatedly in physics, as you will see. These and the uncertainty principle are all important problems in *epistemology,* the part of metaphysics which deals with how we gain knowledge from our surroundings. Physicist Niels Bohr, with his principle of complementarity, expanded the uncertainty principle into a general statement about the limits of our knowledge about an event or thing. By looking at simple things, or at things simply, physicists have accomplished much with physics.

Considerable criticism has been leveled at physics and the other sciences recently because of their *analytical* approach to problems. The laments of Lewis Mumford and others (see the readings suggested at end of Chapters 5 and 10) are just the most recent chapter in a long conflict between two points of view:

> **(1)** The analytical or mechanistic view, which sees the world as a machine which can be understood by taking it apart and studying the pieces.
>
> **(2)** The holistic or organic view, which says the overall picture is best. The whole is more than its parts, and beauty, feeling, and love are destroyed if subdivided.

When Aristotle was first preaching the organic view, Archimedes and Pythagoras concentrated on numbers and analysis. The Enlightenment philosophers who lionized Newton's analytical mechanism were soon followed by the Romantics, rhapsodizing over the organic whole.

Much of physics is indeed based on analysis into parts. It does not try to ask or answer questions about love, hate, or God. It does not deny their existence, but simply says "that's not part of my game". But to stop there would be unfair to physics. In the study of waves, interactions and patterns become important. A rainbow is an entity for physical study but it is a whole. In modern particle physics, which we won't discuss in this book, *groups* play an important role. One subject which *is* embarrassing to physics is life, or more specifically, *intelligence.* The second law of thermodynamics says that in the long run, disorder or entropy will win out over order. A person's mind may organize things but the body more than makes up for this by leaving a trail of disordered materials behind it, says physics. The joker, as we'll see in Chapter 16, lies in one's definition of "order". Through physics quantitative values can be assigned to certain simple patterns in music or art. Paintings cannot be ranked according to their orderliness. Furthermore, there are difficulties in accounting for the fact that certain arrangements of atoms and molecules seem preferred over others and are able to propagate themselves, whether as crystals or as living tissue.

But perhaps the greatest contemporary criticism of physics arises from its connection with a technology which threatens to rule us rather than serve us. For a long time, physicists claimed an innocent detachment from the technological use that their brother engineers made of their scientific discoveries. Science seeks understanding of nature whereas engineering or technology tries to build better devices. The two have a long symbiotic history. Technology helps science. Many of the early discoveries in atomic physics were made possible by better vacuum pumps which could remove disturbing air molecules from the chambers in which the experiments were done. An engineering device, the electronic digital computer, has encouraged physicists to undertake problems they had ignored previously, simply because they would take too long to solve. On the other hand, the radar engineering feats during World War II would not have been possible without the accumulation of electronic know-how by physicists during the 1920's and '30's. If this were *not* true—if there were not promise of practical application of the work done by the physicists, there would be far fewer physics laboratories. The connection between physics and engineering cannot be ignored. Scientists are rapidly coming to realize that they must share the responsibility for what is eventually done with their ideas. Some have dropped research projects which have military applications. Others have become active politically, working on disarmament, environment control, and other social problems. This may constitute the final step in the adjustment of physicists to their role in the mainstream of modern society.

One must recognize the importance of physics in the world of intellect as well as in technology, politics, and weaponry. The study of physics has made us more conscious of the connections between observer and object which Western thought had ignored for centuries. Through physics, we have discovered new ways of seeing space and time. In another direction, contemporary physics has revealed a universe to us, which is far larger and older than we thought it to be even 20 years ago. It is beginning to suggest that the conditions which led to life on earth probably exist at many other places in the universe. It has found beauty and regularity in the world far beyond the wildest imaginings based on common sense observation.

Further Reading

Cassirer, Ernest, "Galileo: A New Science and a New Spirit", *American Scholar 12*, Winter, 1942–43.

Jaki, S. L., Chapter 1, "The World as Organism" and Chapter 2, "The World as Mechanism", *The Relevance of Physics*, University of Chicago Press, 1966.

Polanyi, Michael, "Science and Man's Place in the Universe", p. 54–76 in *Science as a Cultural Force*, H. Woolf, ed., John Hopkins Press, Baltimore, 1964. (The organic can be rationalized on the basis of laws of physics and chemistry, but seldom predicted. Knowing the parts, it's difficult to visualize the whole.)

Siu, R. G. H., "Abstraction", Chapter 8 in *The Tao of Science*, M.I.T. Press, Cambridge, 1957. (Points out the consequences of studying anything in isolation from its surroundings.)

Marcel Duchamp, *Nude Descending a Staircase*, No. 2, 1912. *(Philadelphia Museum of Art: The Louise and Walter Arensberg Collection.)*

MOTION AND FORCE

2

MEASUREMENT

Lord Kelvin, an eminent nineteenth century British scientist and mathematician, said,

> "When you can *measure* what you are speaking about and express it in numbers, you know something about it, but when you cannot measure it in numbers, your knowledge is of a meager and unsatisfactory kind; it may be the beginning of knowledge, but you have scarcely, in your thoughts, advanced to the stage of science".

There are intellectual dangers in such a point of view. Most scientists, as we said in the Introduction, would now admit that there are elements of reality which are not numerical. Nevertheless, when one wants to make an observation public and communicable, numbers are one of the best symbols available.

If a measurement can be applied to a phenomenon or object, that is, if you can determine a ratio between the magnitude of the phenomenon or object and the magnitude of some agreed-upon *standard*, then you are in a position to communicate that observation to someone else. The comparison with the standard is usually indirect. For example, to measure the width of a curtain, you see how many times a ruler must be laid down end to

(*By permission of Johnny Hart and Field Enterprises, Inc.*)

end to make a length equal to the width of the curtain. You can then say, "The curtain is so many rulers wide". Then the ruler itself, whether it is a foot long or a meter long, must at some time in its life be made equal in length to some master ruler which has been compared with an international standard of length. Two observers in different parts of the world can understand each other's curtain measurements only if they both use rulers which have been compared with the same standard. Since these comparisons cannot be made simultaneously, a standard needs to be unchanging, indestructible, readily available, and accurately and easily measurable. Some of the interesting standards which have been used in the history of measurement will be reviewed in the next few sections.

The result of a measurement is always expressed as a *number* and a *unit* (20 centimeters, 5 hours, 1.5 pounds). The unit employed should match in magnitude the object or event being measured. For curtain widths, the foot or the meter serves nicely, but for screw lengths, the inch or centimeter is more appropriate. The unit is defined as being such and such a fraction of, or so many times, the standard agreed upon for the dimension being measured. Let's examine the role of standards, units, and measuring instruments, in measuring the three basic physical dimensions; length, time and mass.

Length

History has recorded many standards of length. The Egyptians built the pyramids using the cubit (20.6 inches or 52.4 cm.) as their unit, and the forearm of the Pharaoh as the standard. Since Pharaohs change, their forearms were somewhat lacking as a standard. In medieval times, the barleycorn (about one-third of an inch long) was a common unit and standard. For uniformity, the corns were supposed to be chosen from the middle of the head, but this still allowed considerable variations. The metric system of units, based upon the earth as a standard of length, was instituted in 1791. The meter, basic unit in the present metric system, was originally defined as one ten-millionth (10^{-7}) of the distance from the earth's North Pole down to the equator, using the quadrant which passed through Paris. Since that distance was in practice impossible to measure with any accuracy, it was replaced as length standard in 1799 by the surveyed distance from Dunkirk to Barcelona. The meter was re-defined as one-millionth of this new distance, and the manufacture of accurate meter sticks, equal in length to those used today, began. More accurate surveys have since shown the Dunkirk-Barcelona distance to be 1,075,039 not 1,000,000 meters. Furthermore, continents warp and change their dimensions with time. Therefore, continents are not only an inconvenient, but a non-permanent standard of length. In 1875 an international convention decided to adopt a man-made meter as the standard of length. Two very accurately positioned scratches were made on a bar of platinum-iridium alloy a distance apart equal, by definition, to one meter. This bar, maintained at constant temperature and humidity to prevent expansion, and supported on two cross rollers to prevent warping, is kept in the International Bureau of Weights and Measures, located in St. Cloud Park, west of Paris.

Figure 2-1
International Bureau of Weights and Measures. (Rear entrance.)

Figure 2-2
*The American Standard Meter being carried into the new administration building of the
National Bureau of Standards. (*National Bureau of Standards photograph.*)*

Various nations have their own national meters, made by duplicating the International Prototype Meter in France. Manufacturers of meter sticks and other measuring devices in this country then had to develop gauge blocks which were measured to be accurate fractions or multiples of the national meter kept in a vault in the National Bureau of Standards (NBS) in Washington. In 1966 the NBS moved to suburban Gaithersburg, Maryland and the national meter had to be moved too. The rulers and meter sticks and tape measures in common use are made to match the gauge blocks mentioned above.

This hierarchy of standards may sound like academic hairsplitting, but it has been prerequisite to the mass production upon which our economy is based. Historians believe that Eli Whitney's main contribution to our society was not the cotton gin but the idea of the interchangeability of parts. By using the measurement standards and machining methods suggested by Whitney, his factory was eventually able to make gun parts so precisely that even if they had been made at various times and places, they could be attached to the rest of a given gun and would function well. (Whitney had obtained a government contract to develop his methods and kept running out of money. After numerous extensions and additional grants, he finally delivered the first guns in 1809 — ten years after the first contract was let. Whitney's guns thus became one of the government's first peacetime defense contract cost over-runs.) With the advent of the twentieth century, length measurements in terms of light waves were becoming far more accurate than the best techniques for mechanically comparing two objects laid side by side. Furthermore, the international and national prototype meters are relatively inaccessible and susceptible to bomb and earthquake damage. Therefore, in 1960 the international General Conference on Weights and Measures re-defined the meter in terms of a less material, more accurately measurable (2 parts in 10^9) distance. The distance is the "wavelength" of the red-orange light emitted by the gas Krypton 86 when "excited" in an electric discharge. The meter is now defined as 1,650,763.63 times the wavelength of that light. Why this particular light? Primarily because the light is bright and sharp in wavelength. It can thus be measured accurately by redefined spectroscopic techniques, and the brightness makes length comparisons with it easier.

Because the light emitted from lasers is so sharp in wavelength, these devices can be used to make even more accurate

Table 2-1   *Measuring Length.*

Typical Object	Unit	Measuring Instrument
Atomic nuclei	Femtometer (10^{-15} meter) or "fermi"	Nuclear collision apparatus
Atoms in solids	Nanometer (10^{-9} meter) Ångstrom (10^{-10} meter)	Electron microscope; field ion microscope
Biological cells	Micron (10^{-6} meter)	Light, electron microscopes
Machined parts	Millimeters (10^{-3} meter)	Micrometer calipers
Dress hems	Centimeters (10^{-2} meter)	Meter stick
Houses; discus throws	Meters	Steel tape
Countryside; highways	Kilometers (10^3 meters)	Surveying equipment
Distance to sun; moon	Light minutes, seconds	Triangulation, using telescopes on different parts of the earth; laser; radar
Distance to stars	Light years (9.45×10^{15} meter)	Triangulation, using telescopes on earth at opposite sides of its orbit around the sun; for the more distant stars, indirect schemes dependent upon knowing intrinsic brightness of stars
Universe	10^{26} meters	Gravitational-cosmological theory.

1 Meter = 39.37"

1" = 2.54 Centimeters

1 Mile = 1610 Meters = 1.61 Kilometers

length measurements (one part in 10^{12} to 10^{14}). Laser light may thus eventually become a source of length standards.[1]

Table 2-1 shows some units and instruments used in length measurement. The powers of 10 notation used in the table is explained in Appendix E, at the end of the book.

Time

The second dimension, time, is far more illusive than length. Poets, philosophers and psychologists have pondered its passage for centuries. We have no dependable intuitive feel for the pure passage of time. Instead, we depend on succession of events or cues whose accumulation tells us time must be passing. With fewer cues, time seems to drag. When we are unconscious, as during a surgical operation, all sensation of time passage disappears. Afterwards, we have the feeling that time has stood still. When we are awake, our body gives us many cues such as pulse, onset of hunger and certain repetitive changes which act as "biological clocks." However, these clocks are all subject to variation.[2] We must search outside of ourselves for phenomena to serve as adequate standards for physical time. In general, a phenomenon which repeats itself regularly is a better time standard than a phenomenon which simply involves accumulation. Hour glasses and water running out of a tap were both fairly good clocks, but not regular enough to be good standards. The hour glass was only as precise as the monk who turned it over after the sand ran through, and there are many things which can prevent water from running out of a tap consistently. People began to look for events in nature which repeated themselves regularly and an early candidate for time standard was "a day," or the time required for the earth to revolve once. By the 1930's it was clear that even the rotation of the earth was irregular. The next step was to use a year, the time required for the earth to travel around the sun once. Years vary in length, however, so in 1956 the year 1900 was taken to be the time standard and a second was defined as a certain fraction of that interval. The year 1900 was a more accurately known standard, but comparisons with it, involving the apparent motion of stars, were bound to be indirect. Therefore, in 1964, the General Conference agreed upon a more accessible and even more accurate time standard. It is the rate of vibration of one kind of wave given off by Cesium 133 gas atoms. The second, everyone's favorite time unit, is now defined as 9,192,631,770 times the period of vibration of Cesium 133 energy waves as measured by an atomic beam clock. The preci-

(1) We speak further of spectroscopy on page 371 and of lasers on page 379 ff.
(2) Cohen, John, "Psychological Time," *Scientific American 212*, November, 1964, p. 116. Offprint No. 489.

Table 2-2   *Measuring Time.*

Time Interval	In Seconds	Measuring Instrument
Short lived nuclear particles	10^{-8}–10^{-10}	Photos of cloud chamber or bubble chamber tracks
Explosions	10^{-6}–10^{-3}	Electronic circuits
Sporting events	10–10^4	Stop watches; phototimers
College semesters	10^7	Calendars
Life of humans	$(2$–$3) \times 10^9$	
Recorded history	2×10^{11}	Radioactive dating of organic materials
Age of human race	6×10^{13} $(2 \times 10^6$ years?$)$	
Age of earth	1.5×10^{17} $(5 \times 10^9$ years$)$	Radioactive dating of rocks
Age of universe	10^{18}?	Expansion of the universe

1 Year = 3.15×10^7 Seconds

sion[3] is 1 part in 10^{12}, an error of one second in 30,000 years! For those who can't set up atomic clocks, the National Bureau of Standards broadcasts a tick every second over its radio station, WWV. These ticks constitute a far better timepiece than the best mechanical clock. If you are interested, you will discover that after you have heard 86,400 ticks, a day will have gone by.

The origin of the meter has been mentioned, but where did the *second* come from? Historically, it came into being as 1/3600 of an hour. The hour itself dates from Egyptian times. One source[4] says that the hour, dividing the day into 24 parts, was

(3) In measurement, "accuracy" and "precision" don't mean the same thing. Accuracy means the apparatus is well tuned up, and gives "correct" results in comparing with a known standard. Precision has to do with repeatability—how well the measurements in a set agree with each other.

(4) Thorndyke, Lynn, *History of Magic and Experimental Sciences*, Columbia University Press, N.Y., 1941, Vol. VI, pg. 353.

"discovered" by observing the urination habits of the royal animals! By 500 B.C. the Babylonians had established a connection between the 12 daylight hours and the Zodiac, which divides the sky into 12 equal parts. Their mathematicians counted in sixties so to obtain more convenient time units, they divided the hour into 60 parts, and each of those parts into 60 smaller portions. By early Roman times the two resulting time units were called *partes minutae primae* (first small division)—hence our name, "minute", and *partes minutae secundie* (second small division) hence our name, "second". Table 2.2 shows some time intervals expressed in terms of seconds, along with instruments used for measuring them.

Mass

Mass, which we will sharply distinguish from weight after we have studied gravitation, is a characteristic of matter which shows up when you try to speed up or slow down an object. The kilogram, the primary mass unit in the metric system, was derived from the meter by setting the density of water at a certain temperature at one "gram" per cubic centimeter. The kilogram was then simply the mass of 10^3 cm^3., or one liter of water. The

Table 2-3 *Measuring Mass.*

Object	Mass, kg	Measuring Instrument
Electron	9×10^{-31}	Cloud or bubble chamber tracks; collisions with other particles
Proton (hydrogen nucleus)	2×10^{-27}	Cloud or bubble chamber tracks; collisions with other particles
DNA molecule	10^{-23}	Centrifuge; electron microscope
A nickel	4.8×10^{-3} (4.8 g)	Weight balance
A pound of butter	.454	
Man	60–120	Bathroom scales
Automobile	$(1–3) \times 10^3$	Large drive-on scales
The earth	6×10^{24}	Gravity theory
The sun	2×10^{30}	Gravity theory and period of earth's orbit around sun;
The universe	10^{53}	star counts; general theory of relativity

1 Kilogram = 2.205 Pounds
454 Grams = 1 Pound
28.4 Grams = 1 Ounce

present international mass standard is a platinum-iridium cylinder kept at the International Bureau of Weights and Measures near Paris. Although the mass standard is a material one and is thus subject to destruction, the pressure to replace it with an atomic standard is not as great as in the case of length and time. This is because the masses of large objects can be compared quite accurately: to within several parts in 10^9.

The Meaning of Terms

From what we have said thus far, objects and events have no inherent or absolute size, duration, or mass. These quantities are expressible only as ratios to the magnitude of carefully selected standards. The ratios constitute public, fully objective, and communicable information. It would be wrong to call the wavelength of the red-orange light from Krypton 86 an "absolute" length.[5] It is simply a steady and accurately measurable length existing in nature which all the world's physicists have agreed to compare with. Having done this, they can then say that all the objects or events observed from that point on are so much longer or shorter, quicker or slower, heavy or lighter than the appropriate standard.

Just as length statements imply that you have laid a measuring instrument down next to an object and noted the location of the object's left end and right end, *all* physical statements depend for their meaning upon an *operation*. According to this, even the idea of constancy has in back of it an unchanging standard which can be referred to. When you measure very large or very small things (such as the distance to the Crab Nebula or the diameter of the nucleus of the atom) the length-comparing operations become very indirect, so that the word "length" loses the definite meaning it has when you are measuring human-sized objects. More is involved than just laying a meter stick down beside the thing you want to measure.

The physicist Percy W. Bridgman (1882–1961) developed these ideas into a philosophical principle called operationalism. It insists that all statements, not just physical ones, require an operational basis for their meaning. This offers some interesting possibilities for separating the rhetoric and double-talk of our age from the meaningful statements. It asks what statements mean in terms of existing or possible human activities. Operationalism goes on to say that the only meaningful questions which can be asked in physics are those which can be answered by doing an experiment. For example, the question, "What color is an electron?" would have no meaning because there is no possible experiment which can answer the question.

(5) In Chapter 8, on relativity, we will look into the physical definition of "absolute".

The Metric System

Tables 2-1 and 2-3 show several conversions between the inch-foot-pound system of units still used in the United States and the meter-kilogram-second (MKS) metric system.[6] From this point on, we will speak primarily in terms of the metric system, which is used almost exclusively in scientific circles. An act of congress put the United States on the metric system officially (but not in practice) in 1866. In 1968, the National Bureau of Standards was allotted 2.5×10^6 for a three-year study of switching from the inch-pound system to the metric system, and in 1971, the Secretary of Commerce presented the report[7] to Congress, suggesting that the switch *should* be made, but over a 10-year period. It will cost American Industry between $10 billion and $40 billion to make the conversion. Screw threads and wrenches will have to be redesigned in terms of centimeters instead of

Table 2-4 *Prefixes used in the Metric System.*

Prefix	Power	Example
atto-	10^{-18}	
femto-	10^{-15}	
pico-	10^{-12}	picosecond
nano-	10^{-9}	nanometer
micro-	10^{-6}	micron (10^{-6} meter)
milli-	10^{-3}	millimeter
centi-	10^{-2}	centimeter
deci-	10^{-1}	decibel (sound intensity)
deca-	10^1	decapod (10-wheeled locomotive; class of 10 legged crustaceans)
hecto-	10^2	hectare (a square piece of land 100 meters on a side)
kilo-	10^3	kilogram
mega-	10^6	megabuck
giga-	10^9	gigavolt
terra-	10^{12}	

(6) It is more complete to call it the MKSA system, to include the basic unit in electrical measurements, the ampere (A) of current.

(7) De Simone, Daniel V (director of study) *A Metric America: a Decision Whose Time has Come.* National Bureau of Standards Special Publication 343, July, 1971.

inches.[8] Housewives will learn to buy butter by the half-kilogram. Motorists will have to realize that a speed limit sign reading "60" means 60 kilometers per hour (36 miles per hour). However, the switch does need to be made eventually if we expect to continue a profitable foreign trade. The switch will cost far less if it is carried out according to a nationally planned program.[7] In 1970, 147 nations were practicing the metric system and only United States, Jamaica, Ceylon, Nigeria, and 5 small African and South American nations remained on inches and pounds.[9]

The main advantage of the metric system lies in the fact that it is decimal like our number system, so that one can convert from one unit to another simply by multiplying or dividing by the proper power of 10 (Appendix E). Prefixes have been selected to show how a unit compares with the basic unit (see Table 2-4).

The Size of Humans

As can be seen from Table 2-1, the universe is thought to be about 10^{26} meters in extent, and the smallest observable distance about 10^{-15} meter. Man is thus unimaginably larger than the nucleus of the atom but even more unimaginably smaller than the universe. These in themselves put a stamp on the descriptions, the scientific explanations, which the physicist as a human being develops. Objects and events which are human-sized in speed, size, or duration will be most familiar and directly related to everyday experience. Classical[10] or Newtonian physics is the name given to the set of explanations which deal with these sorts of phenomena. However, there are also natural phenomena which involve systems that are much larger, smaller, or faster than the things we ordinarily experience. Physicists try to explain these events as well, using the newer principles of relativity and quantum physics. Since these branches of physics are dealing with phenomena that are widely separate from everyday experience, problems arise concerning the operational meaning of the words used. The greater part of this book will deal with classical physics, but we shall consider enough of relativistic and quantum phenomena to show that the physicist,

(8) The switch to the metric system will involve a second usage of the word "standard": one goal of the switch is to *standardize* the sizes of manufactured items. Not only will their dimensions be measured in terms of meters and kilograms, but their sizes will match better with items manufactured elsewhere, and the *number* of sizes will be sharply reduced.

(9) Ritchie-Calder, Lord, "Conversion to the Metric System", *Scientific American 225*, July, 1970, pg. 17. Offprint No. 334.

(10) Classical does not mean Greek or Roman here. We will lump them under "ancient."

like the poet and the theologian, is seeking to communicate a certain perspective on the world by means of symbols.

Beyond this, the forces which act on the surface of the earth have probably led to our being the size that we are. According to F. W. Went,[11] humans could not have been much different from their present size and still survived and developed the culture they have. If humans were even two to three times larger than they presently are, a walk in the earth's gravitational field could be very dangerous. For example, if we were three times taller and had the same proportions, we would weigh $3 \times 3 \times 3 = 27$ times as much. Tripping and falling would involve an object 27 times as heavy falling three times as far, so we would hit the sidewalk with 81 times the impact. But our bones would only be nine times stronger, since their strength depends on their cross sectional area. Therefore, the fall would be nine times as damaging; something like jumping from a third-story window!

On the other hand, humans would also have great difficulties surviving if we were very much smaller than we are. Our volume decreases as the cube of our size; but our surface area only as the square. Therefore the ratio of our surface area to our volume increases as we get smaller. We would avoid the weight problems of the giant, but would have to face surface problems almost as serious. A great deal more energy per unit of volume would be required just to keep our bodies up to normal temperature, since our relatively larger surface area would increase energy losses to our surroundings. (The smallest warm-blooded animal, the shrew, must eat prodigious amounts of food per unit volume to get enough energy to keep warm.) Trying to build a fire would not help. If humans were as small as ants, they could not bring fuel fast enough and close enough to the fire to maintain it. There are no miniature fires. So much energy is lost to the surrounding cool air that the fire must be of certain size before it can keep itself going. The relation between brain size and intelligence is not simple, but certainly an ant-sized brain has limited information storage capabilities. Now, suppose ants *were* as intelligent as humans. They could not use books with pages because surface adhesive forces would make the ant-sized book pages stick together. All things stick together to a certain extent but the ratio of stickiness to weight increases as objects get smaller. They could try carving on rocks, but the inherent graininess of rock would prevent very detailed symbols from being recorded. And without practice in primitive data storage systems, how could ants learn to make and use the microphotographic information recording systems that humans are beginning to use?

(11) "The Size of Man", *American Scientist* 56, no. 4, Winter, 1968, pg. 400.

Ants seem to have survived better than dinosaurs and might yet outlast humans, but they will never write symphonies or do physics.

Summary

Measurement can be one of the most objective—that is, communicable elements in observation. However, this is only true if the measurement is expressed as a number of units, with the unit being comparable with a world-wide standard.

A standard needs to be unchanging, indestructible, readily available and accurately measurable.

International agreements have been reached whereby the standard for length is the wavelength of the red-orange light emitted by Krypton 86; the standard for time is the period of vibration of the Cesium 133 atom; and, the standard for mass is a certain cylinder machined from platinum-iridium alloy and kept in a vault near Paris.

The words used in physics must have an operational basis; in principle, there must be a way to observe or measure the entity which the word stands for.

Many different measurement units have been used in the past but a majority of the world now uses the metric system of units, based on the meter, the second, and the kilogram. Larger and smaller units may be derived from these three by multiplying by the proper power of ten.

The physical forces at work on the surface of the earth have probably played an important part in determining how large we are. Our size, in turn, determines the kinds of experiences we have and also the perspective from which we see and describe the world, scientifically or otherwise.

Further Reading

Born, Max, "Symbol and Reality", Published most recently as an appendix to *Natural Philosophy of Cause and Chance*, Dover Publications, N.Y., 1964. (Why is science mathematical?)

Fraser, J. T. (ed.) *The Voices of Time*, G. Braziller, N.Y., 1966. (A momentous collection of essays relating time to religion, literature, psychology, science and life in general.)

Haldane, J. B. S., "On Being the Right Size", pp. 23–27 in *The Project Physics Reader #1*, Holt-Rinehart-Winston, N.Y., 1970. (Illustrates problems we would have if we were larger or smaller.)

Hawkins, David, "Measurement", Chapter 4 in *The Language of Nature*, W. H. Freeman, San Francisco, 1964. (A sophisticated treatment of measurement as the bridge between nature and knowledge.)

Hudson, George E., "Of Time and the Atom", *Physics Today 18*, August, 1965, p. 34. (A discussion of the atomic basis for time standards.)

Kemeny, John G., *A Philosopher Looks at Science*, D. Van Nostrand, Princeton, 1959. pp. 125–131. (On operationalism.)

Physical Science Study Committee, *College Physics*, Raytheon Education Co., 1968, pp. 35–39. (A discussion of scaling, and why the Lilliputians and the Brobdignabians could not have had the proportions Jonathan Swift said they had.)

Priestley, J. B., *Man and Time*, Aldus Books, in association with W. H. Allen, London, 1964. (A well-illustrated book on ideas concerning time in literature, science and history. Priestley himself was quite fascinated with the passage or non-passage of time in drama.)

Problems

2-1 Express your height in centimeters and your mass in kilograms.

2-2 How many more minutes do you expect to live?

2-3 Express the wavelength of the red-orange light given off by Krypton 86 in Ångstrom units and nanometers.

2-4 Suppose a man is 5′5″ tall and weighs 130 lbs. Show that a man standing 6′5″ tall and having exactly the same proportions would weigh 214 lbs.

2-5 The volume of a ball having a radius r is $4/3 \pi r^3$. The surface area of the same ball is $4\pi r^2$. Show that the ratio of surface to volume is greater for smaller balls.

2-6 If the liquid in the round container on the left is poured into the container on the right, will the liquid come up to level a, b, or c?

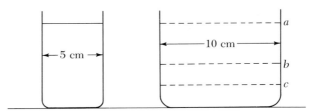

2-7 Suppose pizza pies of equal thickness are baked in two different sized circular pans. Which is the better buy: a 12″ pizza for $1.75 or an 18″ pizza for $4.00?

2-8 Explain in physical terms why it makes sense for humans and other warm-blooded animals to curl up into a ball when sleeping on a cold night.

2-9 A liter is equal to 1.056 liquid quarts. When American grocery stores switch to the metric system will the price of a container of milk go up or down? How much?

3
DESCRIBING MOTION

From the time we are born (or even before), we are fed sensations by our surroundings. We search for patterns in our surroundings and then begin to sense changes in these patterns. As scientists, human beings try to find clear-cut ways of communicating these patterns and changes with other people. One particular sort of pattern change is the motion of an object. A British historian of science, Herbert Butterfield, has said that the description and explanation of motion has been one of man's primary intellectual problems for centuries: the planets in the night sky, the tides, stirred soup rotating in a bowl, rocks rolling downhill (and the rise and fall of the stock market). All these motions raise questions. In the next several chapters we will be looking into the complex evolution of the ways in which people have tried to answer some of the questions. We will begin in this chapter by examining the way physics describes a few simple motions.

Objects

Discussions of motion usually begin with a scheme for describing the change in position of an object with time. They normally assume that everyone participating in the discussion knows first hand what an *object* is, and is prepared to accept certain assumptions about objects which need not even be stated. An integral part of Western thought, dating back at least to Aristotle, 2300 years ago, has been that perceived objects exist and endure even when we can't see them or are looking at something else. The subject and the object are quite separate and distinguishable.[1] This is in contrast to many non-Western cultures in which the viewing self and the rest of the world are more interconnected and co-existent.[2] Psychologists are still not sure how we integrate sensations into a perceived object. Certainly an object becomes "known" as a result of being the center of repeated sense impressions. Insofar as these impressions localize at one place, they constitute a group of invariant sensations and an object develops there.

The ability to see an object as moving requires experience. We learn to "track" an object as it moves even though we may not see it all the time. But this ability seems to come only after we have gone through several stages of mental development. Jean Piaget, the Swiss psychologist, has shown that the concept of "object" itself comes to infants only after they are able to construct spatial concepts. Infants do this by body movements, reaching, and touching. After they have in their minds the idea of a spot in space somewhere in front of them, then they begin to realize that certain spots out there continue to send them visual cues as long as they look. The idea of these objects being able to move comes later. Piaget tells[3] of an experiment in which a screen is set up about three meters in front of an infant. The researcher comes into the room from the left, signals and smiles at the infant, and then walks across the room behind the screen. As the researcher reappears from behind the screen at the right, the infant is surprised! It sees no connection between the disappearing and the reappearing person. If the researcher walks on out of the room to the right, the infant still looks at the left side of the screen, expecting the original object to reappear there. That is the point in space from which the original visual cues came.

As infants grow older, they finally learn to see the motion as we do. What makes all of this interesting is that the adult view might not always be the most valid. Nuclear physicists must unlearn some of the habits they have developed when they begin studying particles which are too small to see directly. They begin to see their movements the way the surprised infant did. A particle, after all, is an object which is confined to a

(1) In Chapter 17, we will discuss some evidence from physics which runs counter to this.

(2) Werner, Heinz, *Comparative Psychology of Mental Development*, International Universities Press, N.Y., 1948, Chap. 4, 5.

(3) Piaget, Jean, "The Child and Modern Physics", *Scientific American 196*, March, 1957, p. 46.

limited region of space. It does not spread all over. It does not exist as a particle if it cannot be localized by observation. If it can't be localized, then it must be spread out, somewhat like a cloud, and it is no longer a particle. Evidence exists that this sort of behavior does occur on the atomic scale. Although we may feel justified in assuming the permanence and "ball-ness" of a baseball which can be seen every time we look at it, it does not follow that a sub-microscopic particle, which we "see" only indirectly, maintains an independent existence when it is not being observed. More of this in Chapter 17. For now, the important thing is to be aware that seeing and interpreting the motion of, an object involves considerable learning.

Locating Objects

In Chapter 2 we said length, mass and time intervals have communicable meaning only insofar as they are compared with, or referred to, a common standard. The same may be said about the position of an object in space. If a number of objects or identifiable points are visible, then one chooses a particular anchor point or *origin,* and lets it become the center of a *frame of reference.* The object is then located in space by giving the directions for getting from the origin to the object. The actual directions will depend upon the particular frame of reference. Many frames can be set up having the same origin. Consider the problem of describing the location of a book on a table in a room. A possible origin might be the southwest corner of the room where the floor meets the intersection of two walls. The book could then be located by saying that it is three meters east, two meters north and one meter above the origin. In this case, the actual system would

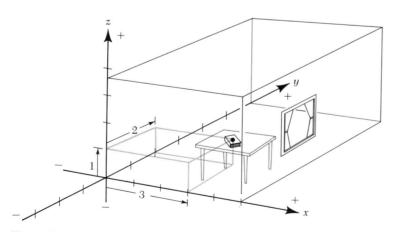

Figure 3-1
Locating objects in a three-dimensional cartesian coordinate system.

resemble three long rods, all arranged at right angles to each other, one oriented east-west (x), one north-south (y), and one up-down (z). Such a set of three perpendicular lines for locating points in three-dimensional space is called a cartesian coordinate system, after the French mathematician, René Déscartes (1596–1650). In the example cited above, the east-west or left-right or x coordinate of the book would be 3, the north-south or y coordinate would be 2, and the up-down or z component would be 1. If the book were restricted to positions solely on the (two-dimensional) floor, only two numbers would be required for specifying how to get from the origin to the book. The problem would be two-dimensional. The position of an ant crawling along the intersection of the floor with the south wall of the room could be specified by just one number; it would be moving in just one dimension. Observers facing the room from other directions might choose different frames of reference having the same origin, in which case the coordinates of the book would be different. For example, someone looking in through the east window might say the book is 2 meters to the right of the origin, 3 meters this side of the origin, and 1 meter above the origin. The book's three coordinates would then be 2, minus 3, and 1, instead of 3, 2, 1, as in the first case.[4] Other origins, all perfectly communicable, could also be chosen and other, non-Cartesian systems for expressing the location of the book in space could be used. If the origin of the coordinate system is *moving*, this will certainly affect the position measurements. The important thing to remember is that objects have no absolute location in space, no physical means for telling you where they are. *They are located only in reference to other objects or arbitrarily chosen points.* We will see subsequently that this is a conclusion from twentieth century physics, not an assumption. The idea that objects have an absolute location in space endured for centuries but eventually did die a slow death.

The Perception of Motion

There are several ways in which we decide that a perceived object is moving. First the retinal image of the object may move across the back of one's eyeballs, or the eyeballs may move to keep the retinal image in the same place. An object which stands still but is seen against a moving reference object or background also seems to move. Psychological testing indicates that when motion relative to an observer and relative to a background are

(4) Note that the distance from the origin to the book is the same in both coordinate systems: $\sqrt{(3)^2 + (2)^2 + (1)^2} = \sqrt{14} = \sqrt{(2)^2 + (-3)^2 + (1)^2}$. Quantities such as this, which measure the same in two different frames of reference are said to be *invariant*.

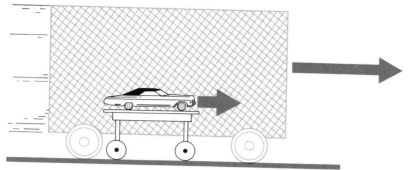

Figure 3-2

made to compete, the latter wins out. That is, we subconsciously feel that the background must be at rest and the object moving, even when the opposite is true. ("Rest" and "moving" are of course meant to refer to the observer.) We look around for the largest thing we can see and assume that it must be at rest. For example, if a large backdrop is pulled toward the right and a table holding a car is moved more slowly to the right, as illustrated in Fig. 3-2, a stationary observer who gets no other visual cues will think the car is moving to the left—for a while.

An interesting psychological experiment[5] shows that we tend to locate objects and to see their motion according to coordinate systems which are *most simply* oriented with respect to the objects themselves. Subjects are placed in a dark room where they see two lights moving against a dark background (Fig. 3-3a). They are then asked to describe what sort of motion is taking place. They usually will say that the two lights are moving toward each other and that both are moving together downward and to the left as shown in Fig. 3-3b.

Figure 3-3a

The motion is thus seen by the subject in a coordinate system consisting of two lines; one oriented so as to be parallel to a line containing the only objects visible, and the other perpendicular to the first one. We mention these tendencies because physicists are humans and these human tendencies helped color mechanics, the science of motion and force, as it developed.

Figure 3-3b

Very soon after children develop the concept of "objectiveness," they grasp the concept of size conservation. Objects don't normally swell up or shrink. Thus, when the retinal image of an object enlarges, we perceive the object to be coming toward us, not just being blown up like a balloon. If we lived in a world where objects quite commonly inflated and deflated, we would perhaps not interpret changes in retinal image size as approach

(5) Wallach, Hans, "The Perception of Motion", *Scientific American* 200, July 1959, pg. 56. Offprint 409.

or departure. If that were the case, we would have a more diffi-cult time learning to duck out of the way of things coming toward us. Avoidance has survival value.

We tend to see motion as smooth, passing through all points of space from start to finish. A good illustration of this is the mo-tion picture, where we are shown separate still pictures in rapid succession. A shutter turns out the light between pictures on the film, but what we think we see is a smooth, continuous move-ment of the figures on the screen.

Various artists have attempted to *represent* motion in their paintings and sculpture for centuries. Blurring, straining or elon-gation of figures can give a feeling of dynamism. Organization of a painting can also make the eye follow from one side to the other or from close-up to mid- and far distances. Recent pop art draws on optical illusions to make the painting appear to shimmer or flow. (See *Current*, by Bridget Riley, which is reproduced on page 266, just preceding Chapter 11.)

Just before World War I, a school of painting called "Futurism" began in Italy. These artists were intensely concerned with de-picting motion as part of an over-all attachment to the idea of mechanism and machines. However, most of their work was more caricature than great art. An example is Giacomo Balla's *Dog On Leash.*

Marcel Duchamp was not a Futurist, but his *"Nude Descend-ing a Staircase"* is one of the best-known representations of

Figure 3-4
Giacomo Balla, *Dog on Leash.* Courtesy George F. Goodyear and Buffalo Fine Arts Academy.

motion in painting, so it was used to introduce this section of the book. We shall soon see that Duchamp's multiple images depict motion in the same strobe-light[6] fashion as physics does.

The Description of Motion

It is one thing to perceive the motion of an object or to bring about the sensation of motion in graphic art, but quite another thing to *describe* motion to someone who is not also looking at the object. The problem is one of analysis; separating out the parts of the observation which can be communicated. The approach of physics to the description of motion is just the reverse of what someone does when they see a motion picture. It "freezes" the motion, records the location of the object relative to the origin, freezes it again a known time later. etc. The analysis is like that of a high-speed repeating camera, or of a world lit by a strobe light.

But before we plunge into the physics of motion, a word of warning and encouragement. The next 15 or 20 pages contain a lot of graphs, equations and algebra. The whole book is not going to be this mathematical. In fact, there is more math in this chapter than in any other. There is really no other way to give you a taste of the things physicists look for or the way they handle their data. So take your time, look for numbered equations, try to express them to yourself in words, and hang in there!

Constant Speed

As a simple example, let us first consider motion in one dimension. A car rolls along a level, straight railroad track. We first lay out a coordinate system or frame of reference along the track with an origin and positive and negative numbers will be taken to the right and left of the origin. Suppose we photograph the car every two seconds and from the pictures, get a record of where the car was as time passed. By "Where the car was" we mean where

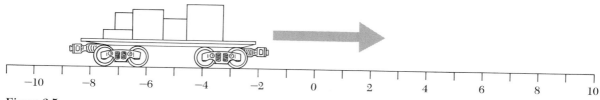

Figure 3-5
Locating a freight car by means of a one-dimensional cartesian coordinate system.

(6) A stroboscopic light is an instrument which emits a very bright but short pulse of light which is repeated at a fast or slow rate.

Table 3-1 *Position-Time
Data For Car Rolling
Along a Level Track.*

Position, x, *in Meters*	Time, t, *in Seconds*
−11.2	0
− 8.1	2
− 4.9	4
− 2.0	6
+ 1.0	8
+ 4.2	10
+ 7.2	12

some reference point on the car was. Let us choose the center of one front wheel as the point which will locate the car

The photographs taken will contain all the information required to describe how the car is moving. But that information is not in the most accessible or communicable form. In order to better see what is going on physicists now put the information through a *data processing* stage. First, they will often make a table of positions and times. Suppose measurement photographs gave the data shown in Table 3-1.

Such a list is not very interesting or revealing. The next step in studying the data would be to make a graph, using the two vari-

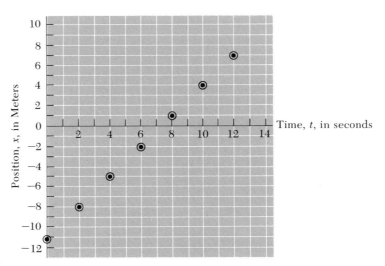

Figure 3-6
Graph of position vs time for a car rolling along a level track.

ables,[7] position and time, as the axes. The plot of position *vs* time would look like Fig. 3-6.

The graph begins to suggest things to us. First, we note that a straight line could be drawn through the data points on the plot. Such a graph is called "linear". The scientist will usually do just that. But in so doing, he is guessing at something he really has not observed. The photographs tell him only where the car was at 2, 4, 6, etc., seconds. By *interpolating* between the data points, that is, by drawing the line, he is assuming that the car rolled smoothly over the distance between photographs. If he happened to be watching the car as well as photographing it, he might feel reasonably safe in drawing the line, but if the object were sub-microscopic and never directly visible, the interpolation would be more risky. In drawing the straight line, the physicist is applying the principle of simplicity.

How fast is the car moving? The way a physicist answers this question is to see where the car is at one instant and where it is a known time interval later. If the initial position is x_i and the initial time is t_i and the final position and time x_f, and t_f, then he or she can define an *average velocity* during the time interval $t_f - t_i$:

$$v_{avg} = \frac{x_f - x_i}{t_f - t_i} = \frac{\Delta x}{\Delta t}, \qquad \text{Equation (3-1)}$$

where Δ, pronounced "delta", means "change in".

In words, Equation (3-1) says that the average velocity of an object is equal to the change of position (displacement) of the object divided by the time interval during which that change of position occurred. For the car we've been looking at, the average velocity between 8 and 10 seconds is: $v_{avg} = (4.2-1)/(10-8) = 3.2$ m/2 sec $= 1.6$ m/sec. If we use other sets of position and time measurements, we find that the average velocity is close to 1.5 m/sec during all the 2 second intervals in which photographs were taken. The average velocity from start to finish is 18.4 m/12 sec or 1.53 m/sec. (If you were to divide 12 seconds by 18.4 meters, what would you get? In words, what kind of physical quantity would your answer represent?)

From your high school geometry you might remember a concept called the *slope* of a graph. It had to do with how fast the line in the graph was going up. Numerically, it was equal to the rise divided by the run: slope = rise/run. In physics, where graphs can represent the changes occurring in all kinds of things, the slope may take on various particular physical meanings. If position is being graphed or plotted vertically *vs* time horizon-

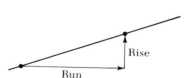

(7) In mathematics or physics, a variable is simply some characteristic of a system which can be measured numerically and which is singled out for attention. Its numerical value may or may not change during the observation.

tally, then the slope represents $\Delta x/\Delta t$ or velocity. In Figure 3-6, where the graph would be nearly a straight line, (has constant slope) then we know the velocity is nearly constant. The car covers almost the same distance during each equal interval of time. It is even possible to make a table of velocity values and a plot of velocity *vs* time by dividing successive changes of position by changes in time (Table 3-2).

Note that the points in the velocity time plot are placed at 1, 3, 5, 7, etc., seconds. This is because the points represent the average velocity during intervals which begin at one time and end 2 seconds later. They are thus placed at the middle of the intervals. The points in Figure 3-7 almost lie on a horizontal line, representing constant velocity.

The last step in the process of data reduction cannot always be reached, but is a goal toward which the physicist works. It is the presentation of the experimental information in the form of an *equation of motion*. It constitutes the limit of brevity and generality, with all the risks that come with them. If the velocity

Table 3-2 *Average Velocity of a Car Rolling Along a Level Track.*

x_f	x_i	Δx	t_f	t_i	Δt	$\dfrac{\Delta x}{\Delta t}$	"avg" t
−8.1	−11.2	3.1	2	0	2	1.55	1
−4.9	− 8.1	3.2	4	2	2	1.60	3
−2.0	− 4.9	2.9	6	4	2	1.45	5
1.0	− 2.0	3.0	8	6	2	1.50	7
4.2	1.0	3.2	10	8	2	1.60	9
7.2	4.2	3.0	12	10	2	1.50	11

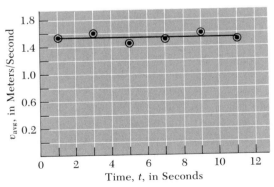

Figure 3-7
Graph of average velocity vs time for a car rolling along a level track.

of the car is constant, then at all times, $v = \Delta x/\Delta t$, or $\Delta x = v\Delta t$, or $x_f - x_i = v(t_f - t_i)$. Suppose we want to express x as a function of t; that is, to see how x depends upon t algebraically. Then we could let x_f and t_f be simply x and t. Suppose we let x_i and t_i be replaced by x_0 and 0. That is, we let the car occupy an initial position x_0 when t is equal to zero. Then $x - x_0 = vt$ or

$$x = x_0 + vt \qquad \text{Equation (3-2)}$$

(To repeat, x_0 is the initial position or position at time $t = 0$ and x is its position at some later time t. The velocity is v.) For our car rolling along the track, $v = 1.53$ m/sec on the average and $x_0 = -11.2$ m, so the equation of motion of the car is $x = -11.2 + 1.53t$.

Not only are x and t allowed to take on values in between those at which observations were actually made, but as far as the equation is concerned, t can increase without limit. The equation suggests that when t is equal to 1000 seconds, for example, the car will be 1519.8 meters down the track toward the right. No concern over the track bending or ending or a wheel falling off the car! *Extrapolation* beyond the last observation is always risky.

Constant Acceleration

Suppose now that the car is put on a track which slants downhill. We release the car from a standing start at -11.0 m and photograph it every 2 seconds as before. A table and graph of position *vs* time are prepared, as shown in Table 3-3 and Fig. 3-8.

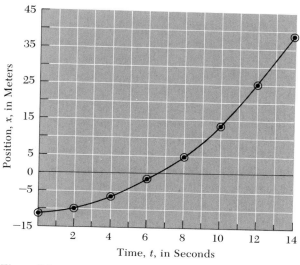

Figure 3-8
Graph of position vs time for a car rolling down a sloping track.

Table 3-3 *Position-Time
Data For a Car Rolling
Down a Sloping Track.*

Position, x, *in meters*	Time, t, *in seconds*
−11.0	0
−10.3	2
− 6.9	4
− 2.1	6
4.9	8
14.4	10
25.0	12
38.8	14

Table 3-4   *Average Velocities For a Car Rolling Down a Sloping
Track.*

x_f	x_i	Δx	t_f	t_i	Δt	$\Delta x/\Delta t$	"avg" t
−10.3	−11.0	0.7	2	0	2	0.35	1
− 6.9	−10.3	3.4	4	2	2	1.7	3
− 2.1	− 6.9	4.8	6	4	2	2.4	5
4.9	− 2.1	7.0	8	6	2	3.5	7
14.4	4.9	9.5	10	8	2	4.8	9
25.0	14.4	10.6	12	10	2	5.3	11
38.8	25.0	13.8	14	12	2	6.9	13

The graph tells us that the car speeds up. It covers a larger distance during each succeeding equal period of time; alternatively,
the slope of the position-time plot is increasing. But *how* is the
velocity increasing? As before we make a table of Δx's and Δt's
so as to calculate the average velocities during successive
time intervals (Table 3-4). We can present these calculations in
the form of a plot, as is done in Fig. 3-9.

The velocity is certainly increasing, but more than that, it is
increasing in linear (straight line) fashion with time. The graph
also goes through the origin. Thus we call it a "proportionality".
(Velocity is equal to a constant times the elapsed time.) Just for
interest's sake, let us see how velocity changes with change of
position: We can do this by first looking up positions in Table

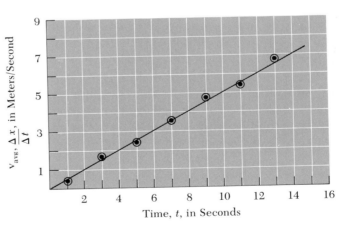

Figure 3-9
*Graph of average velocity vs time for a car rolling down a
sloping track.*

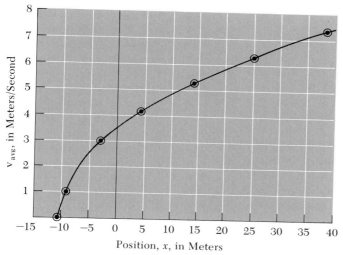

Figure 3-10
Graph of average velocity vs position for a car rolling down a sloping track.

Table 3-5 *Average Velocity and Position For a Car Rolling Down a Sloping Track.*

t, *in sec*	x, *in meters*	v_{avg} *m/sec*
0	−11.0	0
2	−10.3	1.0
4	− 6.9	2.0
6	− 2.1	3.1
8	4.9	4.1
10	14.4	5.2
12	25.0	6.3
14	38.8	7.3

3-3 and matching them with velocity values taken from Figure 3-9 at $t = 0, 2, 4, 6$, etc., making a table (see Table 3-5) and then plotting (see Fig. 3-10).

Although we have looked into the relationship between velocity and distance moved (Fig. 3-10) as well as velocity and time (Fig. 3-9), the $v - t$ plot shows a simpler, linear relationship. In general, we will be more interested in the change of velocity with *time* and will give it a special name: *acceleration.*[8] If the velocity of an object changes by Δv in a time Δt, we say its average acceleration during that time interval is:

$$a_{avg} = \Delta v/\Delta t. \hspace{3cm} \text{Equation (3-3)}$$

Just as the velocity of our car turned out to be numerically equal to the slope of the position-time plot, *acceleration is equal to the slope of the velocity-time plot.* (See if you can show that the acceleration of our car is nearly constant and equal to approximately 0.5 m/sec.²) The m/sec² is read "meters per second per second," which, when you stop to think of it, is how you would have to express the time rate of change of velocity. Our car picks up 0.5 m/sec of velocity every second, no matter what its velocity.

Note: You should realize that the motion of this car constitutes a special case, but one which we'll study pretty thoroughly. The acceleration of an object is *not* always constant.

(8) Galileo Galilei (1554–1642) was one of the first to call attention to acceleration and to show that acceleration is constant for a ball rolling down a ramp. Once he had done this, acceleration and not rate of change of velocity with distance became a basic concept in the science of motion.

The fact that velocity is changing with time makes it difficult to state what the velocity is at a particular instant of time. We calculated a series of average velocities, plotted the points in Figure 3-9, and then happily drew the best straight line we could through the points. The graph then purported to tell us what the velocity of the car must have been at every instant from $t = 0$ to $t = 14$ seconds. But what is the operational basis for saying that the velocity of the car at $t = 6$ seconds, for example, was 3.1 m/sec? How can we *experimentally* measure the velocity of the car at the instant $t = 6$ seconds? The answer is that we can't, exactly.[9] Since a velocity statement has behind it the knowledge of where an object is one time plus knowledge of where it is a little later, we can only *approximate* a measurement of its velocity at a particular instant within the time interval. For example, if we want to approximate what its velocity is at $t = 6$ seconds, we measure its position at 5.99 seconds and at 6.01 seconds. Then we calculate $\Delta x/\Delta t$ and get an average velocity over the .02 second interval centered on $t = 6$ seconds. If that's not good enough, we measure where the object is at 5.9999 seconds and where it is at 6.0001 seconds, etc. We can then *approach* an experimental, not just a graphical basis for stating the instantaneous velocity at $t = 6$ seconds. Even so, graphs might be the best means for visualizing the process we are carrying through. We are trying to find what the slope of the position-time plot is at a particular point on the graph. We do this by taking pairs of points which lie closer and closer to each other but an equal interval before and after $t = 6$ seconds. We draw a line through the two points and determine its slope, $\Delta x/\Delta t$. The closer the two points are, the closer the slope of the line comes to matching the slope of, or the tangent line to the graph at $t = 6$ seconds.

The same thing may be said about instantaneous acceleration. If we wish to know the acceleration at $t = 6$ seconds, we must determine the velocity of the object a very short time before $t = 6$ and a very short time after $t = 6$ seconds. At least three position measurements, then, are required to define an acceleration:

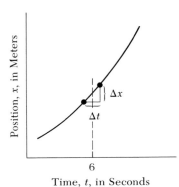

Figure 3-11

The slope of a graph of position vs time.

$$a = \frac{\Delta v}{\Delta t} = \frac{v_f - v_i}{\Delta t} = \frac{\dfrac{\Delta x_f}{\Delta t_f} - \dfrac{\Delta x_i}{\Delta t_i}}{\frac{1}{2}\,\Delta t_i + \frac{1}{2}\,\Delta t_f}$$

Let's try that again: we are saying that the average velocity at a time $\frac{1}{2}\,\Delta t_i$ before 6 seconds is v_i and the average velocity at a time $\frac{1}{2}\,\Delta t_f$ after 6 seconds is v_f. (Remember, we plot the average velocity during a time interval at the middle of the interval.) Then the change in velocity $v_f - v_i$ occurs during a time interval

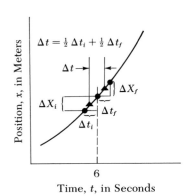

Figure 3-12

Change of slope of a position-time graph.

(9) A speedometer gives velocity readings at various instants of time, but at some time in its history it had to be *calibrated*. That is, numbers were put on it (or on a similar speedometer) by letting the car which carried it cover a measured distance in a measured time and seeing where the needle pointed on the dial.

$\Delta t = \frac{1}{2} \Delta t_i + \frac{1}{2} \Delta t_f$. As Δt gets shorter, we come closer and closer to the *bending* of the $x - t$ plot. If we were to make a separate plot of these velocity *vs* time values, the *slope* of this $v - t$ graph at the time $t = 6$ seconds is the acceleration at that time. Therefore, acceleration can be thought of either as the change in slope of a position-time graph, or, as the slope of a velocity-time graph.

It is interesting to see if we can represent the curve of Figure 3-8 by means of an equation, using the equations for average velocity and for position which we have obtained thus far If acceleration is constant, both average and instantaneous acceleration is given by $a = \Delta v / \Delta t$.

$$\Delta v = v_f - v_i = a\Delta t.$$

If we choose a time interval which begins when velocity is equal to zero, then the velocity at a time Δt later will be $v_f = v = a\Delta t$. A glance at Figure 3-9 shows that *when acceleration is constant*, that is, *velocity increases linearly with time*, then the *average* value of the velocity during any interval is just half-way between the initial and final velocity:[10]

$$v_{avg} = \frac{v_f + v_i}{2} \qquad\qquad \text{Equation (3-4)}$$

This means that if an object starts from rest ($v_i = 0$) and undergoes a constant acceleration to a final velocity (v_f) then its average velocity during that interval will be

$$v_{avg} = \frac{v_f + 0}{2} = \frac{a\Delta t}{2}.$$

But from Equation (3-1), the average velocity

$$v_{avg} = \frac{\Delta x}{\Delta t}.$$

This tells us that an object starting from rest covers a distance

$$\Delta x = (v_{avg})\Delta t$$

$$= \frac{a(\Delta t)^2}{2}$$

in the time interval Δt.

$$\Delta x = x_f - x_i = \frac{a(\Delta t)^2}{2}.$$

(10) This equation, or the ideas in it, dates all the way back to 1350, when it was originated by a group of scholars at Merton College, Oxford. It was called the "mean speed theorem."

If now $x_f = x$, the position after a time Δt, and $x_i = x_0$, the position when $t = 0$, we obtain

$$x = x_0 + \frac{a(\Delta t)^2}{2} \qquad \text{Equation (3-5)}$$

This equation constitutes an expression for the position of any object starting from rest at the point x_0 and undergoing constant acceleration. For our car rolling downhill, $x_0 = -11.0$ m and $a = 0.5$ m/sec². Thus,

$$x = -11.0 + \frac{0.5\ (\Delta t)^2}{2}.$$

To check Equation (3-5), let Δt equal 14 seconds and see if it matches the observed value:

$$x = -11.0 + \frac{0.5\ (14)^2}{2}$$

$$= -11.0 + 49$$

$$= 38 \text{ m}.$$

This agrees fairly well with Table 3-3.

Constant Acceleration with a Running Start

Now suppose the car coasts down the same sloping track, having been launched at some initial velocity. Position and times are

Table 3-6   *Position-Time Data For a Car, With a Running Start, Rolling Down a Sloping Track.*

Position x, in meters	Time, t, in Seconds
−11.0	0
0	2
13.2	4
28.1	6
45.0	8
64.2	10
84.8	12
108.3	14

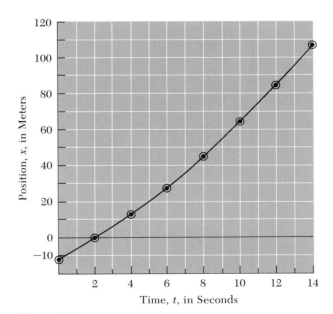

Figure 3-13

Graph of position vs time for a car with a running start rolling down a sloping track.

recorded as previously and presented in tabular and graphic form (Table 3-6, and Fig. 3-13).

The car plainly goes down the track more rapidly than it did when it started from rest, but just how the motion differs remains to be seen. We note immediately that the curve does not start out horizontally, with zero slope. Since the car has velocity right from the start, a curve drawn through the data points should have an initial slope. Average velocities can again be calculated for each of the two-second intervals and the results plotted in Fig. 3-14 as a velocity-time graph.

The straightness of the plot, that is, the constancy of slope, indicates a constant acceleration. Furthermore, comparison of Figures 3-14 and 3-9 shows that the rate of change of velocity is the same along a given sloping track, regardless of the initial velocity

An equation relating position and time for a car with constant acceleration and a running start can be obtained in the same manner as was Equation (3-5) if we simply use the proper expression for the change in velocity: $\Delta v = a \Delta t$. If we take the initial velocity v_i to be v_o and the final velocity $v_f = v$, then $\Delta v = v - v_o = a \Delta t$ or

$$v = v_o + a\Delta t \hspace{4cm} \text{Equation (3-6)}$$

From Equation (3-4),

$$v_{avg} = \frac{v_i + v_f}{2}$$

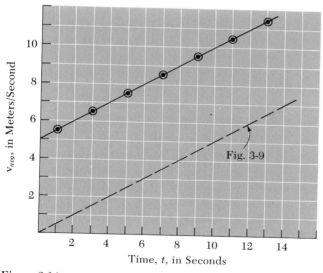

Figure 3-14

Graphs of average velocity vs time for a car rolling down a sloping track. Running start (solid line), and stationary start (dashed line).

But, if v_i equals v_o, the velocity when $t=0$, and v_f equals $v_o + a\Delta t$, then

$$v_{avg} = \frac{v_o + (v_o + a\Delta t)}{2}$$

$$= v_o + \frac{a\Delta t}{2}$$

From Equation (3-1) again,

$$\Delta x = (v_{avg})\,\Delta t,$$

so

$$\Delta x = \left(v_o + \frac{a\Delta t}{2}\right)\Delta t$$

$$= v_o\Delta t + \tfrac{1}{2}a(\Delta t)^2.$$

If we take x_i as being equal to x_o, the position when t equals 0, and x_f equal to x, then Δx equals $x - x_o$. Solving this equation for x and substituting the last expression derived above for Δx yields:

$$x = x_o + v_o\Delta t + \tfrac{1}{2}a(\Delta t)^2 \qquad \text{Equation (3-7)}$$

Note that Equation (3-5) can be obtained from this by letting v_o equal zero, and that Equation (3-2) can be obtained by letting a equal zero. (v_o is equal to v in Equation (3-2) since the initial, final, and average velocities are all the same for the constant velocity case. Δt equals t because t_o is taken to be zero.)

Turning to our specific car again, we have $x_o = -11.0$ m, $v_o = 5$ m/sec and $a = 0.5$ m/sec². Therefore, the equation of motion for the car should be

$$x = -11.0 + 5 \text{ m/sec } (\Delta t) + \frac{0.5 \text{ m/sec}^2\ (\Delta t)^2}{2}.$$

To check the equation, let $t = 8$ seconds, for example. Then

$$x = -11.0 + 5 \text{ m/sec } (8 \text{ sec}) + \left(\frac{0.5 \text{ m/sec}^2\ (8 \text{ sec})^2}{2}\right)$$

$$= -11.0 + 40 + 16$$

$$= 45 \text{ m}$$

This matches Table 3-6 exactly.

In comparing Equations (3-5) and (3-7), note that the term $v_o\Delta t$ increases with time. This means that a car with a running start gains a lead which increases with time assuming that both cars continue to accelerate. In the real world, accelerations are seldom maintained for very long. Sooner or later the object arrives at a physical or legal speed limit.

Expressing results in a general algebraic form enables one to manipulate symbols to obtain new relationships between variables without doing new experiments. The mathematics does not generate any information which was not there before, but rather uncovers and makes more noticeable certain relationships. For example, one useful expression for constant accelerated motion relates velocity, acceleration, and position with the specific time of the observation "eliminated". From Equation (3-6), $(v - v_o)/a = \Delta t$. Using this, we can eliminate Δt from Equation (3-7).

$$x = x_o + v_o \, \Delta t + \tfrac{1}{2}a \, (\Delta t)^2$$

$$= x_o + v_o \left(\frac{v - v_o}{a}\right) + \tfrac{1}{2}a \left(\frac{v - v_o}{a}\right)^2$$

$$= x_o + \frac{v_o v}{a} - \frac{(v_o)^2}{a} + \tfrac{1}{2}a \left(\frac{v^2 - 2vv_o + v_o^2}{a^2}\right)$$

$$= x_o + \frac{v_o v}{a} - \frac{(v_o)^2}{a} + \frac{v^2}{2a} - \frac{vv_o}{a} + \frac{v_o^2}{2a}$$

$$= x_o + \frac{v^2}{2a} - \frac{v_o^2}{2a}$$

Finally, therefore:

$$x = x_o + \frac{v^2 - v_o^2}{2a} \qquad \qquad \text{Equation (3-8)}$$

This relates initial (x_o) and final (x) position, acceleration (a) and initial (v_o) and final (v) velocity in cases where you don't know, or aren't interested in the elapsed time.

If the initial velocity, $v_0 = 0$ and the initial position, $x_o = 0$, then

$$x = (\tfrac{1}{2}a) v^2$$

or

$$v = \sqrt{2ax} \qquad \qquad \text{Equation (3-9)}$$

As an example, and to obtain some practice in manipulating units, consider a drag racer which starts from rest and covers a

quarter mile strip in 10.90 seconds. First, what is its *average
velocity?*

$$v_{avg} = \frac{\Delta x}{\Delta t} = \frac{0.25 \text{ mile}}{10.90 \text{ sec}} = \left(\frac{0.25 \text{ mile}}{10.90 \text{ sec}}\right)\left(\frac{3600 \text{ sec}}{1 \text{ hour}}\right)$$

$$= \frac{(0.25)(3600)}{10.90} \frac{\text{mile}}{\text{hour}}$$

$$= 82.5 \text{ miles/hour}$$

In meters/sec,

$$v_{avg} = 82.5 \text{ miles/hour} = \left(82.5 \frac{\text{mile}}{\text{hour}}\right)\left(\frac{1 \text{ hour}}{3600 \text{ sec}}\right)\left(\frac{1610 \text{ meters}}{1 \text{ mile}}\right)$$

$$= \frac{82.5 \times 1610}{3600} = 36.9 \text{ meters/sec.}$$

(Note that in these changes of units you just keep multiplying
by "one"! 3600 sec./1 hour = 1; 1610 meters/1 mile = 1; etc.)
Assuming the acceleration to be constant (not actually true in
drag racing),[11] what is the final velocity of the racer at the end
of the strip? First, to calculate the acceleration, we use Equation
(3-7), recognizing that initial position and velocity, x_o and t_o,
are zero:

$$x = \tfrac{1}{2}a(\Delta t)^2 \qquad \text{or} \qquad a = \frac{2x}{(\Delta t)^2}.$$

If $x = .25$ mile $= 402.5$ m, then

$$a = \frac{2(402.5 \text{ m})}{(10.9 \text{ sec})^2} = \frac{805 \text{ m}}{119 \text{ sec}^2}$$

$$= 6.76 \text{ m/sec}^2$$

Then from Equation (3-9), $x = v^2/2a$ where v is the value of
velocity when x is 402.5 meters. Re-writing $v^2 = 2ax$, we get

$$v^2 = (2)(6.78 \text{ m/s}^2)(402.5 \text{ m}) = 5450 \text{ m}^2/\text{s}^2$$

Taking the square root,

$$v = \sqrt{5450 \text{ m}^2/\text{sec}^2} = 73.8 \text{ m/sec}$$

In miles per hour,

$$v = 73.8 \left(\frac{\text{mtr}}{\text{sec}}\right)\left(\frac{1 \text{ mile}}{1610 \text{ mtr}}\right)\left(\frac{3600 \text{ sec}}{1 \text{ hour}}\right)$$

$$= 165 \text{ miles/hour.}$$

(11) Fox, Geoffrey T., "On the Physics of Drag Racing", *Amer. Jour. Physics 41*,
March, 1973, pg. 311.

This is the final speed of the car at the end of the strip. We could also have obtained this answer using Equation (3-3):

$$v_{avg} = \frac{v_f + v_i}{2}.$$

We know $v_{avg} = 82.5$ miles/hour and that $v_i = v_o = 0$. Therefore,

$$v_{avg} = \frac{v_f}{2}$$

$$v_f = 2(v_{avg}) = (2)(82.5 \text{ miles/hour})$$

$$= 165 \text{ miles/hour}$$

Free Fall

Let us now take a look at objects which have been dropped or thrown up in the air. We will make two idealizations: first, that the gravitational pull of the earth does not change with height; second, that air drag (resistance to motion of objects through the atmosphere) is small in comparison with the earth's gravitational pull. We will examine these two assumptions more critically later. But for the present we can take confidence in the fact that the assumptions are very nearly justifiable in ordinary laboratory experiments, so long as we don't experiment with feathers, or balls of cork.

"Free fall" is an idealized sort of motion which objects undergo when they fall in the absence of air. The acceleration of *all* objects near the surface of the earth has been found to be 9.8 m/sec² when they fall freely; that is, when gravity is the only influence acting upon them. Chemical composition, color, weight, age, style have no effect. The acceleration due to gravity at the surface of the earth, "g", not only has a *magnitude* of 9.8 m/sec²—but it also has a *direction*: down. If we arbitrarily choose quantities which are directed downward to be negative and those directed upward as positive, then "g" = −9.8 m/sec². (See Appendix G, On Vectors.)

For the sake of illustration of what can be done with an equation of motion, consider a student who throws a physics textbook straight up with a velocity of 15.0 m/sec.

(a) Where is the book after one second? We know it won't have gone up quite 15 meters because gravity is slowing the book as it goes up. Its average speed during the first second will be something less than 15 m/sec. If we let y represent vertical distances, then, from Equation (3-7),

$$y = 15.0 \, \Delta t - [\tfrac{1}{2} (9.8)(\Delta t)^2],$$

since the initial velocity $v_o = 15.0$ m/sec, $a =$ "g" $= -9.8$ m/(sec)^2, and y_o, the initial height, is zero. This initial height could be taken at the student's waist or the level at which the book left his hand. If Δt equals one second,

$$y = (15.0 \text{ m/sec})(1 \text{ sec}) - [\tfrac{1}{2} (9.8 \text{ m/s}^2)(1 \text{ sec})^2]$$
$$= 15.0 \text{ m} - 4.9 \text{ m} = 10.1 \text{ m}$$

above the level from which it was thrown.

(b) How fast is the book traveling at that instant? Its velocity is changed by only 9.8 m/sec every second, so that after one second, it would have some upward-directed velocity left. From Equation (3-6),

$$v = v_o + a\Delta t$$
$$= 15.0 \text{ m/sec} - (9.8 \text{ m/sec}^2)(1 \text{ sec})$$
$$= 5.2 \text{ m/sec}.$$

The velocity is positive, so the book is still traveling upward, as we expected.

(c) How high will the book go? Here we need to think a little. We know that when the book is at the top of its flight there has to be an instant at which the book quits going up and starts coming down. Thus, there must be an instant when the velocity is zero. At the top,

$$v = v_o + a\Delta t$$
$$0 = 15.0 \text{ m/sec} - (9.8 \text{ m/sec}^2)(\Delta t)$$

or

$$\Delta t = \frac{15.0 \text{ m/sec}}{9.8 \text{ m/sec}^2} = 1.53 \text{ seconds.}$$

This is the time required for the book to travel up to the top of its flight. The value of y when Δt is equal to 1.53 sec is

$$y = (15.0 \text{ m/sec})(1.53 \text{ sec}) - \tfrac{1}{2} (9.8 \text{ m/sec}^2)(1.53 \text{ sec})^2$$
$$= 22.9 \text{ m} - 11.4 \text{ m}$$
$$= 11.5 \text{ Meters (at the top)}$$

(d) What is the acceleration at the top of the flight? Not zero! Even though the velocity itself is zero at that instant, it's still *changing*

at the same rate, -9.8 m/sec every second. To check this, let us calculate the velocities, v_1, 0.5 second before 1.53 seconds, and v_2 0.5 second after 1.53 seconds, respectively.

At $t = 1.03$ sec,

$$v_1 = 15 \text{ m/sec} - 9.8 \ (1.03) \text{ m/sec}$$

$$= 15 \text{ m/sec} - 10.1 \text{ m/sec}$$

$$= 4.9 \text{ m/sec}.$$

At $t = 2.03$ sec,

$$v_2 = 15 \text{ m/sec} - 9.8 \ (2.03) \text{ m/sec}$$

$$= 15 \text{ m/sec} - 19.9 \text{ m/sec}$$

$$= -4.9 \text{ m/sec}$$

Thus, during the second between 1.03 and 2.03 seconds, the velocity changes by $\Delta v = v_2 - v_1 = (-4.9 \text{ m/sec}) - (4.9 \text{ m/sec}) = -9.8$ m/sec. The average acceleration over that interval of time is thus -9.8 m/sec^2.

(e) At what time does the book get back down to the level of the student's waist? Here, the overall change of position amounts to zero:

$$y = 0 = 15.0 \text{ m/sec } \Delta t - \tfrac{1}{2} \ (9.8 \text{ m/sec}^2)(\Delta t)^2$$

then

$$15.0 \ \Delta t = \tfrac{1}{2} \ (9.8)(\Delta t)^2$$

Dividing both sides of the equation by Δt, we obtain:

$$\Delta t = \frac{15.0 \text{ m/sec}}{4.9 \text{ m/sec}^2} = 3.06 \text{ sec}$$

Notice that there is symmetry in the rise and fall of the book. It takes the book 1.53 sec to go up and another 1.53 sec to go down.

(f) What is the velocity of the book when it gets back to $y = 0$? We could obtain the answer two ways.

(1) First, at $t = 3.06$,

$$v = (15.0 \text{ m/sec}) - (9.8 \text{ m/sec}^2)(3.06 \text{ sec})$$

$$= 15.0 \text{ m/sec} - 30.0 \text{ m/sec}$$

$$= -15 \text{ m/sec}$$

The rate of travel is the same as at launch time, but the velocity is directed down, not up. We know this from the sign convention. The minus sign indicates downward velocity.

(2) From Equation (3-8), $y = y_o + (v^2 - v_o^2)/2a$. If y and y_o are equal, then $v^2 - v_o^2$ is zero, or $v = \pm v_o$. The mathematics allows for two possibilities, a velocity v directed up, equal to v_o (v_o for short) and a velocity directed down, equal to $-v_o$, but we know physically that the solution applying to the return of the bullet is

(By permission of Johnny Hart and Field Enterprises, Inc.)

$-v_0 = -15.0$ m/sec. Mathematics often indicates multiple correct answers and the physicist must decide which of these applies to the observable world.

Projectile Motion

Suppose an object is thrown out sideways instead of just being dropped. What happens to it? First, if there is no air resistance to its motion, we can assume that the horizontal and vertical motions are independent.[12] That is, traveling sideways has no effect on the downward motion and vice versa. This can be checked by experiment. An object thrown out horizontally takes the same time to reach the ground as one dropped straight down! Of course, the sideways toss *has* to be horizontal. Try it. Because of this fact bombs dropped by an airplane stay directly beneath the plane while they are falling, if the plane does not change its velocity. Motion which combines horizontal travel with free fall (such as that shown in Figure 3-15) is called *projectile* motion. The path followed by a projectile is often called its *trajectory*.

It is not hard to determine the trajectory of an object thrown in a horizontal direction. The horizontal distance it covers is simply its initial horizontal velocity times the elapsed time. It just keeps chugging along at constant horizontal speed. If its initial horizontal speed was 10 m/sec, then after one second it will have traveled 10 meters horizontally, after two seconds, 20 meters, etc. At the same time, the distance of fall increases as (9.8/2) times the *square* of the elapsed time. You should recognize the shape of the path: it is a parabola, a geometrical curve which results when one variable increases as the square of the other.

In studying projectiles and other objects moving in more than one dimension it becomes helpful to make a distinction which we have avoided up to now. *Velocity* represents the rate of change of position in a specified direction with time. It is the

(12) This may or may not seem obvious, but it is a good example of a profound principle which physics takes advantage of. It holds that in many instances the total behavior of a system is a simple sum of several things happening at once.

rate of "getting somewhere" and includes the idea of direction as well as rate of travel. Velocity can be represented by a vector (Appendix G), having components with respect to some coordinate system. *Speed*, which we mention for the first time, is simply a rate of motion. It is the *magnitude* of the velocity vector, but it lacks a specified direction. Suppose a runner in a race goes around a 400 m track in 48 seconds. The runner's average speed would be 400 m divided by 48 sec., or 8.33 m/sec. However, if the race ends at the same spot where it began, the runner's net change of position, and average velocity, would be zero.

In the so-called one-dimensional problems, such as the ones involving the railroad car and flying book, the distinction between speed and velocity can be accounted for with a plus or minus sign. (The thrown book didn't go back and forth, but it did go down as well as up.) Now, having refined the difference between the concept of velocity and speed, we can say that the book had a speed of 15 m/sec., both at the start and at the finish of its flight. However, its velocity changed from 15 m/sec. *up* (or +15 m/sec.) at the start to 15 m/sec. *down* (or −15 m/sec.) at the finish.

The projectile in Figure 3-15 has a constant horizontal component of velocity equal to 10 m/sec. The vertical component of

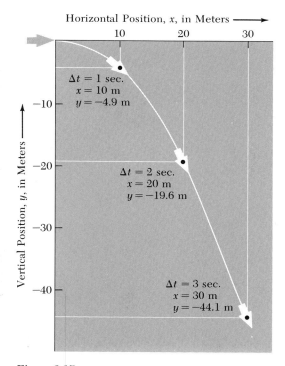

Horizontal Position, x, in Meters

$\Delta t = 1$ sec.
$x = 10$ m
$y = -4.9$ m

$\Delta t = 2$ sec.
$x = 20$ m
$y = -19.6$ m

$\Delta t = 3$ sec.
$x = 30$ m
$y = -44.1$ m

Vertical Position, y, in Meters

Figure 3-15
Graph of position vs time for a projectile thrown horizontally at 10 m/sec.

velocity increases with time, so that the total velocity vector slants downward more and more with time:

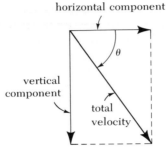

horizontal component

After three seconds the projectile still has a horizontal component of velocity, v_x, equal to $+ 10$ m/sec. (The plus sign tells us that it is moving toward the right.) Its vertical component of velocity is

$$v_y = a\Delta t$$

$$= -9.8 \text{ m/sec}^2 \ (3 \text{ sec})$$

$$= -29.6 \text{ m/sec}.$$

(The minus sign signifies downward motion.)

Using Pythagoras' theorem, we can find the magnitude of the velocity (speed):

$$v = \sqrt{(v_x)^2 + (v_y)^2} = \sqrt{(10)^2 + (-29.6)^2}$$

$$= \sqrt{100 + 876}$$

$$= 31.2 \text{ m/sec}.$$

The direction of the velocity makes an angle θ (pronounced theta) with the horizon. To find θ, we think of the vertical and horizontal components of velocity as making a triangle. Then, using Appendix F, we see that

$$\text{tangent } \theta = \frac{v_y}{v_x} = \frac{29.6}{10}$$

$$= 2.96.$$

Therefore, θ is a little more than $70°$.

Some Examples

(a) Suppose a car crashes through the side of a bridge and lands on the river bottom below. The police investigating the accident find that the front wheels of the car lie a horizontal distance of 25 meters from where the skid marks go through the side of the bridge. What was the speed of the car when it left the bridge? First we need to find out how long the car took to fall five meters from the bridge to the river bed. From Equation (3-7): $y = -5 \text{ m} = \frac{1}{2}(-9.8 \text{ m/sec}^2)(\Delta t)^2$, since $y_0 = 0$ and v_{oy}, the initial vertical velocity (in the $-y$ direction) is zero. Then,

$$(\Delta t)^2 = \frac{-5 \text{ m}}{\frac{1}{2}(-9.8 \text{ m/sec}^2)} = +1.02 \text{ sec}^2$$

$$\Delta t = \sqrt{1.02 \text{ sec}^2}$$

$$= 1.01 \text{ sec}$$

For the horizontal motion, $x = v_{ox}\Delta t$, where v_{ox} is the horizontal velocity in the $+x$ direction as the car leaves the bridge. We assume that this horizontal velocity stays constant until the car lands in the mud.

$$v_{ox} = x/\Delta t = 2.5 \text{ m}/1.01 \text{ sec.}$$

$$= 24.7 \text{ m/sec.} \qquad \text{(What is this in miles per hour?)}$$

(b) Suppose a shot putter heaves the shot up and out at an angle of 45 degrees to the ground. Let's say the shot has a horizontal component of velocity $v_x = +8.5$ m/sec. and an initial vertical component of velocity $v_{oy} = +8.5$ m/sec. as it leaves the athlete's hand. (From Pythagoras' theorem, the speed or magnitude of the total initial velocity will be:[13]

$$v = \sqrt{(v_x)^2 + (v_{oy})^2} = \sqrt{(8.5)^2 + (8.5)^2}$$

$$= 12.03 \text{ m/sec.})$$

Let the shot be two meters above the ground when it leaves the

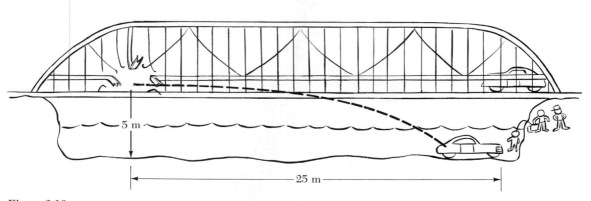

Figure 3-16
Sketch of a car which has crashed through a bridge and landed in a river, 5 m below.

(13) From Appendix G it can be seen that since the shot is launched with a velocity of 12.03 m/sec. at an angle of 45°, the horizontal and vertical components of initial velocity could be found using the cosine and sine:

$$v_{ox} = v_o \cos \theta = 12 \text{ m/sec. } (.707) = 8.5 \text{ m/sec.}$$

$$v_{oy} = v_o \sin \theta = 12 \text{ m/sec. } (.707) = 8.5 \text{ m/sec.}$$

athlete's hand. Question: how far will the shot travel horizontally before it hits the ground?

First, we need to find the time required for the shot to travel to the peak of its trajectory. At this point, the vertical component of velocity, v_y, is zero. Following the same procedure as we did on page 44,

$$v_y = v_{oy} - 9.8 \, \Delta t = 0$$

or

$$v_{oy} = 9.8 \, \Delta t;$$
$$\Delta t = v_{oy}/9.8$$
$$= 8.5/9.8$$
$$= .867 \text{ sec.}$$

Figure 3-17
Shotputter launching shot. (Photo by Chip Gani, *courtesy of Track and Field News.*)

At this instant the shot is above the ground a distance y, that can be found by using Equation (3-7)

$$y = y_o + v_{oy} \Delta t - \tfrac{1}{2} (9.8) (\Delta t)^2$$
$$= 2 \text{ m} + 8.5 \text{ m/sec } (.867) - \tfrac{1}{2} (9.8 \text{ m/sec}^2) (.867)^2$$
$$= 5.68 \text{ m}.$$

We needed to find this distance because we now must find the time in addition to .867 seconds required for the shot to fall to the ground. To find this time, $\Delta t'$, we use the relationship

$$y = 5.68 \text{ m} - \tfrac{1}{2} (9.8 \text{ m/sec}^2)(\Delta t')^2$$

where $\Delta t'$ begins with the shot at the peak of its trajectory. Then we set $y = 0$ and solve for $\Delta t'$:

$$(\Delta t')^2 = \frac{5.68 \text{ m}}{4.9 \text{ m/sec}^2} = 1.16 \text{ sec}^2$$
$$\Delta t' = \sqrt{1.16 \text{ sec}^2} = 1.077 \text{ sec}$$

Now we know that the shot was in the air $.867 + 1.077 = 1.944$ seconds. All this time it was moving horizontally at a constant velocity equal to $v_{ox} = 8.5$ m/sec. Therefore, it must travel a total distance of

$$(8.5 \text{ m/sec}) (1.944) = 16.52 \text{ m}$$
$$= 54.18 \text{ feet}$$

Uniform Circular Motion

Now let us consider a type of motion in which the distinction between speed and velocity is greatest. It involves an object traveling in a circular path at constant speed and is called uniform circular motion. To analyze it properly we need to examine first the ways in which a vector can change. Insofar as we use vectors — arrows — to represent velocity, then what we say will be true of velocity as well.

A vector can increase in length or magnitude:

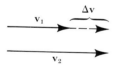

A vector can decrease in length or magnitude:

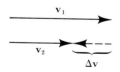

And a vector can also change direction!

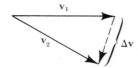

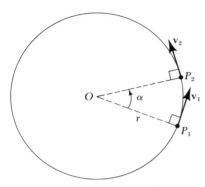

Figure 3-18
*Diagram of an object moving from
point P₁ to point P₂ along a
circular path.*

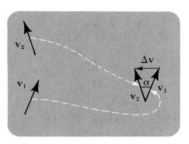

Figure 3-19
*Change in velocity, Δv, from v₁
to v₂.*

Speaking in terms of velocities, we would say in the first two cases that the object speeds up, or slows down, respectively. In the third case, the object keeps the same speed but changes direction. All three cases correspond to real changes in velocity: that is, to *accelerations*. Constant velocity is motion at a constant speed *in a straight line*.

Now, let us consider an object which is forced to travel in a circular path at constant speed. It neither speeds up nor slows down, but it is accelerated inward all the time in order to follow the circular path. Let us find what this acceleration must be. Suppose it travels in a circular path of radius r, centered on O, moving from point P_1 to point P_2 in a time Δt. During this time interval, the speed stays the same but the velocity changes from v_1 to v_2, a change of direction which amounts to an acceleration. (α is pronounced "alpha".)

We can, if we wish, move a vector around in space so long as its length (magnitude), or direction is not changed. This is shown in Figure 3-19. (Also, see page 474 in Appendix G.)

If we do move the two velocity vectors as shown, it is easier to see Δv, the change in velocity between v_1 and v_2. Note that in Figure 3-18 the line OP_1 is perpendicular to v_1 and that line OP_2 is perpendicular to v_2. (Think of OP_1, (or r) and v_1, as forming an arm with a stiff bent elbow.) Then the angle α between v_1 and v_2 is the same as the angle between OP_1 and OP_2. Now try to imagine two triangles. One of them is formed by the two equal velocity vectors v_1 and v_2 plus Δv. The other is formed by OP_1 and OP_2 and a straight line P_1P_2 joining their ends:

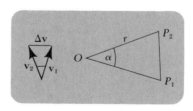

These two triangles have the same proportions. That is, two of the sides of each triangle are equal in length, and, in both triangles, the angle between these sides is α. In the small triangle, $v_1 = v_2 = v$. The two triangles are similar isosceles triangles, and for similar triangles like these, the ratio of two similar sides will be equal. That is, $\Delta v/v$ in the one triangle is equal to P_1P_2/r in the other. Therefore, since $\Delta v/v = (P_1P_2)/r$ then $\Delta v = [v(P_1P_2)]/r$.

The average acceleration during the time Δt during which the velocity changes from v_1 to v_2 is $(v_2 - v_1)/\Delta t = \Delta v/\Delta t$. Then, from above, $a_{avg} = \Delta v/\Delta t = v \, (P_1P_2)/r\Delta t$. Now if Δt becomes very small, so will the angle α. When α is small, the straight chord P_1P_2 comes closer and closer in length to the arc curving from P_1 to P_2. The arc is equal to $r \, \alpha$, so for small Δt,

$$a_{avg} = \frac{\Delta v}{\Delta t} \xrightarrow[\text{approaches}]{} \frac{v \, (r\alpha)}{r\Delta t} = a$$

$r\alpha/\Delta t$ is simply the distance covered in the time Δt, or the speed, v. Therefore,

$$a = \frac{(v) \, (v)}{r} = \frac{v^2}{r} \qquad\qquad \text{Equation (3-10)}$$

We note also that for small Δt and small α, Δv comes very close to being perpendicular to v_1 and v_2. Since they are both perpendicular to r, then the change in velocity (acceleration) is directed toward the *center* of the circular path. It may seem odd at first that the change in velocity is toward the center of the circle, but in fact this is the case. For this reason, the acceleration is called *centripetal* acceleration, (from the Latin meaning, "center seeking"). Without this inward acceleration the object would fly off tangentially in a straight line. It would NOT be thrown radially outward. Even though people riding on a merry-go-round feel as though they must resist a "centrifugal force" trying to throw them outward, this is merely their physical reaction to being accelerated inward when the natural thing would be to travel in a straight line. There is *no* centrifugal force. (At least, there is none when we look at the motion from a well-behaved, non-rotating frame of reference.) If a hammer thrower wants to throw the hammer toward the right, it must be let go of at a, not at b.

An old rural trick is to whirl a full bucket of milk over one's head without spilling the contents. How fast must the bucket travel to keep the milk in it? Since the liquid is falling freely at the top of the circle, it has a downward acceleration of 9.8 m/sec². In order to keep the bucket *around* the milk it must also be accelerated downward at least 9.8 m/sec². This can be done by moving the bucket in a vertical circle in such a way that at the top of the circle, the centripetal acceleration v^2/r is 9.8 m/sec. If $r = 0.7$ m, an average arm length, then

$$v^2 = (9.8 \text{ m/sec}^2) \, (0.7 \text{ m}) = 6.86 \text{ m}^2/\text{sec}^2$$

$$v = \sqrt{6.86 \text{ m}^2/\text{sec.}^2} = 2.62 \text{ m/sec.}$$

It is difficult to keep the bucket moving at a constant speed, since it tends to swoop faster through the lower part of the circle. But, if the bucket could be kept moving in the circular path with constant speed, the time required for one revolution, the

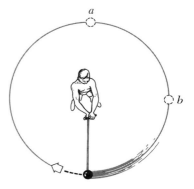

Figure 3-20
Overhead view of a hammer thrower.

Figure 3-21
Photograph of the author swinging a bucket of water over his head.

period of the motion could be calculated. The *period, τ* (tau) is equal to the circumference divided by the speed, or

$$\tau = \frac{2\pi r}{v} = \frac{2\pi \ (0.7 \text{ m})}{2.62 \text{ m/sec}} = 1.68 \text{ sec.}$$

The motion of manned or unmanned satellites around the earth can provide an example of uniform circular motion on a larger scale. It is motion which is particularly interesting because once the satellite is given the requisite speed, it coasts around the earth with no need for further propulsion. The satellite undergoes genuine free fall toward the earth, but is moving sideways with such velocity that its downward curving path just fits around the earth! Figure 3-22 represents a stone being thrown sideways from a mountain top. If it's thrown fast enough, but not too fast, it will fall right on around the earth! It will always be pulled away from straight line motion with a downward acceleration of "g" without ever getting any nearer to the earth. No rockets are needed to propel it. Let us see what sort of sideways speeds are required. Consider a low-altitude satellite. For it to be in orbit, v^2/r must be just equal to the gravitational acceleration. Near the surface of the earth, "g" will be equal to 9.8 m/sec². The radius of the orbit, r, will be close to the radius of the earth, 4000 miles or 6.4×10^6 m. Then from Equation (3-10),

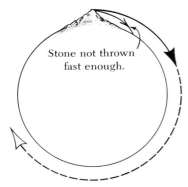

$$\frac{v^2}{r} = g$$

$$v^2 = gr$$

$$= (9.8 \text{ m/sec}^2)(6.4 \times 10^6 \text{ m})$$

$$= 62.7 \times 10^6 \text{ m}^2/\text{sec}^2$$

$$v = \sqrt{62.7 \times 10^6} \text{ m/sec.} = 7.92 \times 10^3 \text{ m/sec}$$
$$= 17{,}700 \text{ miles/hour.}$$

Figure 3-22
A stone thrown horizontally with sufficient speed will become an earth satellite.

Actual satellites orbit at higher altitudes, where the friction of the atmosphere will not burn them up. Here the gravitational pull of the earth is less strong. We shall look into this later when we specifically study gravitation, but for the present, we will simply say that, in order to maintain a circular path, a satellite farther from the earth travels with a speed of less than 17,700 mph.

The Newtonian View of Space and Time

A number of examples of the graphical and algebraic description of motion have been given. What significance do they hold,

(Figure label in image: Stone not thrown fast enough.)

beyond telling us where an accelerated car will be at some time in the future, or how fast you have to swing a full bucket over your head to keep it from spilling? First, they mark a significant change in humanity's ability to understand, predict, and control the external world. As we saw in Footnote 10, some of the ideas date from as early as 1350. However, it was Galileo, rolling balls down ramps and analyzing his results in terms of geometrical diagrams, who gave us most of the key equations displayed in this chapter. Isaac Newton (1642–1727), along with others, was able to see that the moon and the planets obeyed the same equations of motion. As we shall see in the next chapter, Newton also tackled the problem of finding a *cause* for these motions. Galilean-Newtonian physics is still the best means we have for predicting or controlling the movements of objects which are large enough, and which are not travelling too fast. That is, they apply to the realm of everyday experience and fit with the learned, adult view of motion which we mentioned on page 24. We will see later on that when physicists tried to explain atomic phenomena, or the motion of objects at extremely high speed, the Galilean-Newtonian analysis was inadequate.

A specific set of views concerning space and time accompanies the Newtonian analysis of motion. Again, these views match quite well the views we learn culturally. Space is three-dimensional, infinite in extent, and Euclidean; that is, two parallel lines never meet, and if you keep going in one direction you never come back to the place where you started. Empty space is *isotropic*: a physical device will act in the same way no matter how it is tilted, or the direction in which it is pointed is changed. (However, in many places, gravitational pulls do impose special "up" and "down" directions.) Space is continuous: we can put an object anywhere we want to; there are no preferred locations where the object must exist, such as the holes in a Chinese checkerboard. Space is infinitesimally divisible: there is no lower limit as to how small a movement you can inflict on an object. Space contains objects. It pre-exists them and makes possible their movement.

Time, according to Newton, is separate from space and independent of it. It flows continuously, ticking away at the same rate in all frames of reference whether or not events are occurring. Since time is independent of space, we can define a simultaneous "now" which is the same for everyone, no matter how far apart they might be. All motions through space can be recorded in terms of one definite time.

The vast majority of people would not argue with the description of space and time above. It fits their everyday experience. However, it does represent a limited view of the natural world.

In this chapter we have looked at motion the way Galileo and Newton did. However, we did emphasize one "anti-Newtonian" idea: that objects have a position and undergo motion only with respect to—or relative to—other objects. This was a viewpoint which Newton was never able to accept. Even though his equations of motion, which we will discuss in the next two chapters, do not indicate so, he believed that space "stood still". He believed that there existed a frame of reference which was absolutely at rest and that objects *really* moved (or stood still) depending on whether they moved with respect to this frame of reference. The idea of an absolute frame of reference and the hope of detecting motion relative to it endured until the mid-twentieth century, when the views that Albert Finstein (1879–1955) developed in 1905 finally put an end to it. More of this later.

Summary

An object is essentially the center of an unchanging group of sensations in the midst of overall change. Objects are locatable only with respect to a frame of reference and an origin, or to some other objects. They have no absolute position in space. One quantity which does stay the same when referred to various frames of reference is the distance between two points. A quantity which measures the same in all frames of reference is said to be invariant.

Tables, graphs, and in some cases, algebraic equations are steps through which the data relating two physical variables may be processed.

Speed is simply the rate of movement of an object.

Velocity is the change of position in a specified direction divided by the time interval during which that change occurred:

$$v_{avg} = \frac{x_f - x_i}{t_f - t_i} = \frac{\Delta x}{\Delta t} \qquad \text{Equation (3-1)}$$

Velocity is represented mathematically as a vector with a magnitude equal to the speed of the object, and with direction components.

Acceleration is the rate of change of velocity:

$$a_{avg} = \frac{\Delta v}{\Delta t} \qquad \text{Equation (3-3)}$$

The instantaneous velocity of an object at the time t can be seen as being equal to the slope of the object's position-time graph at the time t.

The instantaneous acceleration of an object can be seen as being equal to the slope of the object's velocity-time graph at the time t.

When acceleration is constant:

$$x = x_0 + v_0 \, \Delta t + \tfrac{1}{2} a \, (\Delta t)^2$$

Equation (3-7)

where x is the position after a time Δt, x_0 is the initial position, v_0 is the initial velocity, and a is the constant acceleration. When the velocity is constant, $(a = 0)$, this reduces to Equation (3-2). When initial velocity $v_0 = 0$, this reduces to Equation (3-5).

The change in velocity with time, when acceleration is constant, is given by

$$v = v_0 + a\Delta t.$$

Equation (3-6)

The average velocity during an interval in which the acceleration is constant can be written as

$$v_{avg} = \frac{v_f + v_i}{2}.$$

Equation (3-4)

When acceleration is constant, the relation between v, x, and a can be written

$$x = x_0 + \frac{v_f^2 - v_i^2}{2a}.$$

Equation (3-8)

When an object falls freely near the earth's surface its constant acceleration is $a = -9.8$ m/sec.^{2} The minus signifies "down".

Projectile motion results from two simultaneous but independent motions: horizontal, at constant velocity; and vertical or up-down, under acceleration $a = -9.8$ m/sec.^{2}

Velocity is vectorial in nature, while speed is thought of as the magnitude of the velocity vector. Speed is "how fast an object is moving", whereas velocity is the "rate at which it is getting somewhere". (Away from where it started.) An object will be accelerated if it speeds up, slows down, or changes direction.

In uniform circular motion, the object moves around a circle with a constant speed but with a constantly changing velocity. The rate of change of the velocity points toward the center of the circle, hence it is called centripetal acceleration, and its magnitude is equal to

$$a = \frac{v^2}{r}$$

Equation (3-10)

where v is the speed and r is the radius of the circle.

The period, τ, of a motion, that is, the time required for it to repeat itself, can be found from the equation:

$$\tau = \frac{\text{distance}}{\text{speed}}.$$

The view of space and time which best fits "common sense", and upon which the physics of Galileo and Newton is based, sees space and time as separate and independent. Space is infinite, three-dimensional, Euclidean, and objects move around in it with complete freedom. Time is different, ticking away at an absolute rate which is the same in all frames of reference. However, when physicists tried to understand phenomena occurring at high speeds, or over long distances, they found that a new view of space and time was required.

Further Reading

Čapek, Milič, *The Philosophical Impact of Contemporary Physics*, Van Nostrand-Reinhold, N.Y., 1961. (Part I outlines the concepts of space, time and motion in Newtonian physics.)

Drake, Stillman, "Galileo's Discovery of Free Fall", *Scientific American 228*, May 1973, pg. 84. (Using documents which had not been employed before, Drake tries to recreate the sequence of ideas, mistakes, and discoveries which led to Galileo's distance-time squared law.)

Galilei, Galileo, *Dialogues Concerning Two New Sciences*, Reprinted by Dover Publications, New York, 1954. (Classic studies of cohesion in materials, and accelerated motion.)

Gibson, Eleanor J., "The Development of Perception as an Adaptive Process", *American Scientist 58*, January-February, 1970, pg. 98. (Perception of events in space develops early, while perception of objects shows evolution and greater dependence on learning.)

Goldhaber, Judith, "New Clues in J. F. K. Assassination Photos", reprinted in *Physics and Man*, Robert Karplus, (ed.) W. A. Benjamin, New York, 1970. (An account of how an eminent physicist, Luis Alverez, concluded various things about the John F. Kennedy assassination by examining movie film frame by frame. An application of some of the ideas discussed early in this chapter.)

Holton, Gerald, "Science and the Deallegorization of Motion", in Kepes' *The Nature and Art of Motion*, G. Braziller, New York, 1965. (Aristotle's view of motion was very general: something like moving into potentiality. This excellent article traces the changes brought about by Galileo's specific, numerical notings of motion.)

Packard, Vance, *A Nation of Strangers*, David McKay, 1972. (The effect of mobility on American culture.)

Ward, Ritchie R., "Stress and the Space Age", Chapter 20 in *The Living Clocks*, Alfred A. Knopf, New York, 1971. (Effects of rapid travel over long periods of time.)

Watson, W. H., *Understanding Physics Today*, Cambridge University Press, 1963 (Euclidean space and Newtonian time have played a dominant role in man's adaptation to his surroundings and are even seen in the logic of our language.)

3-1   A car traveling on a turnpike passes mile markers every 55 seconds. What is the car's average speed in miles per hour?

3-2   A motorist drives at a speed of 30 miles per hour for one mile, then at 60 miles per hour for a mile, then at 30 mph for a mile, then at 60 mph for a mile, etc. What is the average speed? A second motorist drives at a speed of 30 mph for one minute, then at 60 mph for one minute, then at 30 mph for one minute, then at 60 mph for one minute, etc. What is the average speed?

3-3   A man drives a car from Upham to Waynefleet, a distance of 60 kilometers, in 1.5 hours. He travels at a constant speed during the trip. Make plots of position, and velocity, *vs* time during his trip. A second motorist also takes 1.5 hours to make the trip, but she reaches Bath, a city of 40 kilometers from Upham, in 45 minutes and then proceeds on to Waynefleet at slower speed. Make plots of distance and speed *vs* time during her trip.

3-4   Two bicyclists are 10 miles apart, each pedaling toward the other at a speed of 10 miles per hour. A fly starts from one cyclist's nose, flies at 20 miles per hour to the other cyclist's nose, then back, etc. until the two cyclists meet each other. How far does the fly fly?

3-5   A dog makes its way across a field, traveling in a straight line. As it enters the field it is trotting at 1 m/sec. The dog holds this speed for 5 seconds and then it stops suddenly to look at an insect on the ground for 15 seconds. Then the dog takes off again, accelerating at a constant rate for 6 seconds, at which time it has a speed of 5 meters per second. It runs at that speed for 10 seconds and then reaches the other side of the field.

   (a) How wide is the field?

   (b) What is the dog's average speed?

   (c) Make plots of position, and velocity, *vs* time, making clear the differences between the various parts of the dog's motion, and achieving some semblance of quantitative accuracy.

3-6   Some students are in a jet airliner flying at an altitude of 40,000 feet, on a clear day over open farmland. They notice that if they hold their heads still and sight past the edge of the window down at the ground, the plane passes from one edge of a farm to the next in 10 seconds. Later, they learn that the farms are square and contain 1,280 acres. (There are 640 acres in a square mile.) What is the velocity of the plane in miles per hour? What assumptions did you make about the path of the plane?

3-7   A 10-story building has 200 workers on each floor. Figure out how many elevators the building must have if no one is to wait more than two minutes on the main floor during the morning rush period. (You will have to make assumptions regarding the capacity of the elevators, their average vertical speed, the distance between floors, and the time required to load and unload passengers.)

3-8   A motorist goes on a 60-mile trip. She drives the first 15 miles at
an average speed of 30 miles per hour. How fast must she travel
over the last 45 miles in order to average 60 miles per hour?

3-9   An athlete runs the 100-meter dash and a plot is made of his
speed as a function of time as shown below.

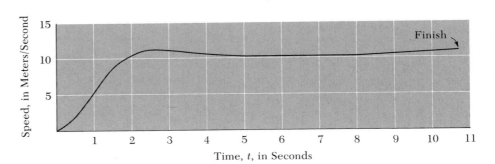

Sketch a graph of the athlete's acceleration as a function of time.
Estimate the average speed over the 100-meter distance.

3-10   The following graphs show plots of position *vs* time for an object
traveling in a straight line. Show, for each part of the motion,
whether the velocity is positive, negative, or zero; and whether
the acceleration is positive, negative, or zero.

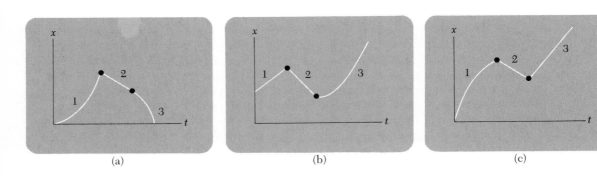

3-11   A body falls freely from rest with a constant acceleration of 9.8
m/sec². (a) Find its speed after falling one second, and; (b) the
distance it falls during the first second. (c) Find the average speed
of fall during the second second, and; (d) how far it falls before
reaching a speed of 9.8 m/sec.

3-12   Suppose a rocket is designed to accelerate uniformly at 3 "g's"
(96.6 ft/sec²) and reach the orbital velocity of 17,800 miles per
hour. How long must the rocket accelerate in order to reach this
velocity?

3-13 A body moving at initially 15 m/sec is given a constant acceleration. In the next two one-second intervals it travels 18 and 24 meters, respectively. Find (a) the acceleration, and; (b) the distance traveled during the third and fourth seconds after the acceleration begins.

3-14 A skier makes her way down a hillside, accelerating uniformly as she goes. At a certain instant her speed is 5 m/sec. Two seconds later her speed is 15 m/sec. What is the skier's acceleration? Another skier travels at a speed of 5 mtr/sec at a particular instant. After traveling an extra 10 meters her speed has increased to 15 m/sec. What is her average acceleration?

3-15 A passenger jet flies at a constant velocity of 660 mph and at an altitude of 35,000 feet. An engine falls off the plane. Neglecting wind resistance, how long does it take the engine to fall to the earth and how far does it land from the point directly below that point at which it fell off? ($g = 9.8$ m/sec.^{2} = 32.2 ft/sec^2).

3-16 An automobile tire is 70 cm. in diameter. Its hubcap is 30 cm. in diameter. One of the wheel nuts works itself loose and rattles around inside the hubcap. Above a certain speed of the car the nut lies against the inside of the hubcap and quits rattling. What is the speed at which this happens? Note: The speed of a point on a wheel increases proportionally to the radius. Assume that the tire does not slip on the pavement.

3-17 A satellite circles the earth at an altitude small in comparison with the earth's radius. How long does it take the satellite to travel once around the earth?

3-18 An astronaut in training is housed in a capsule being rotated in a horizontal circle at the end of a 15 meter boom. How much time is required for the capsule to move around the circle if the centripetal acceleration acting on the astronaut is limited to five "*g*'s"?

3-19 Suppose the pilot of a plane wishes to change direction by 180 degrees, but does not want to undergo more than four "*g*'s" of centripetal acceleration. How long will it take to turn around if with an air speed of 100 m/sec? 300 m/sec?

4

WHY THINGS MOVE

What makes things move? This is a more profound question than simply asking how an object seems to be changing its position. The ways in which ancient, medieval, and modern thinkers have answered the question can give an interesting view into the evolution of Western thought and, in particular, into the birth of modern science.

Aristotelian Physics

The ancient Greek view of the physical world was largely subjective and idealistic. It began with a picture in the mind and speculated outward, beginning with premises and reaching conclusions which might or might not match the perceived world in detail. It was a humanistic, not a naturalistic view.

For Plato (428–347 B.C.), true knowledge was concerned with ideal forms. Since sense data gave knowledge only about specific manifestations of the ideal form, not the forms themselves, experimentation would never match pure speculation as a channel toward truth.

Aristotle (384–322 B.C.), on the other hand, was quite interested in the world of nature and was a careful observer of many things animate and inanimate. Nevertheless, his natural philosophy was far more speculative than experimental. It presented a comprehensive view of the world which was still dominating Western thought almost 2000 years later, and the view was organic.[1]

Every object in Aristotle's world had its appointed place and function, depending on which of the four elements (earth, water, air, and fire) went into its composition. A rock, made mostly of earth and water, had a rightful place close to the center of the earth, which was also the center of the universe. If a rock were somehow carried to a high place, its *natural motion* was to fall straight toward the center of the earth. It moved because of an innate desire and little could be gained by studying in detail how the rock appeared to move. The larger the rock, the greater its desire to reach the earth, thus the faster it *had to* fall. On the other hand, the inclination of objects made primarily of fire and air was to flee the center of the earth, seeking the outer spherical layers where air and fire rightly belonged. Natural motions were thus the result of things trying to get to where they belonged. Aristotle gave his official stamp of approval to the separation of the earth from the heavens by postulating that the planets and stars were made of a fifth substance, "quintessence". The natural motion of objects made out of it was a circular path, not up or down.

Forced motion, such as that of a wagon along a road or an arrow through the air, required the intervention of an outside mover. The greater the push or pull, the greater the velocity. Aristotle's "equation of motion" could be written as

$$v = F/R \hspace{3cm} \text{Equation (4-1)}$$

where F is the external force exerted on the body and R is the resistance to motion offered by friction or the medium through which the object moves. The idea basic to Aristotle's mechanics was that objects not allowed to follow their natural motions would *sit still*. Rest was basic and motion required a cause. This fits most of our everyday observations quite well.

Plato and Aristotle were not the sole contributors to Greek natural philosophy. It is possible now to find in the work of the Stoics and Atomists a great many ideas which anticipate modern science, but the primary Greek influence on science has been Aristotelean. This influence, that is, the way in which his ideas were used and solidified by later admirers, was at least partly reactionary.

(1) *Aristotle's Physics*, translated by Hippocrates G. Apostle, Indiana University Press, Bloomington, 1969.

After the decline of Graeco-Roman culture the seeds of Greek science were kept alive by Moslem scholars. Greek thought was not re-introduced to Western culture to any extent until the discovery and translation of Greek and Arabic treatises in the twelfth and thirteenth centuries. Soon after this, the scholastic philosophers, notably Thomas Aquinas (1225–1274) had reconciled Greek philosophy with Christian theology to produce a theoretical synthesis almost too imposing to question. The Catholic Church of the Middle Ages took in wholeheartedly the Greek confidence in an ordered, rational world, adding to it the concept of God as the final intelligence which had set up the universe and given everything its place and purpose. God was also the unmoved mover, residing in a ninth heaven out beyond the heaven containing the stars. Another part of the Greek inheritance, which we discussed in Chapter 3, was the concept of identity; of being able to distinguish an object from its surroundings. The task of the natural philosopher in medieval times was to show how phenomena fit into the theoretical world picture and explain them on the basis of final cause or function in that world-wide scheme. In general, therefore they did not feel the need to observe phenomena in detail in order to build from *them* a picture based on empirical evidence. On July 4, 1054, an exploding star (supernova) was observed by Japanese and Chinese astronomers (and probably also by cave-dwelling Indians in Western America.) It was bright enough to be seen by day, but was not recorded by a single European skywatcher. Medieval Europeans looked to the church, rather than to the perceived sky itself, to find out about heaven. The climate was therefore right for the acceptance of the rediscovered Aristotelian hierarchical world structure in the thirteenth century. The Church did not forbid experimentation so much as it failed to reward it. It provided a complete and final explanation for *almost* every phenomenon one could hope to observe. To question Aristotle's physics or world model was to question the entire body of Christian theology.

It is too easy now to condemn the medieval church for "holding back science" and generally crushing innovations in thought. It forced people to see the world through stained glass windows, as a marvel of God's handiwork. Nature was seen as a foretaste of heaven, not to be investigated. However, the church did support scholars Roger Bacon (1214–1294), John Duns Scotus (1265–1308), William of Ockham (1300–1349), and others who carried on an almost continuous if subdued critical examination of Aristotle's physics.

Aristotelian physics explained the motion of wagons and heavy furniture very well, but had great difficulty explaining, even qualitatively, the flight of projectiles. For example, how did an arrow keep moving once it left the bow? According to the Aristotelians, "eddys" were somehow set up in the air so that the *air* pushed on the arrow from behind and kept it going. By the 1300's, scholars were wondering what would happen if a string were attached to an arrow. From Equation (4-1), the string, experiencing a smaller resistance, *R*, than the arrow, ought to have a higher velocity than the arrow and precede it through the air. But this was not what was observed. Another problem which arose concerned the possibility of an arrow traveling in a vacuum. Either the arrow would be motionless because the driving force was zero or else it would travel with infinite velocity, because the resistance to motion was zero. Neither of these possibilities seemed valid, so for this and other reasons empty space or vacuum was said not to exist in nature. It might exist out among the planets and stars but a different sort of physics applied there.

It is important to remember that for the most part, Aristotle's physics did fit common sense, and everyday observation, or could be made to. Heavy rocks *do* fall faster than light feathers. If an object falling through the air was observed to speed up as it fell, it could be attributed to increased desire on the part of the object as it neared the earth. Alternatively, we could say that the increasing amount of air above the object as it fell added an increasing forced motion to its natural motion, making it go faster. However, the physical interpretations based on Aristotle's principles were becoming more and more cumbersome. By the end of the fifteenth century, Aristotelian mechanics was being taken far less seriously than the rest of his philosophy. But Western natural philosophy was still speculative rather than experimental.

Empiricism and the Transition to Modern Science

During the fourteenth and fifteenth centuries several trends in natural philosophy began to develop in Western Europe. The first that will be mentioned here is the rise in value assigned to empirical observation in natural philosophy. It has been suggested that the labors of Benedictine monks were instrumental in breaking the barrier between work done by hand and work done with the intellect. In any case, people began investigating the world and the heavens in detail. However, it is important to realize that more than data was required before the Western view of the world could change very much. We have already seen above how observed phenomena can be made to fit an awkward theoretical explanation. An observed fact *by itself* never forced anyone to a conclusion that they did not already have in mind as a possibility. In Barry Stavis' play, "Lamp at Midnight," some of the church fathers refuse to look through Galileo's telescope because it might be a device of the devil, trying to show them things that weren't in the Aristotelian world picture. But Cardinal Bellarmin, a very wise man, looks through the instrument, sees the moons of Jupiter (which Aristotle's heaven had no place for) and says,

> "I have not decided what I have seen. No fact is a pure fact. The aura of attendant consequence can never be severed from the fact itself."

Very slowly, though, empirical evidence assumed prime importance. Hypothesis had to precede the first data collected and interpreted, but it became *the data* in which one placed faith. By the end of the sixteenth century, Johannes Kepler (1571–1630) was able to say "Without sufficient evidence I draw no conclusions."

The second development was careful, formal experimentation, which brings us to Galileo once more. He was a man of many facets and moods. This, plus the transitional age in which he lived, have made him a very popular person to write about, whatever the author's slant happens to be. Even though Galileo was not the first European to experiment, he was the first to persist in the design and refinement of experiments until the observations were conclusive. This is the all-important difference between kitchen science and research. It is just about impossible to conclude *anything* about free fall by merely dropping things and looking. Lacking the timing devices we have now, Galileo was clever enough to dilute and slow down the acceleration by letting the object roll down a ramp. In his early experiments he would time the rolls by letting water run out of a spigot at a controlled rate while the ball rolled, and then weigh the collected water. Only then did he have data which enabled him to infer something about free fall.

It would be wrong to call Galileo the "Father of Empiricism" or any such title. Francis Bacon (1561–1626) might be a better choice for the position because he preached the experimental doctrine more than anyone else,[2] and was the leading critic of Aristotle for not being more experimental. (Whether Bacon did any experiments himself seems doubtful.) Galileo was always guided by a belief in the theoretical formula; the law of motion. His data, be assured, never pointed *exactly* to $x = \frac{1}{2}a(\Delta t)^2$ relationships. Yet he had the conviction that the data would lead to this in an ideal world, free of friction, bumpiness, etc. He was part experimentalist and part theorist. Experimental science, as the Western world has come to know it, is more than experiment.

A third factor in the empirical approach was giving up the grand, teleological explanations of "why", and satisfying oneself with specific description and explanation: "what" and

(2) *The Complete Essays of Francis Bacon*, Washington Square Press, N.Y., 1963.

"how". Francis Bacon said that research into final causes, like a virgin dedicated to God, is barren and produces nothing.

A final factor, one which we will return to, is linked to the one just above. It might be called the mechanical approach. It holds that the cause of an object's motion is not innate, but is external, out where it can be investigated. When the mechanistic point of view reached its peak of popularity early in the nineteenth century, some thought that the whole world could be explained on the basis of interacting particles and forces.

It has been pointed out that the Renaissance and modern science were both born at about the same time; in the hundred years following 1450. Certainly the picture people had of themselves was connected with their curiosity about the world. With the Renaissance, Western Europeans saw themselves as more capable, independent persons. The world in which they lived was no longer just a prelude to heaven. People were free to glorify themselves as well as God and study and question their surroundings. They began to travel and explore. Their painting and music changed, showing more dynamism and sharp observation.[3] Without trying to decide whether science is primarily a Western phenomenon or not, it might be well to pull together our discussion by mentioning Conant's thesis[4] concerning the "birth of science." He says it was the result of the merging of three previously separate human activities:

(1) Theoretical speculation or model building.
(2) Use of deductive logic to eliminate unwarranted ideas.
(3) Orderly experiment to link the theoretical structure to the "real" world.

It is the third of these ingredients whose effects we have just discussed.

Inertia

Having taken a general look at some of the changes which were needed for science to become a serious intellectual activity, we want to look more specifically into the evolution of concepts concerning motion. By the early 1300's, Aristotle's organic view, which attributed an inclination or volition to each moving object, was beginning to die. William of Ockham is best known for his "razor", the principle which says the simpler of two arguments proving the same thing is to be preferred. But he was also interested in mechanics, the science of motion. His theories depersonalized or de-allegorized[5] the motion of objects to the

(3) The background detail in Leonardo da Vinci's *Mona Lisa*, for example, displays his interest in geology.
(4) Conant, J. B., *Science and Common Sense*, Yale University Press, 1951.
(5) Holton, Gerald, "Science and the Deallegorization of Motion", in G. Kepes' *The Nature and Art of Motion*, G. Braziller, New York, 1965.

extent that he saw them moving because of an *impetus* which had been given to them by some initial mover. Impetus was an intermediate agent, still thought to be more in the realm of cause than of effect. With impetus, an object could move without the continual action of an external mover, and it also need not be answering the call to go to some final resting place. By 1340, Jean Buridan, a Parisian scholar, had refined the concept of impetus further, and claimed that it had a numerical value equal to the weight of the moving object times a number that depended on its velocity. The concept of impetus eventually gave way to *inertia*, although nearly three centuries passed before the idea of inertia was fully utilized by scientists. Descartes was one of the first to suggest that *nothing* is required to keep an object moving in a straight line, whether it be on earth or in the heavens. Reasoning like a latter-day Aristotle, he *deduced* the principle of inertia from the general principle that the "motion" of the universe must remain unchanged. Left alone, an object originally moving will keep moving. Galileo's writings show that he had the concept of inertia well in hand. However, Descartes deserves credit for finally stating the *law of inertia* in more or less its present form. (It is usually called "Newton's first law"!) In modern words, Newton's first law says:

> All objects which are not acted upon by a net force remain at rest or in a state of motion at constant speed in a straight line.

We see that *inertia represents "unwillingness" to slow down, as well as speed up, or to turn.* For the first time, it was clearly stated that motion, like rest, can be a fundamental state of being. The basic question in dynamics thus switches from "what causes motion?" to, "what makes objects change their motion?"

Part of Galileo's conceptual genius lay in his ability to visualize friction as a force which could be singled out from the rest of the influences acting on a body and, in principle, eliminated. For instance, if a wagon were without friction, you could get it going with a horse, then unhitch the horse, and the wagon would keep on going forever! The idea of a completely isolated body, experiencing no interactions with the rest of the world, is purely theoretical. (We believe, for example, that all objects exert gravitational forces on each other.) Note, however, that Newton's first law simply says that the *net* force on an object must be zero for a constant velocity to result. There are ways of balancing out pushes and pulls so that their net effect on an object is zero. For example, the downward directed weight of an object, can be balanced by a series of air jets blowing up from beneath it. The object will ride on a thin layer of air, and will experience very little friction when it moves. The hovercraft ferries now in service on the English Channel utilize the same principle, by using downward-blowing fans to form an air bubble to ride on. A rubber skirt around the bottom of the craft helps contain the air.

Figure 4-1
Hovercraft crossing the English Channel. (Courtesy of British Railways Board.)

Mass, the Measure of Inertia

Now what about changing motion? Experience shows that heavy things are harder to get moving than light things. A steel ball is harder to throw than a tennis ball. Likewise, a truck is more difficult to slow down than a small car How can we establish a *measure* of the difficulty of speeding up or slowing down an object? As always, we will measure this *relative to the inertia of some other object.* A simple experiment can be done in the laboratory which, in principle, could be applied to any object you might choose. Let a car, A, be placed on a nearly frictionless "air trough", where it is supported on a layer of air. Give it a velocity v_A to the right. Put another car, B, on the track and give it a velocity v_B to the left:

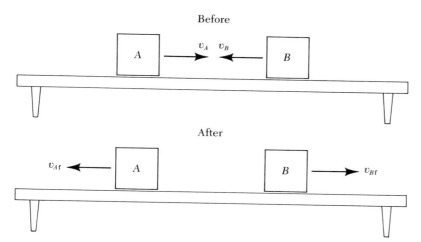

Figure 4-2
Two objects colliding on a frictionless air trough: (a) before (b) after.

Let the two cars collide and rebound. Measure their velocities, v_{Af} and v_{Bf} after the collision. Then their respective changes in velocity, $\Delta v_A = v_{Af} - v_A$ and $\Delta v_B = v_{Bf} - v_B$, can be determined. Since the greater the velocity change an object undergoes, the smaller will be its inertia, we might try the simplest possible way of defining the relative inertia of any pair of objects, A and B. If in a collision two objects undergo velocity changes Δv_A and Δv_B then we can define their relative inertial *masses* with the equation

$$\frac{m_A}{m_B} = -\frac{\Delta v_B}{\Delta v_A}. \hspace{3cm} \text{Equation (4-2)}$$

If Δv_A is twice as great as Δv_B, then the mass of A is half the mass of B. The minus sign simply says that if the velocity of one object is decreased, the velocity of the other increases. To check the validity of this simplest possible relation between masses and velocity changes, we could carry out collisions between various pairs of objects. These objects could be of various shapes, and made of different materials, etc. When these experiments are carried out, it is observed that when an object interacts with various other objects having different masses, its changes in velocity are those which would result only if it had a constant mass. That is, the chemical composition of the objects in collision and other possible variables have no effect on the observed results. A given object possesses a given sluggishness no matter what the experimental conditions:—well, almost. It will be seen later that the mass of an object does depend on its velocity when it travels at speeds approaching that of light. For this reason, mass is *not* defined as the amount of matter in an object. The inertia of a given amount of matter can change. *Mass is simply a numerical statement of the inertia (relative difficulty of changing the velocity) of an object.* The operational basis for the statement is an experiment similar to the one summarized in Equation (4-2).

Two objects needn't bang into each other for their masses to be compared. For example, mass comparisons can be made at a distance by measuring the relative accelerations with which the objects fall toward each other as a result of their mutual gravitational attraction.

It has little value to say that an object has a mass equal to 1.66 times the mass of an object "B" in such and such a physics laboratory. In order to assign a numerical value to the mass of an object, not just a mass ratio, the mass comparison must be made with an agreed-upon standard mass. To obtain a communicable value of the mass of an object, we let the object of unknown mass m collide with a standard kilogram, whose mass we *define* to be one. Suppose that the kilogram is initially at rest on an air trough and that the object with unknown mass approaches it from the left with a velocity of 20 cm/sec:

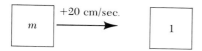

Suppose further that after the collision the object having the unknown mass rebounds with a speed of 10 cm/sec to the left and
the kilogram moves to the right with a speed of 10 cm/sec:

The changes in velocity are:

$$\Delta v_m = -10 - 20 = -30 \text{ cm/sec},$$

$$\Delta v_1 = 10 - 0 = 10 \text{ cm/sec}.$$

Then, from Equation (4-2),

$$\frac{m}{1} = -\frac{\Delta v_1}{\Delta v_m} = -\frac{10 \text{ cm/sec}}{-30 \text{ cm/sec}} = .333$$

So $m = (0.333)1 = 0.333$ kg $= 333$ grams.

Where does inertia arise? Why is matter sluggish? This remains one of the problems in physics which is still not fully
solved. Ernst Mach, (1836–1916) a mathematician, physicist, and
early positivist philosopher, suggested that inertia may not be a
property of the object itself but rather of the whole universe.
Inertia shows up when we try to accelerate an object because in
so doing we are altering the relations between the object and all
the rest of the mass in the universe. This Eastern-sounding
principle has met with considerable criticism since it was proposed in 1893, although it was incorporated by Albert Einstein
into his revolutionary, and highly successful theories of relativity. It is interesting to note that such recent developments in
physics, wherein the interrelatedness of the universe is reappearing, would seem to reiterate Greek and medieval world
views. However, they are based on interaction, not upon Master
Plans. Furthermore, the new interrelatedness follows from numerical analysis, not pure speculation. Finally, Mach's principle
removes almost the last vestige of the Greek organic view of
objects. Not even inertia is innate. It results from external
influences.

Newton's Second Law of Motion

Now suppose we pull steadily on an object whose mass is known
from experiments, such as the ones previously described. It is

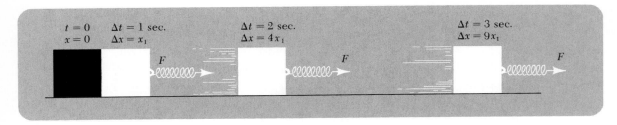

$t = 0$ $\Delta t = 1$ sec.
$x = 0$ $\Delta x = x_1$

$\Delta t = 2$ sec.
$\Delta x = 4x_1$

$\Delta t = 3$ sec.
$\Delta x = 9x_1$

F F F

Figure 4-3
Change of position when a constant force is applied.

not easy to exert a steady pull, but one way to do it is to attach a spring to the object and pull it in such a way as to always keep the stretched out length of the spring the same. Assume that the object is on a frictionless surface, so that the spring pull is the only unbalanced push or pull on the object. What happens?

The object moves, of course. Furthermore, it will be found that the distance the object moves, starting from rest, increases in proportion to the square of the time, as long as we keep pulling steadily. From Chapter 3 we know this means that the acceleration is constant. Thus, the continuous application of a *constant force to an object produces a constant acceleration*, not a constant velocity, as Aristotle thought.

Now suppose we apply to the object forces which increase in regular steps. For example, after attaching one end of the spring to a support we could hang 1 kg, 2 kg, 3 kg, etc. on the other end of the spring and record the amount that it stretches: Δs_1, Δs_2, Δs_3, etc. (Figure 4-4.) Then we could hook one end of the spring onto the object used before and, by pulling horizontally on the other end of the spring, make it stretch by Δs_1, Δs_2, Δs_3, etc. (Figure 4-5.) By measuring the time Δt required for the object to move a given distance, Δx, in each case, we can calculate the respective accelerations:

$$a = \frac{2(\Delta x)}{(\Delta t)^2}$$

When we do this we find that the *acceleration is proportional to the force exerted*.

Finally, if we change the object's total mass by adding a known increment of mass each time, and exert the same pull on the object each time, *the acceleration of the object will be inversely proportional to the mass* (Figure 4-6). We can summarize all of these observations using one simple relationship.

Force is proportional to mass times acceleration.

This general statement summarizes the following experimental findings.

(a) For a given mass, doubling the force will double the acceleration.
(b) For a given acceleration, double the mass requires double the force.
(c) For a given force, doubling the mass will reduce the acceleration by one-half.

The general statement above is one form of *Newton's second law*. It has become such a common part of the vocabulary of physics that physicists seldom stop to wonder and be surprised that the relationship between mass, acceleration, and force should be this simple. Note that the relation constitutes a definition of force, not a definition of mass. We have independent means for measuring mass and acceleration, so force, by Newton's scheme, is defined in terms of them. Of course, one very soon gets the idea that painting a numerical scale next to the spring will let it be used more handily to directly measure applied forces. But before this, the stretch of the spring must be translated into force units by making it accelerate objects of known mass known distances.

Mass, acceleration, and force units can be chosen so that the proportionality above becomes an equality:

$$F_{net} = ma$$

Equation (4-3)

For example, if m is measured in kilograms and a in meters/sec², force is said to be measured in "newtons" (N), a unit coined in the late 1920's to fill out the metric system.

A newton is equivalent to one kg·m/(sec)². It is the force required to impart an acceleration of one m/(sec)² to one kg of mass.

A nickel lying on a table exerts a downward force of about 1/20th newton (5×10^{-2} N). If masses are measured in grams and accelerations in cm/sec², the appropriate force unit is called the "dyne". Since the same nickel exerts a downward force of 5000 dynes, the dyne constitutes quite a small unit of force. Can you show that 1 dyne = 10^{-5} N?

When an object accelerates, we say that a force acts *on it*, not that it "has" or "exerts" a force. *An object is really inert.* It will accelerate in whatever direction you push it, just as long as you really exert a net force on it. "The object" has no desire or volition, and if a net force acts on it, we don't care by what means the external force is being applied. A living being is partly "object". If a person with a mass of 60 kg is placed on a frictionless cart and accelerated, the response of the cart will be exactly the same as that of a cart holding 60 kg of iron. (Assuming, of course, that the person lies still.)

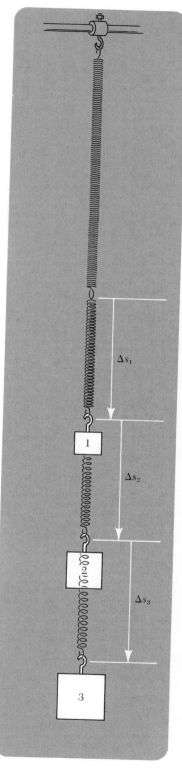

Figure 4-4
Determining the stretch of a spring when various forces are exerted on it.

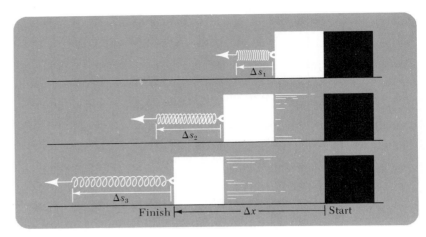

Figure 4-5
Distances moved by an object to which various forces are applied.

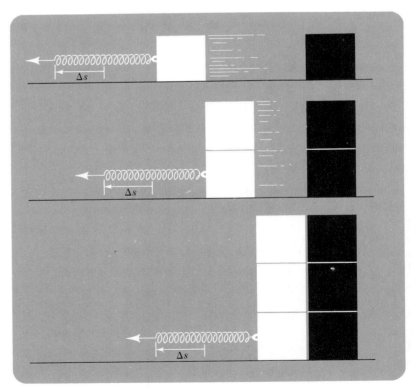

Figure 4-6
Distances various objects move when the same force is applied to each.

Applications of Newton's Laws of Motion

With Newton's laws you can predict the motion of objects under many different conditions. For success in using the laws, you need to practice visualizing cars, people, and planets as simple objects; analogous to boxes or balls, and then account for all the real forces which act on the object.

(a) Suppose a 10 kg block is at rest on a horizontal frictionless table. What force is required to give the block a velocity of 4 cm/sec in 2 seconds? First, the acceleration is

$$a = \frac{\Delta v}{\Delta t} = \frac{4 \text{ cm/sec}}{2 \text{ sec}}$$

$F = ma$

$\qquad = (10 \text{ kg}) \ (2 \text{ cm/sec}^2)$

$\qquad = (10 \text{ kg}) \ (2 \times 10^{-2} \ m/sec^2)$

$\qquad = 0.2 \text{ N.}$

(b) Now suppose a billiard ball, having a mass of 200 grams, moves from left to right across a billiard table with a constant velocity of 10 cm/sec. It bounces off a cushion with a speed of 9 cm/sec after being in contact with the cushion for 0.1 second. What is the average force exerted on the ball by the cushion?

$$\frac{\Delta v}{\Delta t} = \frac{v_f - v_i}{\Delta t}$$

$$\qquad = \frac{-9 \text{ cm/sec} - 10 \text{ cm/sec}}{0.1 \text{ sec}}$$

$$\qquad = -190 \text{ cm/sec}^2$$

Then $a_{avg} = -190 \text{ cm/sec}^2$

$F = ma$

$\qquad = (200 \text{ grams}) \ (-190 \text{ cm/sec}^2)$

$\qquad = -3.8 \times 10^4 \text{ dynes}$

$\qquad = 0.38 \text{ N.}$

By convention velocities and accelerations are defined as being directed toward the right when positive, and toward the left when negative. The minus sign in the answer therefore tells us that the force exerted *by* the cushion *on* the ball is toward the left. The force exerted by the cushion during the 0.1 second is not constant. It starts at a low value and builds up to a peak until the cushion is fully indented by the ball. You might have seen a high-speed photograph (such as Fig. 4-7) of a golf club hitting a golf ball and been surprised at the large amount of deformation of the ball at a time close to the peak force on the ball. The actual force exerted by either a golf club or a billiard table cushion would be a complex function of time. But the average force would produce an effect

$$(F_{net})_{avg} = m \frac{\Delta v}{\Delta t}$$

or

$$(F_{net})_{avg} (\Delta t) = m \Delta v \qquad \text{Equation (4-4)}$$

so the resulting velocity change, Δv is given by

$$\Delta v = \frac{(F_{net})_{avg} (\Delta t)}{m} \qquad \text{Equation (4-5)}$$

Figure 4-7
Golf club hitting a ball. (Courtesy of H. E. Edgerton.)

If follows from Equation (4-4) that the effectiveness of a given force depends on how long it has a chance to act. What matters is the *product* of the average force times the time over which it acts. This product is called the *impulse* of a force.[6]

impulse of a force = (avg. force) (time of action)

It is possible to break your toe by kicking a door in order to close it, or you can close it easily by pushing on it with one finger, for a longer time. In the late 1950's Parry O'Brien, an Olympic shotput champion from California, found that by starting his push on the shot from the very back of the putting circle, he was able to exert his accelerating force over a longer time and distance. This technique enabled him to impart a greater launch velocity to the shot. We know from page 51 that this results in a longer flight path. Unfortunately for him, other, stronger, men began using his technique and O'Brien's world record has now been out-distanced by about eight feet.

Some of the engines that have been proposed for propelling space vehicles on interplanetary flights are not impressive in terms of the force that they are capable of exerting. (Of course, huge engines are required during launching, to give the rocket enough velocity to lift it from the earth, and to push it through the air.) Once the vehicle has attained enough velocity to escape from the earth, and is outside the earth's atmosphere, a fairly small engine will keep it accelerating. Since these smaller engines use little fuel and can keep on exerting their force over a period of years, they will accelerate the space ship to stupendous velocities. (Remember that Δv equals $a\Delta t$, and in this instance Δt can be very large.) Certain space engines called *ion engines*

(6) For the sake of simplicity, science history tends to overlook the contributions of various people and attribute ideas to single individuals; sometimes even to the wrong individual. It's already been pointed out that "Newton's first law" "belongs" to Descartes as much as to anyone else. What is not often realized is that credit for the statement of "Newton's second law" in the form that it is usually written today belongs to Leonhard Euler (1707–1783), a Swiss mathematical physicist. Newton expressed the relation in terms of total impulse [Equation (4-4)], not its rate of change (force). "Newton's second law" was first expressed in its present form [Equation 4-3] by Euler 25 years after Newton's death!

do not burn fuel in the conventional way. Instead, they shoot a beam of tiny electrically charged particles in a backward direction to move the rocket forward.[7] The acceleration itself would be too small to be noticed by any passengers, but if these motors acted for a long enough period of time, they could build up velocities approaching that of light.

Weight

The force which we learn to live with first and which affects us more than any other, except those which hold our bodies together, is weight. *Weight is the name we give to the force which the earth's gravitational pull exerts on objects around it.* If we were moondwellers, then it would be the force exerted by the moon's gravitational pull. An interesting fact is that the weight of an object is proportional to its inertial mass. If a collision shows that one object has twice the mass of another, then it will weigh twice as much. This provides an easy but less basic method for comparing the masses of two objects. You just weigh them. The method is less basic because it depends upon a third instrument, a gravitational field, to make the comparison. It also means the mass comparison must be made at the same place in space, where the given gravitational field has the same strength. There are places in space where an object could be said to be weightless; for example, at a point between the sun and the earth and the moon where the gravitational pull of the earth and moon just balances out the pull of the sun. If the local pull of gravity is zero, then the object is weightless. *But its mass remains the same*: it is just as hard to start or stop as ever. Thus, mass is always with an object whereas *weight* is just the unbalanced net force which acts on an object when it happens to be in a gravitational field. At the surface of the earth a kilogram weighs 9.8 newtons. A gram weighs 980 dynes. More generally, we can say:

weight $= mg$ Equation (4-6)

where m is the mass, and g the gravitational pull or gravitational force per unit mass.

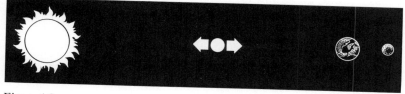

Figure 4-8
Schematic diagram of an object located at a point where the gravitational pull of the earth and the moon balances that of the sun.

(7) Giannini, Gabriel, "Electrical Propulsion in Space," *Scientific American*, *204* March, 1961, pg. 57.

If the only force acting on an object is weight, then it has to accelerate downward, falling freely. What will its acceleration be? By Newton's second law,

$$a = \frac{F_{net}}{m}$$

$$= \frac{mg}{m}$$

$$= g,$$

which is −9.8 m/sec² or −980 cm/sec² at the earth's surface. (One m/sec² is, of course, equivalent to one N/kg.) The acceleration of freely falling objects can now be seen as a result of the gravitational pull of the earth on them.

The earth does *not* pull down on all things with the same force. It exerts a force on objects which is proportional to their mass. In the special case where weight is the *only* force acting, then the acceleration is the same for all objects, regardless of their mass.

Un-free Fall

When objects fall in air, they must plow their way through the air. This is the same as saying that the air exerts a backward drag force on objects which travel through it. Experiment has shown that this force depends only upon the size and shape of the object and its velocity relative to the air. The drag force itself is independent of the weight of the object. The main point is that when real things fall through air, weight is not the only force acting, so that the net force—thus the acceleration—will be less than in a vacuum. Free fall is an idealization which real fall in air approximates when the velocity of fall is not more than a few m/sec (for a steel ball), or a few cm/sec (for a cork ball.)

To illustrate, let us compare the fall of two such small balls having the same shape and diameter and each falling at a speed such that the air drag is 400 dynes. Assume that the cork ball has a mass of one gram and the steel ball a mass of 20 grams. In the absence of air both would have a downward acceleration of 980 cm/sec².

In air, the steel ball experiences an acceleration

$$a = \frac{F_{net}}{m_{steel}} = \frac{-m_{steel}\, g + 400}{m_{steel}} = \frac{-(20 \times 980) + 400 \text{ dynes}}{20 \text{ grams}}$$

$$= \frac{-19{,}600 + 400}{20} = -\frac{19{,}200}{20} \frac{\text{dynes}}{\text{gram}} = -960 \text{ cm/sec}^2$$

Airdrag

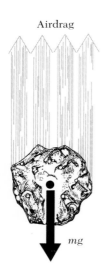

mg

Figure 4-9
An object falling in air is acted upon by two forces: its weight, acting down, and air drag, acting up.

On the other hand, the cork ball experiences an acceleration

$$a = \frac{-(m_{cork})\ g + 400}{m_{cork}} = \frac{-980 + 400}{1}\quad \frac{-580\ \text{dynes}}{\text{gram}}$$

$$= -580\ \text{cm/sec}^2$$

The steel ball hits the ground first.

The air drag increases rapidly with the velocity of a falling object. In time the drag rises to just balance out the downward acting weight force, so that the net force on the object becomes zero. Does this mean that the object stops? By no means! By applying Newton's first law, the motion is seen to continue with a constant velocity. We say that the object has reached its *terminal velocity*. The purpose of a parachute is to reduce an object's terminal velocity. It does this by exerting a large air drag, a drag sufficient to counteract the object's weight at a smaller terminal velocity. Therefore, a person attached to the parachute can land safely. It is interesting to note that physical principles are also applied in the landing process. The safest way to land is with the knees bent. so that the deceleration force is absorbed over a longer period of time. The longer a parachutist takes to stop, the lower the average deceleration force will be. The chance of breaking bones is reduced in proportion to the reduction in deceleration.

Coming to Screeching Halts

Automotive engineers have recently begun to wonder if there is some ultimate speed above which a car might strike another car or a brick wall with no hope of its occupants surviving. People have been killed in accidents where the speed of the car was less than 20 miles per hour, but we are considering the ideal case where the passenger is well seat-belted and in good physical condition. Suppose a car is designed with a two meter projection on the front which can collapse smoothly and completely (like an accordion) over that two meter distance. In this way, the occupants would be decelerated over a 2 meter longer distance than previously possible. From Equation (3-8) the allowable initial velocity, v_0, upon hitting the obstacle would be given by $v_0 = \sqrt{-2a\Delta x}$. (The final velocity, v, in Equation (3-8) is zero, Δx is the distance over which the deceleration occurs, and a is the constant deceleration.) The ultimate speed then depends upon the deceleration which the occupant is able to survive. This is not an easy thing to predict. Astronauts have survived accelerations of eight g's (8×9.8 m/sec^2) for long periods of time during testing in large centrifuges.[8] This is in the "eyeballs out" orien-

(8) Rogers, Terence A., "The Physiological Effects of Acceleration", *Scientific American*, 206 Feb., 1962, pg. 60.

tation, the one for a person facing forward and being stopped. Much higher decelerations have been tolerated by test pilots over shorter periods of time. Men in specially designed "g suits" that have lacings to hold in the viscera (which, of course tend to keep on going) have survived decelerations of 25 g's on rocket sleds which are stopped suddenly. For example; 120 miles per hour to zero in 19 feet! If our car's passengers' heads were held in harnesses; if no loose objects in the car were to become projectiles during the crash; and, if they weren't caught in some metal which was not supposed to fold but did, they *might* survive a head-on collision at a speed

$$v_o = \sqrt{2a\Delta x}$$
$$= \sqrt{2[25(9.8 \text{ m/sec}^2)] \ (2\text{m})}$$
$$= \sqrt{980 \text{ m}^2/\text{sec}^2}$$
$$= 31.3 \text{ m/sec} = 70 \text{ mph}$$

The trouble with most automobile crashes is that the deceleration of the occupants occurs over a much smaller distance and

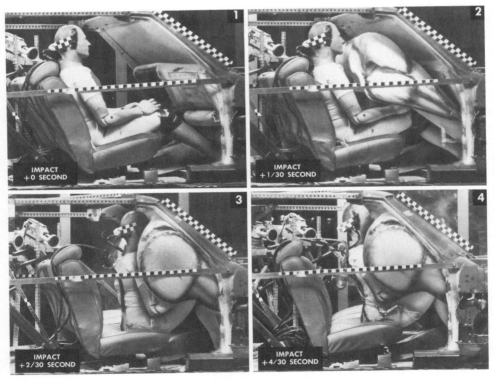

Figure 4-10
Opening of an air bag during an automobile crash. (Courtesy of Fisher Body Division of General Motors.)

time interval. That is, the occupants keep moving until they are stopped *very* suddenly by the dashboard or some other inflexible part of the car. Several automobile manufacturers are experimenting with an expandable air bag which inflates quickly if the car crashes, and acts as a cushion to decelerate the occupant down slowly, over a longer distance. One great difficulty with the concept is that a bag which inflates by mistake is pretty dangerous, tqo. Thus far, the air bags have not been sufficiently dependable in their performance to warrant mass production. Some experts[9] feel that a dashboard filled with collapsible tubes may be a better way to protect the participants in automobile crashes.

Life's Ups and Downs

Newton's second law is also helpful in understanding phenomena that take place in elevators as they start down or up. The same sensations occur to the occupants of airplanes at the beginning or end of a dive. Human beings are not normally in a state of free fall and they dislike anything approaching that state. Normally our bodies are unaccelerated, and the muscles supporting our viscera must exert upward forces which just balance the weight of the organ—stomach or whatever—that is being held up. The muscular force, F, upward just equals the weight, mg, acting down (Figure 4-11). Now suppose the person suffers a downward acceleration. He or she might be riding in an airplane which suddenly begins to lose altitude. The force exerted by the muscles, F, is now less than before, and the change can make the person very uncomfortable. This is partly physiological, but largely psychological, since we learn early that downward acceleration is usually followed by a sudden and painful stop.[10]

The change in muscular force mentioned above can be verified by standing on a bathroom scale in an elevator which is accelerating up or down. During downward accelerations the scale will read less, as in Figure 4-11b. And when the elevator starts up, the scale reading (F) will be higher. To produce an upward acceleration requires an extra force. Whether the force is supplied by a scale from below or a rope from above, it must increase. When the elevator reaches a constant velocity, up or down, then the force exerted by the scale is exactly equal to your weight.

(9) Perrone, Nicholas, *Technology Review*, 73 May, 1971.

(10) Prolonged free fall, or life in places where weight is less than at the earth's surface does have a physiological effect on humans. There seems to be a decrease in the calcium content of bone in humans who have spent time in space. Evidently, relieving a bone of its normal compressive force changes its composition. Astronauts in the later Apollo flights to the moon were fed a diet high in calcium content to compensate for this.

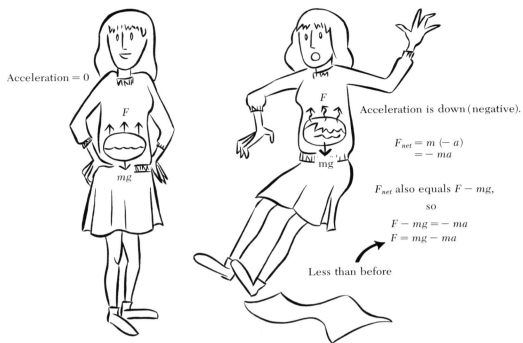

Acceleration = 0

F

mg

Acceleration is down (negative).

$$F_{net} = m(-a)$$
$$= -ma$$

F_{net} also equals $F - mg$,

so

$$F - mg = -ma$$
$$F = mg - ma$$

Less than before

F = Force exerted by muscles holding up stomach.

mg = Weight of stomach.

$F_{net} = F - mg$

a = 0, so $F_{net} = 0$

Thus:

$F - mg = 0$

$F = mg$

Figure 4-11a
Forces acting on a person's stomach under normal conditions.

Figure 4-11b
Forces acting on the stomach while a person is accelerated downward.

A problem: Suppose someone is standing on a bathroom scale. He crouches down, lowering his waist by 30 cm (about a foot.) Then he jumps up, rising to a height, h, such that his waist is 30 cm higher than when he is standing normally. What does the scale read while he is jumping?

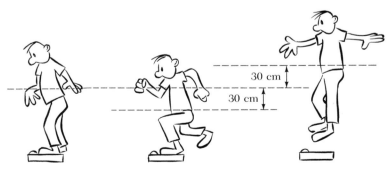

30 cm

30 cm

Figure 4-12 *Bathroom athlete.*

First, if he rises 30 cm he must have jumped with a vertical velocity

$$v = \sqrt{2gh}$$

$$= \sqrt{2(9.8 \text{ m/sec}^2) \ (0.3 \text{ m})} = \sqrt{5.88 \text{ m}^2/\text{sec}^2}$$

$$= 2.42 \text{ m/sec}.$$

To attain this velocity during an upward spring of 30 cm required an acceleration

$$a = v^2/2(.3 \text{ m})$$

$$= \frac{5.88 \text{ m}^2/\text{sec}^2}{.6 \text{ m}} = 9.8 \text{ m/sec}^2,$$

where we assume that he manages to accelerate smoothly up through the 30 cm. He accelerates upward at 9.8 m/sec², so the net force acting on him must be

$$F_{net} = m(9.8 \text{ m/sec}^2).$$

But at the same time, $F_{net} = F - mg$. F is the upward force exerted by the scale and $mg = m(9.8 \text{ m/sec}^2)$ represents his weight, a force acting down. Then

$$F - m(9.8) = m(9.8)$$

$$F = 2 \ m(9.8).$$

Whatever the man's weight is, the scale reads twice that amount.

Friction

A force that affects our everyday lives as much as does weight, is *friction*. Friction is the force which must be overcome in order to make two surfaces slide over one another. Even rolling friction is largely reducible to small-scale sliding. Friction can be annoying either by its *presence*; as in stuck zippers, bicycle tires with insufficient air in them, jars that require a wrench in order to open them, and keys that jam in locks. A lack of friction can be equally annoying; as in slipping on icy sidewalks, having books slide out from under one's arm, and blouses and shirts not staying tucked in. In a word, we learn to live with and expect a certain amount of friction. When friction is higher or lower than the expected amount, trouble can develop. In many cases the friction between two sliding surfaces does not depend upon their sliding speed, once they are started. Furthermore, the frictional force between two surfaces, that is, the force required to just keep them

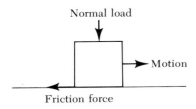

Normal load

Motion

Friction force

sliding, is often a constant fraction of the force holding the surfaces against each other. (Keep in mind that this is a rough, empirical relationship, not in the same class with $F = ma$.) This fraction is called the *coefficient of friction*, μ:

$$F_{fric} = \mu \text{ (normal load)}$$

$$\mu = \frac{F_{fric}}{\text{normal load}}$$

For many pairs of sliding surfaces, μ is between 0.2 and 0.6. Teflon has a much lower coefficient of friction than this when sliding against other surfaces, and for this reason it is now widely used on sliding surfaces.

For a problem involving friction, visualize a 15 kg block sitting on a rough surface. If we push or pull horizontally on the block with a force of 40 newtons, how far will it move in three seconds? We draw an arrow representing the frictional force at the bottom of the block because that's where the friction acts. To find the frictional force we multiply the normal force by the coefficient of friction. And what is the normal force? Just the weight of the block sitting on the surface. Therefore the frictional force

$$F_{fric} = \mu m g$$
$$= -0.2 \ (15 \text{ kgm}) \ (9.8 \text{ m/sec}^2)$$
$$= -29.4 \text{ N}.$$

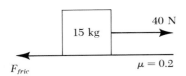

15 kg

40 N

$\mu = 0.2$

F_{fric}

Figure 4-13
A diagram of the forces acting on a 15 kg object being dragged to the right over a rough floor.

(It is negative because it has been chosen as being directed toward the left in the diagram.) To apply Newton's second law, we must first find the net force.

$$F_{net} = F - F_{fric} = 40 - 29.4 = 10.6 \text{ N}$$

Thus

$$10.6 \text{ N} = ma$$

$$a = \frac{10.6 \text{ N}}{m}$$

$$= \frac{10.6 \text{ N}}{15 \text{ kg}} = 0.707 \text{ m/sec}^2.$$

For constant acceleration,

$$\Delta x = \tfrac{1}{2} a (\Delta t)^2$$
$$= \tfrac{1}{2} (0.707) \, (3)^2 = 3.18 \ m.$$

Suppose we push to the right on the block with a force less than 29.4 newtons. The block will sit still, with the frictional resistance just counteracting the push. However, the frictional force cannot *exceed* 29.4 newtons. If the block is pushed harder than this it will accelerate. If the force on the block diminishes back to exactly 29.4 newtons again the block will cease accelerating and will travel at a constant speed. After all, the net force on it has now become zero. If the applied force drops below 29.4 newtons, the net force will be directed toward the left, since, during *sliding* the frictional force remains at 29.4 newtons regardless of sideways pushes. The block will decelerate and come to rest unless the force of the push is increased to 29.4 newtons again.

Forces as Vectors

Forces, just like acceleration and velocity, can be represented by vectors. That means that the effect of a force—or a set of forces—depends not only on their magnitude but also on the direction in which they act. The expression $F_{net} = ma$ ought to be written $\mathbf{F}_{net} = m\mathbf{a}$ to remind us that the net force and the acceleration are both vectorial, and that they are in the same direction. *The acceleration of an object is in the direction of the net applied force.*

Suppose two people are trying to get a piano off a frictionless platform. The piano has a mass of 200 kg. One person pushes "east" with a force of 300 N. Another person pulls "south" with a force of 200 N on a rope tied to the piano. What happens to the piano?

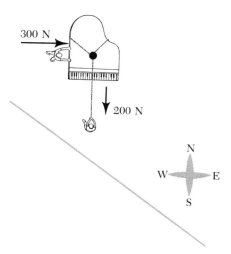

Figure 4-14
Two forces acting on a piano.

First, the net force acting on the piano is the vector sum of the two applied forces. (For information about vector sums, see Appendix G.)

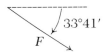

It is a force with an eastward directed component of 300 N and a southward directed component of 200 N. This net force slants at angle θ to the south of east. θ is the angle whose tangent is 200/300, or 0.666. From the table of trigonometric functions in Appendix F, θ must lie between 30° and 35°. (It is actually 33°41', to the nearest minute.) The magnitude (or strength) of the net force can be found by using Pythagoras' theorem:

$$\text{(magnitude of } F_{net})^2 = (300 \text{ newtons})^2 + (200 \text{ newtons})^2$$

$$= 130{,}000 \text{ (newtons)}^2$$

$$\text{magnitude of } F_{net} = \sqrt{130{,}000} \text{ N}$$

$$= 3.6 \times 10^2 \text{ N} = 360 \text{ N}$$

Therefore, F_{net} is a force of 360 N which slants to the south at an angle 33°41' below east:

$$\overline{} \diagdown \diagup 33°41'$$
$$F$$

The acceleration of the piano will be

$$a = \frac{F_{net}}{m} = \frac{360 \text{ N}}{200 \text{ kg}} = 1.8 \text{ m/sec}^2.$$

in a direction 33°41' south of east.

Suppose now that you want to make something *stand still.* That is, you want to apply a set of forces which will just balance each other out. Let's imagine that a girl goes out on a river in a rowboat. She loses her oars and soon after realizes that she is drifting downstream toward a waterfall. She sees people standing on both banks of the river and quickly finds two ropes in the bottom of the boat. One she throws to a child on the near bank and the other she throws to our hero, Tom Terrific, on the far bank. About this time, everyone realizes that the rope on which the child is pulling is really fishline and that a tension of more than 200 N will break it. She ties her two rope ends to the back of her boat so that her hands are free to do a few quick slide rule calculations. She wants to figure where TT should stand and how

hard he should pull. If the child pulls in a direction at right angles to the bank, what must Tom do to keep the boat motionless, assuming that the rushing water exerts a force of 450 N on the boat? First she draws a diagram on the flyleaf of a book she has with her. She wants to find out what T and θ are. To keep the boat from moving, the forces on the boat must add to zero upstream and downstream, *and also from one bank to the other.* If she lets the x axis be along the direction of water flow and the y axis perpendicular to it, then she can set about making the x and the y forces (or, the force components of T in the x and y directions) add to zero.

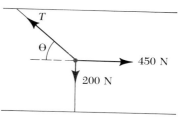

Figure 4-15
Forces acting on the boat as it is being held stationary by two ropes plus water drag. Tom must pull with a force having magnitude T in a direction making an angle θ with the riverbank.

x components:

$$-T_x + 450 \text{ N} = 0$$

y components:

$$T_y - 200 \text{ N} = 0$$

$+$ means up, or to the right (downstream).
$-$ means down, or to the left (upstream).

The 2 components of T are then

$$T_x = 450 \text{ N} \qquad \text{and} \qquad T_y = 200 \text{ N}$$

From Pythagoras' theorem, the magnitude of T is

$$T = \sqrt{(450 \text{ N})^2 + (200 \text{ N})^2} = \sqrt{242{,}500 \text{ N}^2}$$

$$= 492 \text{ N}$$

The direction can be found since we know

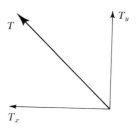

Figure 4-16
The force T has a component T_x upstream and a component T_y across the stream.

$$\tan \theta = \frac{T_y}{T_x} = \frac{200 \text{ N}}{450 \text{ N}} = 0.445$$

On the back of her slide rule, the girl finds θ to be about 24 degrees.

Tom adjusts his pull on the rope accordingly, and pretty soon the girl climbs out of the boat and walks ashore.

Some Force-ful Comments

We have seen that the word "force" is quite narrowly defined in the physicist's description of the world. Given the pushes and pulls exerted by the external world on an object, we can add them together as vectors to find their resultant effect; that is, the acceleration. The tiniest push will move the largest object if the tiny push is really a *net* force. In most everyday experiences, frictional forces make large things stand still until a force sufficient to overcome them is applied.

But, if the John Hancock tower were supported on a frictionless airpad, a single person could get it moving by just leaning against it for a couple of minutes. (Actually, the building would begin to move right away. It would just take awhile for the movement to be noticeable. This is in marked contrast to pushing on things which are locked in place due to friction. In that case nothing moves until the push exceeds the frictional force.)

The paradox of the "irresistible force" encountering an "immovable object" is not a part of the physicist's lore, but it enjoys an amazing vitality in the world of non-scientific thought. Neither an irresistible force nor an immovable object has yet been observed, and to the physicist, speculation as to whether the force or the object would win is fruitless. However, the following information could be added to the debate: large objects are not as immovable as they appear to be. When an outfielder crashes off the wall while chasing a fly ball, the wall gives a little, then straightens out again, just like a cushion on a billiard table, or a trampoline turned on its side.

Mechanism and Determinism

Although the principles which Galileo, Newton, and others have enunciated were put to fairly humble use in this chapter, they constitute part of a magnificent system for predicting the future positions of objects (such as planets) once forces, initial positions, and velocities are given. The scheme was so successful that in the late eighteenth century, Pierre Simon Laplace, the French mathematician and astronomer, claimed that if he knew just the positions and velocities of every object in the universe, he could predict exactly the future of the universe. One could well ask how he could ever get all that information, but the principle remains: *mechanistic determinism*. The future is determined by the present, and so is the past, for that matter. Once you have an equation of motion, such as Equation (3-7), you can insert negative, as well as positive, values of time in order to find where the object *had* to be in the past as well as where it *has* to be in the future. Thus, in *"Newtonian" determinism*, past and future are part of one unrolling tapestry. A motion picture of the orbiting planets could be shown backwards and no one would know the difference.

At the peak of its success in the late eighteenth century the principle of mechanism had spread from physics and astronomy to other fields of thought. It was based on two beliefs:

(1) *All* phenomena of nature are explainable in terms of basic particles exerting forces on each other and combining in various ways. These phenomena included economics, politics, and sociology, as well as individual human behavior. The world is a machine. Particles occupy space, and have no internal structure or secrets. Everything is externalized and exposed to analysis.

(2) Natural laws, discoverable by experiment and human reason, rule everything. The future is determined by, and predictable from, these laws.

Adam Smith in economics, Richard Bentley in theology, David Hartley in early psychology, and John Locke and Thomas Jefferson in political theory: all were greatly influenced by the Newtonian spirit of law, reason, and predictability. It would be a mistake to conclude that Newton was the cause of it all.[11] Possibly he and all the rest were simply reacting in their various ways to a spirit of the times, an age now known as the Enlightenment.

Mechanism Versus Organism

Europe and fledgling America embraced Newtonianism. It constituted a basic theme in the Enlightenment. The spirit of the age, a time when Newton's principles were explaining the motions of the planets so well, was caught by Joseph Addison (1672–1719) in his "Ode":

> The spacious firmament on high
> With all the blue ethereal sky
> And spangled heav'n, a shining frame
> Their great Original proclaim
> · · ·
> In reason's ear they all rejoice,
> And utter forth a glorious voice,
> Forever singing as they shine,
> "The hand that made us is divine".

(11) I. B. Cohen, in *Isaac Newton's Principia*, Harvard University Press, Cambridge, 1971, points out that Newton's main publication was rejected at first by scientists on the Continent. It finally won recognition, largely through the efforts of Voltaire (François Marie Arouet, 1694–1778), who was one of Newton's most enthusiastic supporters.

However, by the early 1800's, a reaction which we call Romanticism had set in. Wordsworth, Byron, Tennyson, and others took a look at the mechanistic rationalism which had developed during the 1700's and cried out for a return to green fields and human feelings.

In Tennyson's poem, "In Memorium", he deplores the stark, numerical view of nature which Newtonianism can bring:

> And all the phantom, Nature, stands —
> With all the music in her tone,
> A hollow echo of my own, —
> A hollow form with empty bank,
> And shall I take a thing so blind,
> Embrace her as my natural good,
> Or crush her, like a vice of blood,
> Upon the threshold of the mind?

It's possible to see this change as one cycle, or episode, in the continuing conflict between mechanism and organism as world views. This was mentioned in Chapter 1 and will be returned to again. The popularity of mechanism as a philosophical viewpoint has decreased in the twentieth century. A human being can be seen as a machine in some respects,[12] but there are just too many things about persons and groups which get lost when you subdivide into smaller units. Mechanism cannot predict or recognize "gestalt" characteristics; that is, things which emerge from the whole, making it more than the sum of its parts. The behavior of the whole can be rationalized on the basis of the elements that make it up, but it's very difficult to go the other way. And though the behavior of the component parts of an organism may be predictable (mechanically determined), the most important characteristics of the organism itself may not.

One critic of our society[13] feels, however, that in spite of the decline of mechanism as an intellectual concept, we have hung on to it at a more every-day level. Our factories, armies, vast governments, computers; our multiplication, distribution, and consumption of uniform products, he says, all point to a tendency for human beings to become parts of a worldwide "megamachine". He attributes the start of this process to Galileo, for deciding to look at the world numerically more than 350 years ago!

Physics itself has moved away from the mechanistic point of view since the mid-1800's. Much of the rest of this book will reflect that trend. The next chapter discusses the main achievement of Newtonian mechanism: a theory of planetary motion.

(12) Woolridge, Dean E., *Mechanical Man: The Physical Basis of Intelligent Life*, McGraw-Hill, N.Y., 1968.

(13) Mumford, Lewis, *The Pentagon of Power*, Harcourt Brace Jovanovich, N.Y., 1970. Reprinted as "Reflections" in the *New Yorker*, Oct. 10, 17, 24, 31, 1970. See also the critical review by Gerald Holton, *New York Times Book Review Section*, Dec. 13, 1970.

But after that it will be downhill for mechanism as a philosophical principle. In the three chapters where momentum and energy are discussed one's attention is drawn to the whole rather than to the interacting parts. The concepts of relativity theory erase the line between matter and energy. In Chapters 11 through 14 the discussion of field theory and waves concentrates on disturbances (not particles) moving through space and interacting with each other. Chapters 15 and 16, on thermal phenomena, *do* utilize a particle model for matter, but all pretense of rigid determinism is abandoned. The elements in the system are too small and numerous to handle or observe, so they are treated statistically, using probabilities. In Chapter 17 atomic phenomena are investigated, and it is found that there *is* interaction between the observer and the observed. With this, the unspoken premise of mechanistic determinism (that a person can be an objective, detached observer) is wiped out.

Summary

Aristotelian organic physics dominated the West for nearly 2000 years. It used experimental evidence, but only in an auxiliary sense. The important element was the rational structure. Every object had its appointed place or purpose, depending upon what it was made of. Motion was either *natural*, in which case the object innately moved toward a proper resting place, or it was *forced*, in which case an outside mover made it go. An Aristotelian equation of motion would have been

$$v = \frac{F}{R} \qquad \text{Equation (4-1)}$$

where v is the speed, F, the external force, and R, the resistance to motion offered by friction or by the medium through which the object moves. Motion itself required a cause. The heavens and earth were different; they consisted of different materials obeying different laws of physics.

Although the transition was less sharp than was previously thought, the fifteenth and sixteenth centuries saw a vast reorientation in Western thought. Mechanics, the science of motion and force, was an important part of the change. Empirical evidence became more important. Acceleration, rather than velocity, was seen as requiring a cause. The concept of inert objects, complete victims of their immediate surroundings and neighbors, developed. Mass became the quantitative measure of the inertia of an object, its relative ease or difficulty in changing velocity. The relative inertial mass of two objects is given by

$$\frac{m_A}{m_B} = -\frac{\Delta v_B}{\Delta v_A} \qquad \text{Equation (4-2)}$$

where Δv_B and Δv_A are the changes in velocity suffered respectively by objects B and A during their interaction.

The cause of inertia is not easy to find. One theory suggests that inertia arises because when you accelerate an object, you are changing its relations with every other mass in the universe.

Newton's laws of motion stated

(1) (The law of inertia.) All objects not acted upon by a net force remain at rest or in a state of motion at constant speed in a straight line.

(2) $F_{net} = ma$. Equation (4-3)

The "newton" is that force which will impart an acceleration of one m/sec² to an object having a mass of one kg.

The impulse of a force equals F_{avg} (Δt).

Weight is the name we give to the force exerted on an object by gravity:

weight $= mg$. Equation (4-6)

When more than one force acts on an object they must be added together to find F_{net}. This can involve vector addition, component by component. When this addition has been accomplished, then Equation (4-3) can be used to calculate the motion of the object.

In auto safety devices, deceleration during crashes is reduced by making Δt as long as possible.

Friction is the force required to just keep two surfaces sliding past each other and was probably the key phenomenon behind Aristotelian physics. In some cases friction is nearly independent of sliding velocity and is directly proportional to the normal load holding the two sliding surfaces together. In such cases,

$F_{fric} = \mu$ (normal load),

where μ equals the coefficient of friction.

Newton's success in explaining and predicting the movements of inert objects was an important factor in the rationalism which swept Europe during the eighteenth century age of Enlightenment. By the nineteenth century a Romantic reaction had set in. Mechanism, in the sense that it reduces a system to basic building blocks, is still an important element in physics. However, physicists have also focused considerable attention (in the last 100 years) on phenomena in which the whole must be looked at; and, where the behavior of individual objects is not fully predictable.

Further Reading

Andrade, A. N. daC., *Isaac Newton: His Life and Work*, Macmillan, N.Y., 1954. (Straightforward biography.)

Bronowski, Jacob, "The Nature of Newton's Insight", from *The Common Sense of Science*, Harvard University Press, Cambridge, 1953. (How Newton connected geometry and physics. The world is not a pattern but clockwork.)

Clark, Harry H., "The Influence of Science on American Ideas From 1775 to 1809", *Trans. Wisconsin Acad of Science, Art and Letters*, 35, 1944, page 312. (An introduction to the religious, political, and social interpretation of Newtonianism in the early days of the United States.)

Cohen, I. B., *Copernicus to Newton: The Birth of a New Science*, Doubleday Anchor Book, 1960. (Interesting non-mathematical account of how ideas such as inertia and gravity developed.)

Dijksterhuis, E. J., *The Mechanization of the World Picture*, The Clarendon Press, Oxford, 1961, section IV. (An excellent account of 17th century mechanics.)

Durant, Will and Durant, Ariel, *The Age of Voltaire*, Simon and Schuster, N.Y., 1965. (Shows the influence of Newtonianism in 18th century Europe.)

Feuer, Lewis S., *The Scientific Intellectual*, Basic Books, N.Y., 1963. (Chapters on intellectuals during various ages. Discussion of whether the Protestant ethic contributed to the growth of science.)

Hall, E. W., "Romanticism and Science", pp. 185–236 in *Modern Science and Human Values*, Van Nostrand, New York, 1956. (After the Enlightenment embraced science and reason, Romanticism rejected them.)

Hook, Sidney, (ed.), *Determinism and Freedom in the Age of Modern Science*, Macmillan, Collier Books, New York, 1961. (A wide selection of readings examining determinism in philosophy, science, law, ethics, and responsibility.)

Jaki, S. L., Chap. 1, "The World as Organism" and Chap. 2, "The World as Mechanism", in *The Relevance of Physics*, University of Chicago Press, 1966.

Kearney, Hugh, *Science and Change: 1500–1700*, World Universities Library, McGraw-Hill Book Co., N.Y., 1971. (A paperback description of the Scientific Revolution. Valuable because it points out the importance of magic as well as Organism and Mechanism as principles during this period of change.)

Keynes, John Maynard, "Newton the Man", reprinted in *Physics and Man*, Robert Karplus, editor, W. A. Benjamin, N.Y., 1970, pp. 22–29. (Some interesting conjectures on Newton's personality.)

Kuenne, Robert E., "The Economy: Mechanism or Organism?" Reprinted in *Physics and Man*, Robert Karplus, editor, W. A. Benjamin, N.Y., 1970, pp. 176–180. (Discusses the holistic versus the analytic approach in an economic setting.)

Guerlac, Henry, "Three 18th Century Social Philosophers: Scientific Influences on Their Thought", *Daedalus* 87, pp. 3–24, 1958. (Voltaire, Montesquieu, and Baron d'Holbach are discussed, and the author concludes that influences besides Newtonian science gave rise to the Enlightenment.)

Nicholson, Marjorie H., *Science and Imagination*, Cornell U. Press, Ithaca, 1956. (Influence of Newtonian physics on literature.)

Randall, J. H., Jr., "The Newtonian World Machine", Chap. XI in *Making of the Modern Mind*, Houghton Mifflin, Boston, 1926. (Still one of the definitive pieces on Newton's world picture.)

Sambursky, S., *The Physical World of the Greeks*, Collier Books, N.Y., 1962. (A physicist looks into Greek science to see what it gave to modern science. He sees the Stoics and Epicureans, rather than Aristotle and Plato, as really affecting the flavor of modern science.)

Schroeer, Dietrich, "The Failure of Greek Science", pp. 60–72 in *Physics and its Fifth Dimension: Society*, Addison-Wesley, Reading, Mass., 1972. (Uses Sambursky and other sources to list reasons why Greek science got no further.)

Whitehead, A. N., "The Century of Genius", in *Science and the Modern World*, Mentor Books, 1948. (One of several excellent chapters in a book on evolution of Western science. Suggests that the Christian faith in reason, and in an orderly world, was basic to the development of science.)

Problems

4-1 How would Aristotelian physics explain the following observations?
   (a) A wagon being pushed along a sidewalk stops when you quit pushing.
   (b) A ball rolled across the floor keeps going awhile before it comes to rest.
   (c) A helium-filled balloon must be tied down to keep it from rising.
   (d) A hot-air balloon rises but later falls.
   (e) A ball thrown up continues to rise after it leaves the thrower's hand.

4-2 What is the difference between the questions "how" and "why"? Does finding a cause answer the first or the second question? What does externalization of cause have to do with answering "how" *vs* answering "why"?

4-3 A mass of 300 grams experiences an acceleration of 95 cm/sec². What net force must be acting on the mass, in dynes, and in newtons?

4-4 A 20 kg object is accelerated by a net force until it reaches a velocity of 10 m/sec. Then the force is removed. What will the object do?

4-5 A fully loaded truck is able to accelerate up to 60 miles per hour in one minute. When emptied of its load the total mass of the truck is half as great. How long will it now take the truck to reach 60 mph, assuming that all the friction, drag, and propulsion forces are the same as before?

4-6 A 500 gram block rests on a horizontal table. Find the acceleration of the block when one-half, one, two and four N of force are applied horizontally, if the frictional drag is one N.

4-7 Describe numerically the behavior of a 10 kg suitcase sitting on the surface of the earth if a rope exerting an upward force of 100 N is tied to it. Where is the suitcase after five seconds?

4-8 Charles Atlas, a well-known strong man of the 1930's, used to push a railroad car along a track unassisted. Suppose that he exerted a horizontal force of 3000 newtons on a car with a mass of 100,000 kg and just managed to keep it moving at a constant speed of 20 cm/sec. If an equally strong man then assisted Atlas in pushing the car what would the velocity of the car have been 10 seconds later?

4-9 An object having a mass of two kg moves along a straight line. At $t = 0$ its position is given by $x = 0$. The graph below shows the net force acting on the object as a function of time. (Positive force means to the right and negative force means to the left.)

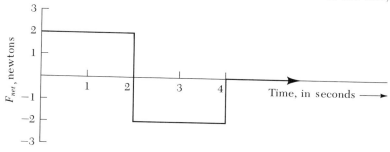

Make graphs of the velocity of the object vs t for $v_o = 0$, $v_o = +4$ m/sec and $v_o = -4$ m/sec.

4-10 What is the average decelerating force on a three gram bullet which burrows 2 cm into a wooden wall after hitting it at a speed of 100 m/sec?

4-11 What is your weight in newtons? What is the mass of a nickel? How many nickels are in a pound of mass?

4-12 Explain how it is that you can slowly lift a stone tied to the end of a string, but that, if you try to accelerate the stone upward rapidly, the string will break. Use a force diagram.

4-13 When you push on a heavy object, how can you tell if it's friction or inertia you are overcoming?

4-14 A 70 kgm man stands on a bathroom scale in an elevator. What will the scales read (in newtons) if the elevator is:
(a) moving down at 9.8 m/sec,
(b) is standing still but is accelerated upward at 1.96 m/sec²,
(c) is accelerated downward at 1.96 m/sec² ?

4-15 Nylon ropes used for mountain climbing are strong but also have the important property of being elastic. They will gradually stop a fall which would break a mountaineer's back if he or she were arrested by a rope having little stretch. A free fall of 20 meters represents a possible fall of a leader climbing 10 meters above his last piton (anchoring pin). Suppose a climber weighs 784 N (176 lbs.). If a three g deceleration is tolerable, how much must a rope be able to stretch and what must be its minimum breaking strength to save the climber in a 20 m fall? Remember to allow for the static weight of the climber hanging on the rope.

4-16 Consider a pilot of mass m who wishes to fly vertical loop-the-loops. (a) What must be the centripetal acceleration of the plane at the top of the loop to keep the pilot from falling out of the plane? In what direction is the centripetal force at that instant? (b) If the inverted plane at the top of the loop experiences a centripetal acceleration of two g, is a force of $1mg$, $2mg$, or $3mg$ felt against the seat of the pilot's pants? (c) What must be the centripetal acceleration of the plane at the top of the loop to keep any extra blood from rushing to the upside-down pilot's head?

4-17 Two people are trying to drag a heavy tree stump across a frozen lake. One person pulls east with a force of 100 N and the other pulls with a force of 150 N in a direction 45° west of south. If the mass of the tree is 200 kg, but the lake is frictionless, find the direction and magnitude of the stump's acceleration.

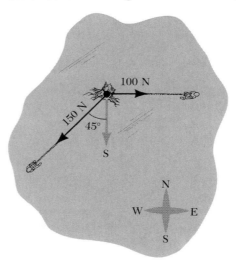

4-18 What is the tension in the string connecting the two masses in the system shown?

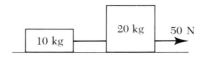

GRAVITY: THE CREATION OF SCIENTIFIC THEORY

Much of what has been discussed thus far has concerned the *observation* and *description* of physical phenomena, and the motion of objects in particular. Even Newton's laws of motion can be seen as summaries of observations of how objects behave in response to influences from the rest of the world. Physicists, however, try to go further than this. They try to *explain* phenomena. Although the boundary between description and explanation is indistinct, scientific *law* can be thought of as correlating or *describing* phenomena, and *theory* as attempting to *explain* them. The purposes of this chapter will be: (1) to trace the development of theories explaining the apparent motion of the planets; (2) to see how the theory of gravitation arose as part of this development; and (3) to discuss more generally the process which leads to scientific theories.

Early Astronomy

Early neolithic peoples noticed that various objects moved across the day and night sky in various ways. The sun seemed to rise and set, and in so doing, gave them a basic division in their lives, the day. The moon also rose and set, and changed its shape, but it took a longer time—defined as a month, or $27\frac{1}{3}$ days—to repeat these changes of shape. The rest of the glowing objects in the night sky underwent more subtle movements from night to night, but save for about 5 of them, the glowing objects all seemed to move together. These objects, moving across the sky in seemingly fixed arrangements relative to each other, were called stars. The five more independently moving objects came to be called planets (meaning "wanderers") and were given names after the gods. All of them were great sources of wonder, and objects of worship, on the chance that their movements might have something to do with the growth of crops or with one's luck in general. By the time of the "golden age" of the ancient Greeks, 400–500 years before the Christian era, the Western belief in a world that was orderly and free of transient spooks was already evident. The spots of light which moved across the night sky were seen as glowing *objects* that followed certain *paths* through space. They were not beings which died and were reborn each night. But the idea of a path itself raises

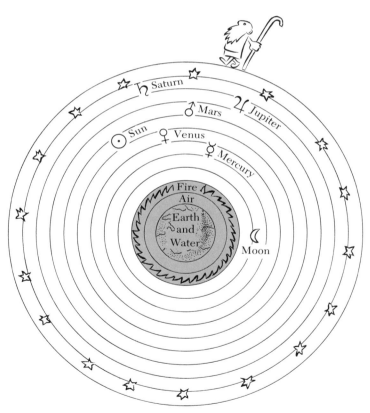

Figure 5-1
Sketch of Aristotle's model of the universe.

three questions: what *are* the shapes of the paths followed by the Sun, Moon, planets, and stars; what keeps these objects on their paths; and, what propels these objects around in their paths? The answers which have been proposed to these questions will give us an interesting means for looking into some of the changes in Western thought since ancient times.

Aristotle's universe was constructed in layers, with the earth at the center; as illustrated in Figure 5-1. The earth itself was layered, with the earth, water, air, and fire tending to occupy successive regions as one went out from the center of the earth. Beginning with the moon and progressing outward through the orbits of Mercury, Venus, the Sun, Mars, Jupiter and Saturn, the objects were made of quintessence, as we saw in the previous chapter. According to Aristotle, the planets were attached to transparent spheres whose centers were located at the center of the earth. The *natural* motion of the spheres was to rotate, so no continued push was needed to keep them going. Two complexities in the behavior of the planets had been noted. First, they occasionally underwent *retrograde motion* relative to the stars, as shown in Figure 5-2.

This backing up was explained by letting the various spheres turn about axes which slanted and were attached to other rotating spheres something like a complex set of gimbals. Toward the end of its reign, the Aristotelian model required at least 55 interrotating spheres! The second complexity, which the model simply overlooked, was that the planets shone brighter and dimmer and the sun grew larger and smaller as time passed. (The sun and planets were thought to be at fixed distances from the earth.) The stars moved together on the 8th sphere or orbit or heaven, and God, the unmoved mover, resided in the 9th.

The Ptolemaic model (after Claudius Ptolemy, a Greek astronomer living in Alexandria during the second century A.D.) altered Aristotle's model in several ways, to account for more

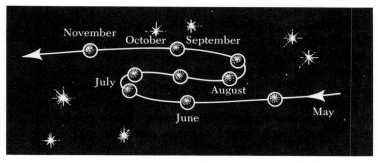

Figure 5-2
Apparent motion of Mars as seen from the earth against a background of stars during a six-month period. Each spot on the path corresponds to the position of the planet at the same time of the night.

detailed observation of the motion of the planets. It kept the stationary earth in the middle. However, Ptolemy recognized that the distance between the earth and the sun and planets must change as they moved around in their orbits. The transparent spheres were discarded. Ptolemy postulated that retrograde motion was the result of *epicyclic* motion. If the earth is seen as being stationary then the motion of Mars can be explained by a counter-clockwise motion of Mars in a small rotating circle whose center also moves counter-clockwise, but along a larger circle, the deferent, which has the earth at its center. By choosing the radii of the two circles correctly and by selecting the proper velocities around the deferent and epicycle, the planet in Ptolemy's model could be made to undergo retrograde motion. Other complexities in planetary motion also began to be detected. For example, all of the postulated deferent circles were not quite centered on the earth. Several circles, in different planes, also had to be added to the circles on the deferent. It was an ingenious model which did work quite well, but it had grown very cumbersome. By the middle of the sixteenth century it required 34 circles to explain the motion of the moon, sun, and the five known planets!

From its earliest days, the Aristotelian-Ptolemaic model had rivals. Aristarchus of Samos had proposed a model of the universe in 280 B.C. which put the sun at the center! If this were true and the earth were moving, then the distant stars ought to have appeared to change position relative to each other. How-

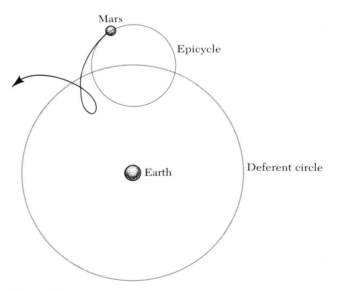

Figure 5-3
Ptolemaic explanation of the retrograde motion of Mars.

ever, the apparent relative motion of the stars was too small to be detected with the instruments which were available then, and the theory did not gain enduring support. Aristarchus' model gave way to Aristotle's. During the fifteenth century Nicholas of Cusa and others had suggested that various planets, and even the earth, might really be moving around the sun. However, the imposing synthesis of Aristotelian physics and medieval theology, which was mentioned earlier, helped preserve the Ptolemaic model, which called for a fixed earth at the center of the universe.

The Copernican Universe

Nicolaus Copernicus' work, *De Revolutionibus Orbum Coelestium*, published in 1543, is often used as a prime example of "Renaissance man" throwing off his medieval shackles. Out of it comes the "Copernican Revolution", wherein the earth (and humanity with it) is removed from the center of the universe and made to rotate around the sun along with the other planets. Given the tight connection between natural philosophy and theology which existed up to the sixteenth century, a revolution *did* have to take place before the Copernican model was fully accepted, more than 100 years after it was proposed. (We shouldn't forget that another barrier to the acceptance of the theory was that it went against common sense. Anyone can tell the earth is standing still just by looking out the window. What's to be gained from an explanation that doesn't fit everyday experience?) However, it would be misleading to call Copernicus himself, or his model, "revolutionary". Copernicus was neither a sharp observer nor bold imaginer. As Herbert Butterfield points out[1] Copernicus was primarily a corrector of the Ptolemaic theory, and his reasons for correcting were conservative. He realized that the Ptolemaic model, with all of its circles with centers rolling on circles, called for a great deal of motion which was *not* centered on the earth. The ancient philosophical principle which said that the earth was the center of the universe and that the heavenly bodies traveled in perfect circular motion was thus already being violated. Since the sun was the noblest of the celestial objects, why not put *it* at the center of things? Copernicus worked out several models one of which had the sun at the center, with Mercury, Venus, Earth, Mars, Jupiter and Saturn traveling around it in successively larger concentric circular orbits. In so doing, Copernicus was able to account for most of the movements of the planets in a much simpler fashion. In this sense, his theory was a

(1) "The Conservatism of Copernicus," Chapter 2 in *Origins of Modern Science*. G. Bell, London, 1949.

better one: it explained just as much as the Ptolemaic theory, but required fewer postulates. It is important to realize that Copernicus did not *prove* that the earth moves around the sun. He merely said that the apparent planetary motions in the sky are easier to understand if we suppose that we are on a moving platform. The retrograde motion of Mars which was shown in Figure 5-2 can now be visualized in a diagram by drawing lines from the earth through Mars as these planets occupy various positions in their respective orbits around the sun.

Copernicus still believed that the planets were kept in their orbits by being attached to crystalline orbs; and, that the orbs rotated freely and naturally, thereby moving the planets around in their orbits.

The way in which Copernicus responded to certain criticisms tends to confirm the notion that his mind still worked mainly with the medieval system of ideas. For example, he said that the earth moved, but dragged the air and clouds with it, because they both contained earth which tended to remain with its mother planet. He also said that a falling stone was not left behind by the moving earth because its natural place was the earth's center, and so it moved naturally in that direction.

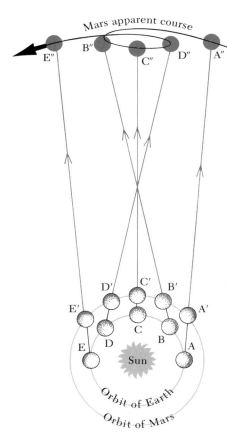

Figure 5-4

Copernican explanation of the retrograde motion of Mars. As the earth moves around its orbit, observations of Mars are made at times corresponding to positions A, B, C, etc. Mars moves at the same time, occupying positions A', B', C', etc. As observers look out at the night sky from earth, they locate Mars along the sightlines AA", BB", CC", etc. These sightlines change direction as a result of the relative motion of the two planets, so Mars appears to "back up" when the earth is between points B and D in its orbit.

The observations of the Danish astronomer Tycho Brahe (1546–1601) represent a distinct advance over previous work. Handsomely financed by the King of Denmark, he built a fine observatory equipped with large sighting instruments, as shown in Figure 5-5, (no telescopes yet). These instruments enabled him to record the positions of planets and stars with far more accuracy than had been possible before. Tycho collected very accu-

Figure 5-5

Tycho Brahe directing an assistant during some observations. The assistant is measuring the angular position of the star by sighting across the instrument and through a hole in the wall. (From Sedgwick, Tyler and Bigelow, *A Short History of Science*, Macmillan, N.Y., 1939.)

rate data, and previously unnoticed details of their motions began to emerge. Whenever observations were being made in his Uraniborg observatory he and his assistants would dress formally, wearing the sixteenth century equivalent of tails, so as to give proper gravity to what was being done.

Johannes Kepler, to whom we referred early in Chapter 4, was a German professor, part-time astrologer, and a deeply religious man. He was associated with Tycho for a short time. He examined and correlated Tycho's astronomical data, and after years of work, derived three empirical laws from them:

(1) The planets do not move in circles, or epicycles, but in ellipses, with the sun located at one focus of the ellipse. (These ellipses come very close to being circles.)

(2) A line drawn from the sun to the moving planet will sweep out equal areas in equal times. (In Fig 5-6 these areas are indicated by shading.) The planet moves more slowly when it is farther from the sun (as in traveling from 1 to 2) than when it is closer to the sun (as in traveling from 3 to 4).

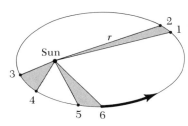

Figure 5-6

(3) In comparing the orbital periods of various planets, they are found to depend upon the average distance from the sun to the planet in the following way:

$$(\tau)^2 = k \ (r)^3 \qquad\qquad \text{Equation (5-1)}$$

In this expression, τ is the *period* of the planet (the time required to go around the orbit once), r is the average distance from the sun to the planet and k is a constant.

The description of the orbits as ellipses came only after great effort on Kepler's part.[2] For 2000 years the circle had been the epitome of symmetry, simplicity, and of all the aesthetic principles which still guide modern science. Several things might have helped Kepler make the mental transition from circles to ellipses. One was the geometrical realization that circles are after all only special kinds of ellipses. Another is that by using

(2) Kepler's labors were made somewhat less arduous by his use of the newly available logarithms, invented by the Scots mathematician, John Napier (1550–1617).

the ellipse, he could make the sun a genuine center of influence. As the evidence accumulated, it became clear to Kepler that the sun was not really at the center of Copernicus' planetary orbits. Still another possibility is that since the oval had made its appearance in European architecture and painting by the late 1500's, the charmed life of the circle was already in decline. The next factor is probably the one which would have the most appeal to a contemporary physicist. Kepler noted that when the planets travel in elliptical orbits, such directly observable quantities as speed, and distance from the sun are not constant. However, if one looks beneath these two changing quantities, as Kepler did in his second law, one finds a quantity which *is* constant; the area swept out by the radius vector per day, or per month. The further we go in our study of physics, the more important it will become to see what it is in a system that stays the same as time passes.

Whatever Kepler's reasons were for his elliptical vision, we need to keep in mind that, for Kepler, the data were given first rank and the details of the model followed. The data were not quite forced to fit pre-conceived ideas.

Heavenly Forces

The organic belief that objects moved as they do because of innate tendencies died a slow death. As we saw in Chapter 4, it gave way eventually to the idea that motion results from external causes: forces. However, the planets presented special problems. What force could be acting on the planets? How did this force act through space in order to pull or push on the planets?

William Gilbert (1540–1603), a physician who attended to Elizabeth I and James I, played an important part in the development of ideas concerning the transmission of forces with his work on magnets. He gave graphic evidence that magnetic attractions (and repulsions) were *mutual,* and that their intensity depends upon the size of the magnets used. Gilbert also began to investigate how the intensity of attraction depended on the separation between magnets. He even suggested a model of the solar system in which magnetism not only attracted the planets toward the sun, but also propelled them in their orbits. His model was not taken seriously, but it left its mark.

Kepler, with Gilbert's magnetic forces in the back of his mind, attempted to *explain* the orbital shapes by means of a single force emanating from the sun. He mistakenly assumed that the attractive force was confined to the plane of the sun and orbits, and deduced that the force must decrease as $1/r$, where r is the distance from the sun to the planet in question. Since the concept of inertia was still in the future, Kepler also needed a force to pull the planets around their orbits. To get this, he said that as the sun rotated, its magnetic field rotated with it, pulling the planets along. Note that he thought in terms of a force emanating from an object; the sun. The force surged through space and affected the planets externally. They did not propel themselves innately or as the result of having to stay on some surface in space. As Kepler said in a letter to D. Fabricius in 1607, ". . . . you use circles, I use forces." This places Kepler's physics in the modern, rather than in the medieval world.

Nevertheless, Kepler's style of physics was transitional, showing elements of the old world as well as the new. When compared with modern scientific papers, his writings have an odd ring.[3] They wander about, mixing astronomy, astrology, geometry, prayer, poetry, theology, and long accounts of ideas that didn't work out. He sees the sun not only as the center of force in the universe, but as a great mystical center, the temple of God. The actual relative sizes of the planetary orbits, Kepler thought, were an expression of a heavenly numerical harmony. It is possible to see the goal of Kepler's work as alternating between two philosophical interpretations of reality.[4] One said that the physically real world is the world of phenomena as explained by mechanical principles. The other stated that the physically real world is a world of mathematically expressed harmonies. God set it up and gave human beings the intelligence to find the harmonies in the midst of overall chaos.

Galileo, just seven years older than Kepler, played an important, but less direct, role in the evolution of astronomy. He was repelled by Kepler's ellipses and preferred epicycles.[5] He had a general idea of gravity but did not yet see it as emanating from massive bodies. However, his work with telescopes was showing many interesting new things about the heavens. The Milky Way was resolved into individual points of light. The planets were seen in greater detail, but the stars were no larger than they had appeared to the unaided eye. And as noted in the previous chapter, small moons could be seen traveling around Jupiter. (This was another mark against Aristotle's world, which said there should be just seven objects in the sky besides the stars: the sun, moon and five planets.) All of this evidence, along with his work on motion and inertia, put Galileo in the important position of helping to overthrow Aristotelian physics and thus, aiding acceptance of the Copernican model.[6]

(3) Translations of some of Kepler's and Copernicus' works are available in Volume 16 of *Great Books of the Western World*, Encyclopaedia Britannica, University of Chicago Press, 1952.

(4) Holton, Gerald, "Johannes Kepler's Universe: Its Physics and Metaphysics", *Amer. J. Physics* 22, May, 1954, pp. 340–351.

(5) Panofsky, Erwin, *Galileo as a Critic of the Arts*, Martinius Nyhoff, The Hague, 1954.

(6) Galilei, Galileo, *A Dialogue Concerning the Two Chief World Systems: Ptolemaic and Copernican.* Translated by Stillman Drake, University of California Press, Berkeley, 1967.

Descartes also applied his considerable rational powers to the problem of explaining the motion of the planets. He felt convinced that the physical world must ultimately obey a set of geometrical axioms, with all observable phenomena following as necessary deductions. The axioms were seen as prior to or even truer than sense experience. We have already said that he deduced the law of inertia and said that *no* force is required to push the planets around, once they have been started. However, he did not believe in gravity as the mechanism for pulling the planets into their orbits around the sun. He was repelled by *action-at-a-distance* ideas, wherein influences are transmitted across empty space with no mechanical connections between objects. Instead, he postulated a giant vortex which whirled around the sun.[7] The planets were swept around by the fluid in the vortex and thus followed their elliptical paths. Descartes' model competed for attention for awhile, but eventually it gave way to Newton's because the idea of the vortex presented more problems than it answered.

The Triumph of Newton

Newton was an immensely successful synthesizer and clarifier of ideas. He was able to merge empirical fact and thought, geometry and mechanics, heaven and earth. As a result, he emerges as a giant in the history of science, in much the same way as Aristotle did 2000 years previously.[8] By this we mean that the influence of his thought was so great that in some fields, such as optics, certain discoveries were ignored for over a century because they ran counter to Newton's theories—or at least, his disciples thought they did. But let us examine some of Newton's contributions to planetary motion, and with it, his theory of gravity.

First, he said that the planets were carried around their orbits by inertia, as did Descrates. Inertia was a principle applying to celestial as well as to mundane objects.

He next pointed out that although the velocities of the planets might be directed any which way at a given time, their *accelerations* were always directed toward the sun. The force causing the accelerations was set up by the sun. He called it *gravity*. (Naming it, of course, did not tell what it was or what caused it.)

Perhaps taking a cue from Gilbert's work on magnets, he postulated that *the gravitational force exerted by an object such as*

(7) *Le Monde*, 1633. Published posthumously in 1664.

(8) It appears that he was a vain but reluctant giant. His *Principia* (*The Mathematical Principles of Natural Philosophy*, translated by Andrew Matte, David Adee, N.Y., 1846.) was published in 1687 only after his friend, the astronomer Edmund Halley, led him to believe that his rival, Robert Hooke, might scoop Newton on the gravitational theory. The *Principia* set forth his laws of motion, his planetary model, and his theory of gravity, along with his basic scheme for scientific reasoning. Newton was extremely reluctant to give credit in the *Principia* to Kepler, Francis Bacon, Hooke, Sir Christopher Wren, Gilbert, and others whose ideas on gravity and planetary motion preceded his own.

the sun is proportional to the mass of the object. (If the mass doubles, the force doubles.) Later on, he could tell from the accelerations of the planets in orbit that *the gravitational force also is proportional to the mass of the planet being pulled by the sun.*

From an analysis made by Christian Huygens (1629–1695) Newton knew that the acceleration required to keep a body of velocity, v, moving on a curved path of radius, r, is v^2/r (page 53). From his own second law, he could then set up a relation between the centripetal force, F_c, furnished by the gravitational field, and the acceleration of the planet: $F_c = m_p v^2/r$. In this relation m_p is the mass of the planet, v is its velocity, and r is the distance from the sun to the planet.

Another major problem was to calculate the way in which the gravitational force varied with distance from the sun. An important mathematical realization enabled Newton to get where he wanted to go. This was the discovery that, as far as its external gravitational effects are concerned, a spherical object can be replaced by a point mass at its center, as shown in Figure 5-7. Although some parts of the sun are closer to the planet and some are farther away, the sum of their effects is the same as if all the parts were concentrated at the center.[9] Once Newton could show this mathematically, then "distance from" began to have some clear meaning. From this point, Newton could then find out how the gravitational pull of the sun must depend on distance by using Kepler's third law. His method can be most simply shown if we assume that the orbit of the planet is circular (has a constant

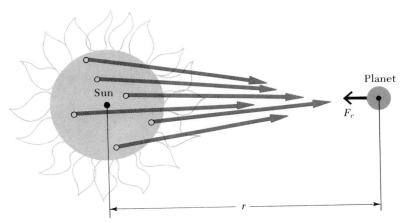

Figure 5-7
All parts of the sun attract the planet gravitationally. The overall effect is the same as if the mass of the sun were concentrated at its center.

(9) This was, perhaps, Newton's most original contribution. Hooke, Ismael Boulliau, and others had suggested that the sun's gravitational force dies off as the square of the distance from the sun, but they were not able to find a satisfactory mathematical demonstration of their theories.

radius r) and make use of Kepler's third law (Equation 5-1, page 104).

For a planet with a mass, m_p, traveling in a circular orbit with a velocity, v, the period,

$$\tau = \frac{2\pi r}{v} \qquad \text{Equation (5-2)}$$

Solving the equation for centripetal force, $F_c = m_p v^2/r$, for v,

$$F_c r = m_p v^2,$$

$$\frac{F_c r}{m_p} = v^2$$

so:

$$v = \sqrt{\frac{F_c r}{m_p}}. \qquad \text{Equation (5-3)}$$

Solving Equation (5-2) for v,

$$\tau v = 2\pi r,$$

$$v = \frac{2\pi r}{\tau}$$

Substituting the value of v from Equation (5-3) gives

$$\sqrt{\frac{F_c r}{m_p}} = \frac{2\pi r}{\tau}.$$

Solving this equation for τ;

$$(\sqrt{F_c r})\tau = (\sqrt{m_p})2\pi r,$$

$$\tau = \frac{(\sqrt{m_p})2\pi r}{\sqrt{F_c r}}.$$

Squaring both sides of the equation to remove the radicals gives

$$\tau^2 = \frac{m_p 4\pi^2 r^2}{F_c r},$$

Performing the indicated division by r gives:

$$\tau^2 = \frac{m_p 4\pi^2 r}{F_c},$$

When solved for F_c, the centripetal force, this equation becomes

$$F_c = \frac{m_p 4\pi^2 r}{\tau^2} \qquad \text{Equation (5-4)}$$

Since Kepler's third law showed Newton that $\tau^2 = kr^3$, Equation (5-4) becomes:

$$F_c = \frac{4\pi^2 r m_p}{kr^3}.$$

This is equivalent to saying:

$$F_c = \left(\frac{4\pi^2 r m_p}{k} \right) \frac{1}{r^3} = \left(\frac{4\pi^2 m_p}{k} \right) \left(\frac{1}{r^2} \right).$$

Since π is a constant, and by our original statement of the problem, m_p is also a constant, as is k, the quantity $4\pi^2 m_p/k$ can be replaced by a single constant, becoming:

$$F_c = (\text{constant}) \frac{1}{r^2}$$

The centripetal force (provided by gravity) must then decrease as the square of the distance between the sun and the planet. This result was one of the first accomplishments in "theoretical" physics and is a good example of what modern physicists spend a great deal of time doing: combining empirical observations with mathematical manipulations to deduce a conclusion which can be tested empirically.

Universal Gravitation: Newton's Leap of Faith

This would have been accomplishment enough. However, Newton went further and made the daring inductive inference that *every* object in the universe attracts every other one with a force which is proportional to both of their masses. If the two masses, m_a and m_b, are point masses (that is, if they are far apart) or if they are spherical in shape, then the gravitational force F_g between them is inversely proportional to the square of the distance r_{ab} between their centers.

$$F_g \propto \frac{m_a m_b}{(r_{ab})^2}$$

(Doubling either mass will double the force. Doubling the distance between them will decrease the force by a factor of four.) If we multiply both sides of the equation by the proper constant, the proportionality becomes an equality:

$$F_g = \frac{(\text{constant}) m_a m_b}{(r_{ab})^2} \qquad\qquad \text{Equation (5-5)}$$

By suggesting the universality of gravity, Newton heralded the end of the ancient idea that earth and the heavens were subject to different laws of physics. His *universal law of gravitation* was not a proof of anything, but a plucky induction from specific evidence to the general. It is a simple but potent idea which takes us far beyond the falling objects discussed in Chapter 3. It will never be *proved* correct because we will never be able to measure the gravitational force between every pair of objects in the universe. However, it seems to fit all the cases seen thus far.

Years before he had polished his theory and found proper mathematical support for it, Newton had shown his ability to extend a single idea into a broader context, and in so doing find explanations for heavenly phenomena in terms of the mundane. Whether or not Newton was actually set musing at a specific time by watching an apple fall from a tree, he did know that the apple was accelerated downward at about 9.8 m/(sec)². Newton then suggested (without proof, at the time) that the moon, 60 times as far from the center of the earth as the apple, should be pulled on and accelerated toward the earth $(1/60)^2$, or $1/3600$, as much. Therefore, the centripetal acceleration of the moon, v^2/r, must be equal to 9.8 m/sec²/3600.

$$\frac{v^2}{r} = \frac{9.8 \text{ m/(sec)}^2}{3600}$$

$$v^2 = \frac{9.8 \text{ m/(sec)}^2 r}{3600}$$

so

$$v = \sqrt{\frac{9.8 \text{ m/(sec)}^2 r}{3600}}$$

Substituting this expression for v in the equation for the period, τ, of the moon,

$$\tau = \frac{2\pi r}{v}$$

$$= \frac{2\pi r}{\sqrt{\dfrac{9.8 \text{ m/(sec)}^2 r}{3600}}}$$

$$= 2\pi r \sqrt{\frac{3600}{9.8 \text{ m/(sec)}^2 r}}$$

$$= 2\pi \sqrt{\frac{3600 r}{9.8 \text{ m/(sec)}^2}}$$

If r, the radius of the moon's orbit, is 240,000 miles (3.86×10^8 m) then:

$$\tau = 2\pi \sqrt{\frac{(3600)(3.86 \times 10^8 \text{ m})}{9.8 \text{ m/(sec)}^2}}$$

$$= 2\pi \sqrt{\frac{(3.6 \times 10^3)(3.86 \times 10^8 \text{ m})}{9.8 \text{ m}/(\sec)^2}}$$

$$= 2\pi \sqrt{\frac{13.9 \times 10^{11} \text{ m}}{9.8 \text{ m}/(\sec)^2}}$$

$$= 2\pi \sqrt{13.9 \times 10^{10} \ (\sec)^2}$$

$$= 2\pi \ (3.77 \times 10^5 \sec)$$

$$= 2\pi \ (3.77 \times 10^5 \sec) \left(\frac{1 \text{ day}}{8.64 \times 10^4 \sec} \right)$$

$$= \frac{(2)(3.14)(3.77)(10^1 \text{ days})}{8.64}$$

$$= 27.4 \text{ days.}$$

This is quite close to the observed time required for the moon to go around the earth once!

What is the gravitational force between two objects really equal to, numerically? Equation (5-5) does not help much because it is merely a proportionality. Henry Cavendish (1731–1810) finally answered this question in 1797 by designing a delicate balance which could measure the tiny gravitational pull between two pairs of lead balls. The large balls were positioned so as to twist the balance first in one direction (1) and then in the other (2). Cavendish could measure the masses of the lead balls, the distance between their centers, and the force required to twist the balance through the observed angle. With this informa-

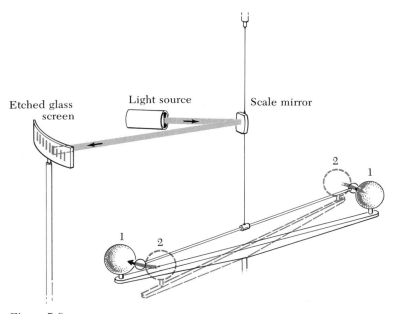

Figure 5-8
A diagram of the Cavendish balance, used to measure the gravitational force between objects of known mass.

tion, he could calculate the universal gravitational constant G, and thereby determine the gravitational attraction between any two masses:

$$F_G = G \frac{m_a m_b}{(r_{ab})^2},$$ Equation (5-6)

where $G = 6.67 \times 10^{-11}$ N·m²/kg².

If m_a and m_b are measured in kilograms and r_{ab} in meters, then G determines the gravitational force (in newtons) between the two masses. Two one-kilogram masses one meter apart experience a gravitational attraction of 6.67×10^{-11} N. Attractions between humans, at least the gravitational type, are pretty weak!

Suppose we consider a mass m_a which is situated in a region where a gravitational field of influence has been set up by a mass m_b. The force acting on m_a can be expressed as

$$F_g = \frac{G m_a m_b}{(r_{ab})^2} = m_a \left[\frac{G m_b}{(r_{ab})^2} \right]$$

As various masses are placed at the same point, various forces will act on them. But the force per unit mass, $F_g/m_a = G m_b/(r_{ab})^2$, remains the same. It depends only on the location and the fact that m_b is a certain distance away.

The ratio, of the gravitational force acting on an object to the mass of that object, is called the gravitational *field intensity* at that point in space. The symbol "g" is used to represent it.

Let us look into the intensity of the earth's gravitational field and into the claim put forward by Cavendish after his experiment that he had "weighed the earth". In chapter 4, we state that "weight" is the name given to the gravitational force exerted on an object. Since an object having a mass of one kilogram at the earth's surface weighs 9.8 newtons the gravitational field intensity at the earth's surface is 9.8 newtons per kilogram. But it is also equal, by Newton's law of gravity, to $G m_b/(r_{ab})^2$, where m_b is the mass of the earth and r_{ab} is the radius of the earth (4000 miles or 6.44×10^6 m). (At a distance of 4000 miles from the surface of the earth, or 8000 miles from its center, the gravitational field intensity would be one-fourth as great, or 2.45 m/sec².) At the earth's surface,

$$\frac{G m_b}{r_{ab}^2} = 9.8 \text{ N/kg}.$$

$$r_{ab} = 6.44 \times 10^6 \text{ m, so}$$

$$m_b = \frac{(r_{ab})^2 \ (9.8 \text{ N/kg})}{G}$$

$$= \frac{(6.44 \times 10^6 \text{ m})^2 \ (9.8 \text{ N/kg})}{6.67 \times 10^{-11} \text{ N·m}^2/\text{kg}^2}$$

$$= 6 \times 10^{24} \text{ kg}!$$

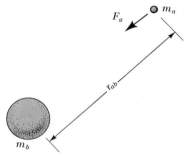

Figure 5-9

If the gravitational force acting on the mass m_a is divided by m_a, a number is obtained which depends only upon m_b and the distance r_{ab}. It does not depend on m_a.

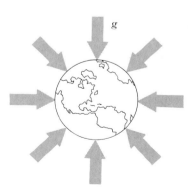

If the power of 10 notation does not allow the visualization of such a big number, then perhaps a six followed by 24 zeros would be more impressive.

While we're at it, let's weigh the sun! The earth is at an average distance of 93×10^6 miles from the sun, and takes one year (3.15×10^7 seconds) to complete one revolution in its orbit. Then the earth's velocity, v_e, must be

$$v_e = \frac{2\pi r}{\tau}$$

$$= \frac{2\pi (9.3 \times 10^7 \text{ miles}) \ (1.61 \times 10^3 \text{ m/mile})}{3.15 \times 10^7 \text{ sec}}$$

$$= 2.97 \times 10^4 \text{ m/sec}$$

$$= 18.5 \text{ miles/sec}.$$

The centripetal force acting on the earth must be

$$\frac{mv^2}{r} = \frac{(6 \times 10^{24} \text{ kg}) (2.97 \times 10^4 \text{ m/sec})^2}{(9.3 \times 10^7) (1.61 \times 10^3 \text{ m})}$$

$$= 3.6 \times 10^{22} \text{ N}$$

To visualize the magnitude of this force, imagine that it is being supplied by a steel cable having a strength of two billion newtons per square meter. The cable would have to be 3000 miles thick!

To exert such a force, the sun's gravitational field intensity, g_s, at the distance of the earth's orbit would have to be:

$$g_s = \frac{F_g}{m}$$

$$= \frac{3.6 \times 10^{22} \text{ N}}{6 \times 10^{24} \text{ kg}}$$

$$= 5.9 \times 10^{-3} \text{ N/kg}.$$

Figure 5-10
The gravitational force, F_g, attracting the earth toward the sun is enormous.

We are thus being pulled on by the sun about 1/1650th as hard as we are being pulled on by the earth. (5.9×10^{-3} N/kg is about 1/1650th of 9.8 N/kg.) The mass, m_s, of the sun can now be found by substitution in the equation $g_s = m_s \, G/(r_{ab})^2$, where $r_{ab} = 9.3 \times 10^7$ miles.

$$m_s = \frac{g_s \, r_{ab}^2}{G}$$

$$= \frac{(5.9 \times 10^{-3} \text{ N/kg}) (9.3 \times 10^6) (1.61 \times 10^3 \text{ m})^2}{6.67 \times 10^{-11} \text{ N} \cdot \text{m}^2/\text{kg}^2}$$

$$= 2 \times 10^{30} \text{ kg}.$$

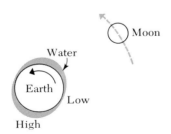

Figure 5-11
As the earth rotates relative to the moon, it experiences approximately two high tides, and two low tides, per day.

This is about 333,000 times the mass of the earth! When the masses of all the sun's planets are added together, the total is still a small fraction of one percent of the solar mass!

The moon's gravitational pull on the earth is about 1/200th that of the sun. But surprisingly, the moon has a larger effect on the tides than does the sun. This is because there is a greater fractional difference between the distances from the moon to the near and far parts of the earth than there is for the sun. Since the sun is about 370 times farther away than the moon all parts of the earth appear to be about the same distance away. However, in the case of the moon, the pull on the ocean waters on the near side of the earth is larger by three percent than its pull on the ocean waters on the far side. Water is thus pulled away from the earth on the near side and the earth is pulled away from the water on the far side, producing a hump of water on both sides of the earth. Local topography may alter the timing and the magnitude of the hump, but a high tide generally reaches a given earth shoreline twice in a little more than one daily rotation of the earth.[10] The earth's crust itself undergoes smaller tidal motions and the overall drag of the rotating earth keeps the tides slightly ahead of the moon's pull.

When a pair of objects attract one another, the force is mutual, in that both are accelerated toward each other. The more massive of the pair is, of course, accelerated less. Nevertheless, the earth should not be thought of as moving rigidly around its orbit like a train on a track, with the moon performing loop-the-loops around it. The earth and the moon both rotate around a common point between their centers. The earth has 81.3 times the mass of the moon. For reasons that will be understood better after the next chapter, the common center of rotation is out 1/81.3 of the distance from the earth to the moon (measured from the center of the earth). Since the moon is about 60 earth radii from the earth, this means that the earth is wobbling about a point about 3/4 earth radii from its center, as well as moving in its orbit around the sun, and spinning around once daily!

Further Successes of Newtonian Mechanics

The Newtonian analysis of planetary motion was not only simpler than Descartes' and Kepler's, but it was more complete, and it put people in the position of being able to predict where heavenly objects ought to be in the future. It has had repeated successes, including the correct prediction, on the basis of Edmund Halley's observations of its orbit in 1682, that Halley's comet should re-appear in 1759. It has come back again on schedule twice since then, and is predicted to return in 1986.

The number of observed planets had stood at six (five plus the earth) since ancient times. The seventh one, Uranus, was found in 1781. That is, it was recognized as a planet after having

(10) Goldreich, Peter, "Tides and the Earth-Moon System", *Scientific American* 226, April, 1972, pg. 42.

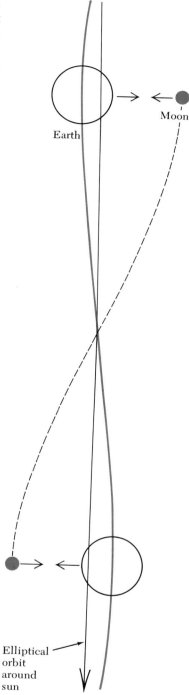

Figure 5-12

The earth and the moon rotate about each other, so that the earth wobbles back and forth from its elliptical path around the sun.

been observed off and on for 90 years.[11] After this, the orbit of Uranus was calculated, using Newton's laws and theory of gravity. However, the agreement with observation was not good. Later on a Frenchman, Joseph Leverrier (1811–1877), decided that another planet, as yet unobserved, was tugging on Uranus, altering its orbit. He calculated where the new planet should be. It was found there in 1846, a year later, and was named Neptune. By 1909, it was clear to astronomers that something was disturbing *Neptune's* orbit. The location of the disturbance was pinned down in 1930, when Pluto, the ninth planet, was finally observed by telescope.

The moon flights were a notable application of Newtonian theory. Before the manned moon flights the Lunar Orbiter series of mapping satellites was put in orbit around the moon. These lunar satellites did not follow the exact orbit predicted for them. It was deduced that this was because the moon has massive concentrations of heavy crustal materials in certain places, so that its gravitational field is irregular. As the Lunar Orbiters traveled around the moon, they entered certain regions where the gravitational pull was greater than expected, so that the ship was drawn in closer than calculated. Computers compiled these alterations and helped to locate these "Mascons", so that, when the first manned ships made the trip, they could be given a boost of power at the needed times to bring them back to their correct orbits.

The Process of Science

Science textbooks often begin with a discussion of "The Scientific Method." In this text, such a discussion has been intentionally withheld until now for two reasons. First, a good historical example or two is usually needed, in order to give sense to the statements that are made. Secondly, beginning this text with a discussion of "The Scientific Method" would have made it appear to be a more formal process than it really is. If the reader comes to the end of this chapter with the realization that doing physics is a fairly informal, sometimes non-logical or creative process, then the discussion will have been a success.

Figure 5-13 is an attempt to relate some of the elements involved in doing an experimental science. It is important to realize first that physicists do not go into the laboratory with blank minds. They have plenty of pre-conceptions about what is going to be "interesting" to look for. These preconceptions and expectations are bound together into one or more *hypotheses* or working theories. The hypotheses have in them ideas as to what is relevant to the investigation and also what is the best

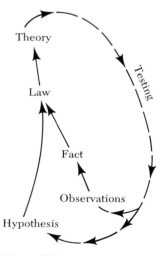

Figure 5-13
The loop of science.

(11) Doig, Peter, A *Concise History of Astronomy*, London, 1950, pp. 115–116.

conceptual language in which to express the questions. For example, in studying the heavens, the Greeks had already developed and agreed upon the concept that planets and stars were permanent bodies. They also had the idea of a path or orbit, and the belief that motion has some cause. As soon as tablets on astronomy were being carved, a set of rules had already been established according to which stars and planets were discussed. Many of these rules are still in use. The point of all this is to insist that scientific thinking does not proceed in an objective vacuum. It proceeds according to certain slants, or biases.

Observation

What the physicist decides to look at is thus a product of hypothesis. It has already been mentioned in Chapter 1 that abstraction is an important part of observation. That is, a great deal of information concerning a system is always ignored by scientists. They concentrate on those portions of systems which can, hopefully, be measured numerically. In this way, scientists unravel certain trends from the tangle of all that is going on, but they can also miss certain truths which are visible only when the whole system is considered.

Chapter 2 contains a discussion of how measurement is communicated as a ratio, comparing the measured quantity with a standard; and how the meaning of the terms communicated must have an operational basis.

The *intellectual* importance of experimentation might have been understated by emphasizing the idea that science begins with hypothesis, as well as by drawing on examples from astronomy, where experiments can't readily be done. Experiments do not just verify or fail to support hypotheses. They are, by themselves, a rich source of new ideas and problems to be solved. Experiment not only keeps the loop process sketched in Figure 5-13 tied to the perceived world, but it helps guide reason in the most productive directions.

Fact

One of the products of observation is *fact*. Our society probably attaches too much importance to fact. One is supposed to stop arguing when presented with "facts." However, facts *can* be prejudiced as a result of all of the circumstances which can make observations selective. As the philosopher Norbert R. Hanson said,[12] "Facts tend to be theory-laden".

(12) *Patterns of Discovery*, Cambridge University Press, 1958.

Quite often the "facts" in a scientific investigation can only be seen after long experience in *trying* to see something a certain way: an astronomer learns to find movements in a single spot of light against a background of spots; a geologist finds facts in a rock which the untrained observer cannot see. Facts can be defined as experimentally verifiable statements about concepts which have been chosen for attention. Much of "scientific fact" are simply those observations which fit well into the research program already underway; into the conceptual scheme. This recalls Cardinal Bellarmine's statement (pg. 66) against the existence of pure fact. The spots he saw around Jupiter when he looked through Galileo's telescope were a long way from being moons, in his mind.

In spite of all this, the most interesting and influential facts are often those which *don't* fit in with the rest. It is these which might lead eventually to changes in science, whether it be Aristotelian, Newtonian, or Einsteinian.

Physical Law

A physical *law* is a concise statement summarizing a large number of observations. It expresses a regularity in the behavior of a physical system. It might be expressed in words or in mathematical symbols. It often relates two observables: Kepler's laws relate position of planets and time; Galileo's laws of motion relate acceleration, initial velocity, distance of movement, and time; Newton's law of gravity connects mass, distance, and gravitational interaction.

Physical laws are not just general sets of facts. Carl Hempel,[13] a philosopher of science, points out that the statement, "All rocks in this box contain iron" would not be regarded as a physical law because the evidence and the scope of prediction coincide. On the other hand, a statement of the type, "If a rock is put into this box, it will contain iron" could be a law. The statement has some prediction implicit in it. It has an applicability beyond what has already been observed.

A physical law can imply that a solid, deterministic relationship exists among observed quantities, as in Newton's law of gravity, or it can imply statistical or probabilistic relationships, as is the case for the thermal phenomena to be discussed in Chapters 15 and 16. For example, in the latter case, chance fluctuations limit one to saying that a hot and a cold object placed together will *probably* come to the same final temperature. Some claim that all laws of physics are laws of probability. The seeming certainty of the laws which describe the motion of the planets arises, then, from the large size of the systems

---

(13) *Philosophy of Natural Science*, Prentice-Hall, Englewood Cliffs, N.J., 1966.

involved. The chance fluctuations which loom large in atomic (and in some thermal) phenomena, are small enough to neglect in planet-sized systems.

Scientific Theory and Explanation

The distinction between physical law and theory is not sharp. A scientific statement can be both law and theory. However, there are important differences in the function and structure of law and theory, as well as the role of the scientist in formulating them.

Insofar as a statement is law, it *describes* a relationship between more or less observable quantities.

Insofar as a statement is theory it is a creative idea which seeks to *explain* laws. It is quite often couched in terms of abstract, non-observable quantities.

If the man on the street were to be asked what is the difference between law and theory, he'd often say that law is something science is sure about but that theory is tentative and man-made. By now, the reader should be convinced that both are man-made. A law is perhaps more certain, since it is meant to summarize what has been observed. However, science is more than naming, classifying, and describing. If its purpose is to explain, or to increase understanding, then law will always be a preliminary to theory, with theory paramount.

If scientific theories are meant to explain, then what is explanation? The answer to this question depends upon one's epistemological point of view, but several things can be said about explanations:

(1) Logically, one explains a law when it is shown to be deducible, along with other laws, from a general premise (theory).

(2) Explanation can involve the re-expression of a statement in a more acceptable form: possibly in a more familiar form, in the sense of being connected to other knowledge. (As an explainer you try to get the explainees to nod their heads.)

(3) A more positivistic or operationalistic (pg. 17) approach would say that explaining is the expressing of complicated facts into as few and simple statements as possible. Nobel Laureate Eugene P. Wigner, a theoretical physicist, has said,[14]

Explanation . . . is the establishment of a few simple principles which *describe* the properties of what is to be explained. If we *understand* something, its behavior—that is, the events which it presents—should not produce any surprises for us.

(14) "Events, Laws of Nature, and Invariance Principles", *Science*, *145*, pg. 995, 4 September, 1964 © by the American Association for the Advancement of Science.

Hempel says that *scientific* explanation must in addition meet the criteria of *relevance* and *testability*. Consider the following statement by Francesco Sizi (an astronomer contemporary with Galileo) concerning the moons of Jupiter that Galileo had seen with his telescope:

> There are seven windows in the head, two nostrils, two ears, two eyes and a mouth; so in the heavens there are two favorable stars, two unpropitious, two luminaries, and Mercury alone undecided and indifferent. From which and many other similar phenomena of nature such as the seven metals, etc., which it were tedious to enumerate, we gather that the number of planets is necessarily seven . . . Moreover, the satellites of Jupiter are invisible to the naked eye and therefore can have no influence on the earth and therefore would be useless and therefore do not exist.[15]

This quotation sheds further light on the episode we have already discussed on pages 66 and 118, and at the same time is a good example of an unscientific explanation in Hempel's terms. Theorizing in terms of the human head is not relevant or helpful in deciding whether Jupiter has satellites or not. In the way it is stated, it is not open to a test of its validity. That is, the allowed means of testing are closely circumscribed.

A second, more recent, example of an explanation which fails the criterion of testability is quoted by Hempel. It concerns a theory[16] which suggests that the gravitational attraction between bodies is a manifestation of appetites or natural tendencies closely related to love! No conceivable observation or experiment can confirm *or* deny the assertion he makes. Therefore, it might be true, but it is not scientific.

We've seen two examples of non-scientific explanation. Suppose now we set out to explain scientifically how it is that water pipes sometimes burst in an unheated house during the Winter. We first employ an experimental law: water expands when it freezes. If the pipe is full of water when freezing occurs, and the metal cannot stretch enough to accommodate the ice, the pipe will burst. Although the expansion of water is not theory but empirical law it still is a good explanation because it leads, by deduction, to other phenomena which can be observed, such as ice floating on the *top* of frozen lakes, rather than sinking to the bottom. (If it didn't, fish would have trouble surviving the winter in lakes and rivers.) The next step is to explain how it is that water expands when it freezes. This can be done by postulating water to be made up of many tiny, separate molecules. If the model is sufficiently detailed, it will tell how the water mole-

(15) From Holton, Gerald, *Introduction to Concepts and Theories in Physical Sciences*, Addison-Wesley, 1973.
(16) O'Brien, J. F., *The Thomist 21*, 1958, pp. 184–193.

cules arrange themselves less densely in ice than in the liquid state. A theoretical model of water would have been applied. Wetness, waves, and transparency—all the perceived qualities of water—have disappeared and in their place is an imaginary network of particles and forces. It does not fit common sense, but it is better at predicting the behavior of water than common sense is.

In the same way, Kepler's Laws can be "explained" by seeing them as consequences of Newton's theory of gravitation; along with pendulum motion, the tides, the free fall of objects, the motion of the moon, the floating of ice (can you see how?), and other phenomena. As we saw, Newton's theory also was used to predict the existence of Neptune and Pluto before they were observed directly. All of these phenomena were tied together—thus were explained—as being examples of a more general theoretical phenomenon.

However, one always reaches a stopping point in scientific explanation which is short of answering the final question "why". To answer the question "why" is to prove that some phenomenon *must* happen. Why should two masses attract each other? Newton was not as willing to enter this metaphysical realm as was Kepler. Newton said,

> To us it is enough that gravity does really exist and act according to the laws which we have explained, and abundantly serves to account for all the motions of the celestial body and our sea.

A theory, then, is a few statements which are physically testable, which leads to specific phenomena by deduction and which can be used to predict phenomena or laws not yet observed. The theory might be *formal*, in which case it consists merely of a mathematical statement, such as $F_g = Gm_a m_b/(r_{ab})^2$. On the other hand, it might take the form of a *model*. A model is usually a miniature, or a magnified, or a simplified structure which is analogous to something that can be directly perceived. Ptolemy's, Copernicus', and Kepler's solar systems were all models. The kinetic molecular model indicates that a solid is something like billions of closely packed tiny billiard balls bouncing around in a box. A model helps make phenomena familiar, but it can also mislead. It can engender a false familiarity. Molecules have many characteristics *unlike* billiard balls.

In the last half-century, there has been a marked movement *away* from models in physics, especially in the physics of atoms. The way in which we see human sized objects; their length, position, solidity, locatability, and other properties, all become inappropriate when we are trying to describe and predict the behavior of atomic sized objects. Atomic models are thus giving way to mathematical statements of the probabilities of various events occurring. Writers have expressed concern over this increasing use of mathematical abstraction in describing nature, and often look back wistfully to the time when physical theories were easier for the layman and scientists in other specialties to understand. So far, the special and general theories of relativity, and quantum mechanics (the study of the motion and energy of atomic, nuclear, and sub-nuclear sized systems), represent the physicists' main contribution to the physical description of nature in the twentieth century. They are both very mathematical in form. In answer to those who criticize this trend toward abstraction, Philipp Frank, a philosopher of science, has written:[17]

> . . . at the time of its discovery Newton's theory of motion was regarded as an abstract, merely mathematical theory; in our day, however, it is often cited as an example of an intuitive theory, especially when philosophers wish to establish the abstract character of the theory of relativity and quantum mechanics by contrasting them with an intuitive theory. In truth, however, the alleged difference between Newtonian and relativistic mechanics depends only on the undeniable fact that the difficult and complicated calculations and deductions required for understanding Einstein's theory of relativity are not required for understanding the phenomena of everyday experience. For these phenomena can be formulated with the help of the more simple Newtonian theory of mechanics. Hence the statement that a theory like Einstein's is abstract and nonintuitive simply means that it is more complex than is necessary for a theory that need describe only the facts of daily experience.

Whether theories are expressed in mathematical symbols or not, they are typically not expressed in terms of directly observable quantities. This is because scientists have realized that the uniformities between natural phenomena are not between the phenomena themselves but between the elements into which the phenomena can be dissected. The acceleration of the moon, for example, becomes comprehensible only when it is analyzed into accelerations toward the earth, the sun, and, to a lesser extent, the other planets. None of these partial accelerations has ever been observed. They are theoretical constructions.

Theories are built of ideas such as mass, molecules, and waves. None are directly observable. The consistencies of natural reality seem to lie at an invisible level. The philosopher and humanist José Ortega Y Gasset said,[18]

> In order to discover reality we must for a moment lay aside the facts that surge about us and remain alone with our minds. Then, on our

(17) *Modern Science and Its Philosophy*, George Braziller, N.Y., 1955, pg. 150.
(18) In "Galileo And His Effect on History", Chapter 1 in *Man and Crisis*, translated by Mildred Adams, Norton, N.Y., 1962.

own risk and account . . . we construct an imaginary reality, a pure invention of our own; then, following in solitude the guidance of our own personal imagining, we find what aspect, what visible shapes, in short, what facts would produce that imaginary reality. It is then that we come out of our imaginative solitude . . . and compare those facts which the imagined reality would produce with the actual facts which surround us . . .

Thus, the elements in this new reality must still connect operationally with the visible world. The "tendencies" which controlled the motion of objects in the medieval world view were not empirically testable, whereas mass, molecules, and waves all have an empirical basis.

The Creative and Aesthetic Side of Theorizing

The word "imagining" which is so important in the quotation from Ortega y Gasset points out the most significant element in the making of scientific theory: creativity. There is no logical formula for getting from a set of empirical laws to a theory which can explain them. Hunches, inspirations, accidents, and dreams have all played important parts in the creation of scientific theory. The higher one goes up the ladder from observation to theory, the less appropriate is the word "discovery." The electron, the molecule, and the light wave were "invented" as much as they were "discovered".

In trying to explain the origin of creative ideas, Arthur Koestler, writes[19] of a process which he calls "bisociation", in which two previously unconnected sets of experiences accidentally connect for the first time. The genius is the person whose mind constantly juggles sets of experiences into various combinations, some of which "click" and become important new ideas, insights, or suggestions for further thought. Koestler also sees *humor* as arising from the same source: a surprise juxtaposition of events which don't normally go together can make us laugh.

The doing of contemporary physics is also creative in the sense that physicists not only try to understand and describe nature, but also, they add to it. Many of the materials which scientists study, and the conditions under which the studies are made: low temperature, high magnetic fields, various vibration frequencies, vacuum, etc.—do not exist on earth, so far as we know. Later, due to widespread technological applications, some of these materials or conditions can become a familiar part of (man-made) nature. The transistor, radioactive isotopes, and antibiotic drugs are examples of this process.

(19) *The Act of Creation*, Dell Publishing Co., N.Y., 1964.

When a piece of scientific work is presented to the world in the form of a publication, it is always shaped and polished to fit, like a brick into a wall, alongside the contributions of other scientists. In so doing, scientists make their results and conclusions appear as though they were almost logically inevitable. In truth, their work was probably full of wrong turns, blind alleys, and bad guesses. Gerald Holton,[20] calls the actual process "private science" and the paper which is published, "public science." Textbooks usually make science appear to be entirely of the logical, public variety when it really is not.

The choice between rival scientific theories is often made on an aesthetic basis: that the simplest, or the most elegant, is best. Simple does not mean easy, in terms of familiarity, but rather that it involves as few statements and assumptions as possible. We have already seen that Copernicus' model for the solar system eventually won out over Ptolemy's partly because it required fewer postulates. Descartes' vortex theory for explaining the attraction of the planets toward the sun was dropped in favor of Newton's gravitational theory because it required too much postulation about the nature of the invisible fluid whirling around the vortex. Three centuries later, Einstein said,

> Our experience justifies us in believing that nature is the realization of the simplest conceivable mathematical ideas.

He believed in simplicity in a very personal way, much as did Mohandas Gandhi. During his life, Einstein gradually eliminated socks, sleeves, shaving soap, and haircuts as unnecessary encumbrances.

On the other hand, the philosopher Alfred North Whitehead warned, "Seek simplicity and mistrust it." The simplicity of physics is partly a result of the selection and abstraction in the original observations made by the physicist. The "exact" sciences are those with few variables, while the "inexact" sciences are those with many. Throughout the structure of physics one can see examples of the effort that has been made to simplify— to find a few static, enduring concepts in the midst of change. However, the multiplication of the "fundamental particles" of matter from three in 1900 to more than 80 by 1970 made some scientists[21] wonder if physics will continue to be basically simple. Will some future synthesis restore a simpler structure to high-energy particle physics, as the Copernican theory did for the Ptolemaic theory of planetary motion, or will physics, like the biological world, become more complex with time?

(20) "On the Duality and Growth of Science," *American Scientist 41*, 1953, pp. 89–99.
(21) Cocconi, G., "The Role of Complexity in Nature," pp. 81–87, in *The Evolution of Particle Physics*, M. Conversi, ed., Academic Press, N.Y., 1970.

Scientific Logic, Mathematics
and Truth

125

Scientific Logic,
Mathematics
and Truth

In hopes of counterbalancing the popular opinion that science and scientists are coldly logical and nothing else, the creative, accidental elements in the explanation of the physical world have been emphasized here. But the doing of physics as a science does, of course, involve orderly thought, and some of this thought involves elements of formal logic. The logic of the sciences is not really any different from the logic that a person uses in finding a lost book or solving a lot of other everyday problems. The care with which terms are defined is greater in the sciences, but the "method" is the same.

In the stating of physical law, and in the creation of physical theory, physicists use *inductive inference*. In this process, they extrapolate from a specific set of observed relationships to a statement claiming that the observed relationship is generally true. Newton found that an attractive force between the sun and each planet, decreasing as the square of the distance from the sun to the planet, can produce an orbit which matches those actually followed by the planets. He then made the general inductive statement that all objects in the universe attract each other with a similar gravitational force. He claimed that what is true in our part of the universe is true everywhere, and what is true now was true in the past, and will be in the future. It is not difficult to see the logical weakness of induction. Its very purpose is to make claims about phenomena not yet observed, and to permit groping from the known to the unknown.

Once scientists have arrived at their premises (or theory) by an inductive inference, then the laws which first engendered the inductive theory, as well as other laws, or phenomena not yet observed, can be arrived at *deductively*. That is, they can be seen as specific conclusions that one would arrive at, starting with the general theory as a premise. The word "deduction" is perhaps too strong to use here, because the specific conclusions drawn from general theoretical premises are not inevitable, necessary, or proven, in the strict logical sense. Logic is primarily a set of rules arranging statements and checking on the consistency of the statements involved. Science on the other hand is interested in relating descriptive statements and *nature*, which does not constitute a logically closed system. There is always something you don't know. There is always the possibility that other theoretical premises could work just as well as the ones you have found. Newton's theory of gravitation can be used to deduce the period of the moon in its revolution around the earth. The fact that the calculated period does match the observed period does not prove that the theory of gravitation is correct. It merely helps to confirm and add support to the theory

The philosopher, Karl Popper, has pointed out that disproof, rather than proof, is crucial in the sciences. Science almost never proves anything to be true. But wrong statements which do not match empirical results can be eliminated. As we saw on page 120 a statement is scientific only if it *is* disprovable. If there is no empirical test which will show a statement to be false, then it is not scientific. A scientific statement has to stick its neck out and make a commitment about physical phenomena which can be tested for falsity.

The logical deductive process is at best only one of the physicist's steps in predicting new phenomena from "established" theory. They learn from experience which mathematical answers to keep and which to discard. Mathematics itself is an empty game as far as revealing new things about the world is concerned. However, the use of mathematics can make mistakes in thought and logic more visible, help uncover things that were there all the time, or help give scientists new frames of reference. In a sense, all the information Newton needed for his law of gravity was in the data that Kepler used. But Newton's mathematics showed him that he could assume the sun and the planets to be point masses. From this, he then proceeded to show that the gravitational force must decrease as the square of the distance.

The Importance of Consensus in Science

It is only partly true to say that in science, "Truth is the result of an experiment." If a theory has been confirmed by many experiments, and fits smoothly into the network of other physical ideas, and especially if a lot of scientists believe in it,[22] no single contradictory experimental result will overthrow the theory. An interesting historical example of this is the Michelson-Morley experiment, which, as is shown in Chapter 8, indicated that a beam of light sent out from the earth travels through space at the same speed in all directions even though the earth itself is moving through space at a high speed. This invariance in the speed of the propagation of light became one of the two postulates upon which Einstein based his special theory of relativity in 1905. Within 10 years after the publication of this theory, its use had made so many other things fall into place that it had become the backbone of a new physics more general than Newton's. But one man, Dayton C. Miller, decided to repeat Michelson and Morley's experiments after he noticed in their data some instances in which there *was* an apparent effect

(22) Ziman, John, *Public Knowledge*, Cambridge University Press, 1968 (Says science at the core is concerned with achieving a public consensus.)

of the Earth's motion in its orbit on the measured speed of light. He gave lectures and published papers on the results he found.[23] But the world of physicists did not pay much attention to him. Special relativity was just too beautiful to throw away. Subsequently, Robert S. Shankland and several colleagues took a third look at the experiments,[24] re-studying Miller's notebooks to determine the conditions under which the work was done. They found that during many of the experiments (done in a small shack on Mount Wilson in California, between 1921 and 1926) the sun shown strongly on one side of the building. They calculated the difference in temperature between the two walls of the shack which would result, how much this would affect the velocity of light in air, and how much the warm side of the apparatus would expand. When they took into account these effects, the dependence of the velocity of light on motion of the earth through space disappeared, and the special theory of relativity was confirmed once again! This is an instance in which a belief in simplicity on the part of the physics community was not shaken by counter-indications and was eventually verified.

Thomas S. Kuhn, in a historical look at science,[25] decides that it changes more typically by revolution than evolution. This happens, he says, because the scientific community not only demands nearly unanimous votes, but at a given time sees the world from a particular perspective; thinks in terms of a particular *paradigm*. If evidence which does not agree with the paradigm shows up, it has no effect at first. But if it continues to collect, counter-theories get born, and in time the whole science community switches to one of them. The old theory is not altered or reinterpreted, he says. It is replaced. Physical truth thus changes by jumps. If it doesn't, it's not science, according to Kuhn.

Summary

The Ptolemaic solar system model expanded on Aristotle's model and endured until the seventeenth century. It claimed that the earth was at the center of the solar system (and the entire universe as well). In order to explain the variations in the apparent motions of the sun and planets, they were said to travel on revolving circles whose centers were always located on other circles. In all, up to 34 circles were required, the number increasing in step with the accuracy of observations.

Copernicus proposed a new model in 1543 which put the sun at the center. Fewer assumptions were required to make the

(23) *Reviews of Modern Physics*, Vol. 5, pg. 203, 1933.
(24) *Reviews of Modern Physics* 27, pg. 167, 1955.
(25) *The Structure of Scientific Revolutions*, University of Chicago Press, 1962.

model work. Early in the seventeenth century, Kepler took advantage of much more accurate data and stated his three empirical equations of planetary motion (pg. 103). They entailed the use of ellipses, a departure from the circles which had dominated Western science and art until late in the sixteenth century.

Late in the seventeenth century Newton saw the motions of the planets around the sun as just one example of his second equation of motion. He said that the centripetal force necessary to pull the planets into their orbits around the sun was provided by a gravitational force:

$$F_g = \frac{Gm_a m_b}{(r_{ab})^2},$$

Equation (5-6)

where m_a and m_b are the masses of the two objects and r_{ab} is the distance between their centers.

Gravitational field intensity at a point where a mass m is located is $F_g/m = g$. For the earth's gravitational field, g is 9.8 N/kg at the earth's surface, and directed down.

Science is a changing combination of observation, empirical law-making, theorizing, and testing. It uses fact, but fact itself presupposes certain methods of observation and attitudes toward what is most important to look for. A theory or model is a creative effort at explaining phenomena which laws merely describe. It is a few statements which are physically testable, which leads to explanations of specific phenomena by deduction, and which can be used to predict phenomena or laws not yet observed. Aesthetic principles, such as simplicity, are often used to determine the choice between competing theories. Since science consists of people trying to understand a physical world which they have not seen all of, it does not try to prove things about the world. A theory concerning the behavior of a system in the world can be verified by repeated observation, but scientists never arrive at a point of proving that a theory is correct, to the exclusion of all others. Style and consensus also play a part in science.

Further Reading

Bernstein, Jeremy, "To Find a Planet: On Neptune." pp. 167–174, in *A Comprehensible World*, Random House, N.Y., 1967. (Good prose account of the finding of the new planet.)

Born, Max, "On the Meaning of Physical Theories," pp. 13–30, in *Physics in My Generation*, Springer Verlag, N.Y., 1969. (A piece of good writing by a profound thinker.)

Bradbury, Ray, et al, *Mars and the Mind of Man*, Harper and Row, N.Y., 1973. (A discussion among five scientists and writers on the scientific investigation of a planet which has been the source of much metaphor.)

Butterfield, Herbert, *Origins of Modern Science*, G. Bell, London, 1949. (Chapter 4 "The Downfall of Aristotle and Ptolemy"; Chapter 8, "History of the Modern Theory of Gravitation"; and, Chapter 2, cited in footnote (1) of this chapter, are valuable and well written.)

Campbell, Norman R., "The Explanation of Laws," in *What is Science?*, Methuen & Co., London, 1921. (The difference between theories and laws.)

Cassirer, Ernst, "Science," Chapter XI in *Essay on Man*, Yale University Press, New Haven, 1944. (Closely written symbolic philosophical view. Says science is humanity's highest cultural achievement.)

Chesterton, G. K., "The Logic of Elfland," reprinted in *Great Essays in Science*, Martin Gardner, ed., Washington Square Press, N.Y., 1957, pp. 78–87. (A charming spoof of scientific law and the idea of cause and chance.)

Cohen, I. Bernard, Introduction to Newton's *Principia*, Harvard University Press, Cambridge, 1971. (Discusses some of the complexities of deciding just who did "invent" the law of gravity.)

Cranberg, Lawrence, "Law—Scientific or Juridical," *American Scientist*, 56, pp. 244–253, 1968. (Shows that many of the differences between these two types of law are more traditional and rhetorical than real.)

Diederich, Mary E., "The Context of Inquiry in Physics," *American Journal of Physics*, 40, March 1972, pg. 449. (What are the means for choosing between two rival theories?)

Feynman, Richard P., *The Character of Physical Law*, British Broadcasting Company, Cox and Wyman, London, 1965. (Chattily written, uses gravity as the main example.)

Gingerich, Owen, "Copernicus and Tycho", *Scientific American 229*, Dec. 1973, pg. 87. (Copernicus' and Tycho's writings show that the Ptolemaic solar system wasn't as complex as recently thought. Tycho carefully studied Copernicus, but developed a geocentric model.)

Holton, Gerald and Roller, Duane H. D., *Foundations of Modern Physical Science*, Addison-Wesley, Reading, Mass., 1958, Chapters 6–12. (Very well written history of planetary theory.)

Koestler, Arthur, *The Sleepwalkers: A History of Man's Changing Vision of the Universe*, Macmillan, N.Y., 1959. (Centered on the work of Kepler but goes on to lament the lack of communication between scientists and non-scientists.)

Kuhn, T. S., *The Copernican Revolution: Planetary Astronomy in the Development of Western Thought*, Random House Vintage Book, New York, 1957. (A thorough, but compact, discussion.)

Lanczos, Cornelius, "Albert Einstein and the Role of Theory in Contemporary Physics", *American Scientist*, 47, March 1959, pp. 41–59. (A lucid historical account of how Galileo, Newton, and Einstein each went about explaining physical reality.)

Nagel, Ernest, *The Structure of Science: Problems in the Logic of Scientific Explanation*, Harcourt Brace Jovanovich, N.Y., 1961. (First 6 chapters give a thorough discussion of law and theory.)

Park, David, "Gravity," Chapter 4 in *Contemporary Physics*, Harcourt, Brace & World, N.Y., 1964. (Non-mathematical. Goes into the principle of equivalence.)

White, Lynn, Jr., *"Science, Scientists and Politics,"* Published by The Center For Study of Democratic Institutions. (Says the climate of society must be right before science can flourish. Suggests that the appearance of ovals in European architecture helped pave the way for Kepler's planetary ellipses.)

Problems

5-1 What does it mean to "calculate an orbit"?

5-2 What are some differences between legal and scientific law? After you write down your answer, read Lawrence Cranberg's article (listed in Further Reading.)

5-3 Have music and painting become more complex or more simple during historical times, or have they undergone cyclic changes? Suggest some reasons for your answer.

5-4 Does a detective usually deduce "who done it"? If your answer is no, give an example of a type of case in which a detective really can deduce who the culprit is.

5-5 J. Ziman (as referred to in footnote (22)) claims that scientific knowledge which is kept secret is no longer science. Does this match your view of science? What *is* public knowledge?

5-6 R. Buckminster Fuller, architect and philosopher, says that our self-centeredness—our parochialism—shows up in our words. For example, we say "upstairs" and "downstairs" instead of "outstairs" and "instairs." Explain the difference in outlook revealed in the two sets of words. Do either of them eliminate our parochialism entirely? Can you devise any less parochial expressions than these?

5-7 Find the ratio of the value of g on the surface of Mars to its value on the surface of the earth, given that the radius of Mars is 1/2, and its mass is 1/8, that of the earth's.

5-8 The gravitational field intensity on Neptune is the same as on earth, but the diameter of Neptune is four times as great. What can you conclude about the material of which Neptune is made?

5-9 If the distance from the earth to the moon is 60 earth radii, where will a space ship experience equal and opposite gravitational forces directed toward both the moon, and the earth? (From page 115, the mass of the earth is approximately 81 times that of the moon.)

5-10 If a satellite is to serve as a space platform that hovers above a fixed point on the earth's equator, show that it must be placed in orbit about 22,000 miles above the earth's surface.

5-11 Suppose the earth were four times as far from the sun as it is now. By what factor would the gravitational force of the sun on the earth be reduced? By what factor would the acceleration of the earth toward the sun be changed? By what factor would the speed of the earth around its orbit change? How long would the new year be?

5-12 If the earth were scaled up by a factor of two, maintaining the same mass per unit volume, how would g be affected? If at the same time the molecules of the atmosphere doubled in diameter and the distance between them doubled, how would atmospheric pressure at sea level be affected? (Atmospheric pressure, measured as a force per unit of area (m²) can be understood as the weight of the air sitting on top of each square meter of the earth's surface.)

5-13 The planet Jupiter has 12 moons. Of the four discovered by Galileo in 1610, the closest to the planet has an orbit whose radius is 4.2×10^5 km, and a period of revolution of 1.77 days. The others have orbits of radius 6.7×10^5 km, 10.7×10^5 km, and 18.8×10^5 km. What are their periods of revolution around Jupiter?

5-14 Imagine (à la 2001) a wheel-shaped space station which rotates about an axis like a wheel as it moves through space. Let the radius of the ship be 100 meters. Suppose the passengers, from earth, are seated in the ship with their feet pointing away from the center of the wheel, on the outer most portion of the wheel. How fast must the space station rotate to make the passengers feel at home?

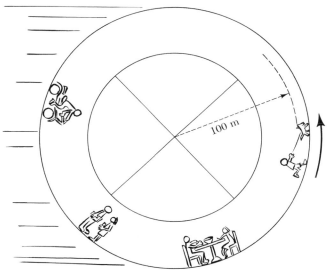

5-15 On page 54 it was shown that a satellite must have a horizontal velocity of about 8×10^3 m/sec, or 18,000 miles per hour to travel in a circular orbit around the earth. How does the required velocity change if the earth is made smaller, assuming that the mass per unit volume of the planet (5.52×10^3 kg/m³) stays the same? Show that if astronauts landed on an asteroid having the same density as the earth, and a radius of 4.06 km (2.4 miles) they could launch themselves into a circular orbit simply by running along the surface at 5 m/sec.

5-16 The average radius of the earth's orbit is about 93 million miles. If the average radius of Venus' orbit is 68 million miles and the average radius of Pluto's orbit is 3.66 billion miles, how long is a "year" on Venus and on Pluto?

M. C. Escher, *Day and Night*. (Courtesy of the Escher Foundation, Haags Gemeentemuseum, The Hague, and Vorpal Gallery, San Francisco.)

INVARIANCE

6

MOMENTUM AND
NEWTON'S THIRD LAW

Up to now change and motion have been the main points of our discussion and the examples we have considered. Now, we will begin looking for things which do not change with time or when frames of reference are changed. In the next three chapters principles of *conservation* are emphasized; principles which have become very important in modern physical thought. We will view conservation as a particular sort of *invariance principle*, and invariance principles will be viewed as examples of certain symmetry concepts with which physicists view their world.

Symmetry of Objects

At first thought, one sometimes views symmetry as some kind of balance: the same on the left as on the right. However, one can also see symmetry in such things as a four-bladed fan (Figure 6-1), in that a 90°, 180°, or 270° rotation about an axis passing through its hub does not visibly change the fan. Ceramic tiles (Figure 6-2) have symmetry, when a pattern is repeated at specific intervals, with each similar pattern having the same group of patterns surrounding it.

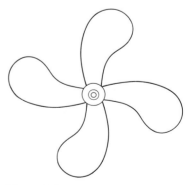

Figure 6-1
Any of the four fan blades could point up. You would not be able to tell the difference.

Figure 6-2
Seventeenth century Delft tiles.

In each of these cases, we could do something to an object which leaves it unchanged. There was an operation that could be performed, after which the object would appear to be unchanged. For example, one of the tiles, a "balanced" object, could be replaced with its *mirror* image and nothing would be different. That is, we could reflect its image in a mirror, so that its left side becomes its right side and no one would be the wiser. In the case of the whole pattern of tiles the operation would consist of a *translation* or a sliding movement of the pattern two, four, or six, etc., tiles to the left, right, or up or down. If you were to move a huge wall of tiles—one whose edges were too far away to be seen—in any of these ways, then the pattern would appear not to be altered. If the fan were *rotated* 90, 180, or 270 degrees in either direction, with no one watching, no one could tell the difference afterwards. A symmetrical object is thus one which can be subjected to certain operations and remain invariant. The objects we chose to look at possessed *mirror, translational,* or *rotational symmetry.* There are other symmetry operations which are more complicated.

From this discussion a unifying idea begins to emerge: symmetry properties always bring with them a simplification. This simplification is probably the root of our desire to find symmetries in the first place. If we can see a complex pattern as a "mere" repetition of some elementary pattern, we usually feel that we understand it better. But with symmetry also comes a hiding or limiting of knowledge; we can't tell when the tiles have been moved or the propeller turned.

Human beings themselves are superficially symmetrical — particularly if they part their hair in the middle. Internally, it's a different story, with the heart, stomach, and other organs located asymmetrically.

Symmetry and Invariance in Physical Laws

Laws, or descriptions of nature, can also possess symmetry.[1] Consider Newton's law of gravitation: $F_g = Gm_am_b/(r_{ab})^2$. The law states that the attraction between two masses depends only upon the distance between their centers, r_{ab}, and the masses of each, m_a and m_b. It makes no mention of the absolute position of either mass or whether one is above, below, to the left, or right of the other. The law may thus be said to possess *translational* and *rotational* symmetry. As long as the distance between the masses remains the same, we could move them around at will without affecting the force between them. Rotational symmetry does not imply that systems are unaffected while they are being spun around. It means rather that their behavior should be the same whether they are "straight up and down" or "tilted sideways." The quotation marks result from the knowledge that up, down, and sideways are different only because we live in a gravitational field which points in a certain direction. The law of gravity says nothing about the *velocity* of either m_a or m_b. Whether they are moving or at rest in the observer's coordinate system, the attraction between them will be the same. It is therefore "symmetrical" with respect to velocity.

When speaking of a law in this way; that is, where different observers (or different coordinate systems) come into play, it is usual to substitute another word for symmetry. The word is *invariance*. A phenomenon, law, or quantity is said to be invariant if it is the same in two different coordinate systems or frames of reference. This concept was encountered early in Chapter 3 (page 26), and now it is seen in a more general way. Physicists, in their quest for consistency and simplicity try to phrase natural laws in terms which will have the same form in different frames of reference. We will return to this subject in chapter 8.

(1) Wigner, Eugene P., "Events, Laws of Nature, and Invariance Principles," *Science, 145*, pg. 995, 4 September, 1964.

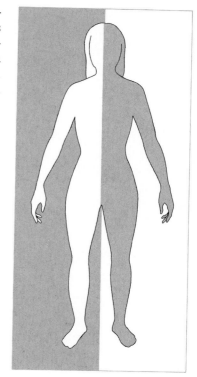

Figure 6-3
Humans, like most animals, have a plane of near-symmetry passing through them.

A different sort of "invariance" called *conservation*, involving time rather than switches in frame of reference, is investigated in this, and the next chapter. According to this principle, in the midst of the overall change within a system, certain quantities associated with physical systems can be singled out and found to have a constant numerical value. Physicists thrive on examining systems to find combinations of variables, which are invariant, or conserved, in the midst of overall change. When physicists are lucky enough to find such a combination, they give it a name and try to use it for understanding and predicting the behavior of systems.

The Conservation of Momentum

Recall Equation (4-2), which represented the operational definition of the relative masses of two objects:

$$\frac{m_a}{m_b} = -\frac{\Delta v_b}{\Delta v_a}.$$

It can be rewritten as

$$m_a \Delta v_a = -m_b \Delta v_b.$$

Or, from the definition for Δv, as

$$m_a(v_{af} - v_a) = -m_b(v_{bf} - v_b)$$

where v_{af} and v_{bf} are the final velocities of masses a and b, respectively, after the collision. Rearranging, we get

$$m_a v_{af} + m_b v_{bf} = m_a v_a + m_b v_b$$

or

$$[(m_a v_a) + (m_b v_b)]_{final} = [(m_a v_a) + (m_b v_b)]_{initial} \quad \text{Equation (6-1)}$$

From Equation (6-1) you can see that, although the velocities of individual objects in the system may change in complex fashion, the sum of mass times velocity for all the objects in the system gives a quantity which is the same before and after a collision. (Remember: You multiply the mass times the velocity of each object and *then* sum them up.) We call this quantity the *momentum* of the system. The momentum of a single object is simply its mass times its velocity.

In this example, the constant value obtained, or the conservation of momentum, is a direct consequence of the way in which we have defined mass. However, experiment has shown that the same principle applies to systems with more than two objects in them, which operate in more than the one dimension used here, and in which the masses or even numbers of the individual objects in the system might vary. Thus, momentum is a basic quantity which is very useful in describing the behavior of a physical system. It is a conceptual invariant quantity which remains the same in the midst of individual flux.

Only one idealization need be kept in mind: The system must be isolated, so that it experiences no interaction with its surroundings. The necessity of this becomes clear if you see the whole system as one object. If no external forces act on an object, and if its mass does not change, Newton's laws indicate that the velocity of the object is constant, as is its momentum. *Conservation of momentum, then, is simply the application of Newton's first law to a cluster of objects.*[2]

Let us illustrate with two examples:

(a) A man stands in a canoe near a dock. He jumps from the canoe onto the dock with a horizontal velocity of 2 m/sec. The canoe recoils in the opposite direction. Question: if we assume that the water exerts no drag force on the canoe, what is its recoil velocity?

First, notice that the system of man plus canoe has zero momentum before the man jumps. When he does jump to the right, the momentum of the whole system remains zero, so the canoe must recoil. Mathematically, both before and after the great leap forward:

$$m_c v_c + m_m v_m = 0$$

Solving for v_c yields the expression:

$$v_c = \frac{-m_m v_m}{m_c} \qquad \text{Equation (6-2)}$$

v_m = velocity of man

v_c = final velocity of canoe

Figure 6-4

The man moves in one direction, the canoe in the other, but the momentum of the whole system remains at zero. (If you try this next summer, be careful. Remember that you're used to jumping from a fixed location.)

(2) We say "simply", yet several decades elapsed between Descartes' enunciation of the law of inertia and a statement of the conservation of momentum of a system during a collision. It was actually worked out in 1668 by Huygens, John Wallis, and Sir Christopher Wren after some urging from the Royal Society of London.

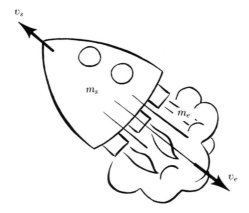

Figure 6-5
*The greater the momentum of the exhaust
gases ($m_e v_e$) in one direction, the greater
the momentum of the ship ($m_s v_s$) in the
opposite direction.*

If the mass of the man is 75 kg and the mass of the canoe is 50 kg (how
much does the canoe weigh?) then the canoe recoils with a velocity
of -3 m/sec:

$$v_c = \frac{-(75 \text{ kg})(2 \text{ m/sec})}{50 \text{ kg}} = -3 \text{ m/sec}$$

From Equation (6-2), you can see that fairly light persons can impart
considerable velocity to a canoe if they jump fast enough. This is the
same principle upon which rocket ships operate. Exhaust gases shoot
from the rear of the rocket at high speed. The rocket picks up momentum in the opposite direction, which results in the momentum of the
system remaining the same. The effectiveness of the ion engines
mentioned on page 76 results from the extremely high velocities
of the charged particles shooting out the rear of the space ship.
(b) A batter illegally bunts a thrown ball by tossing his bat at it. The
mass of the bat is 1.0 kg and its velocity before striking the ball is
-6 m/sec. The mass of the ball is 0.15 kg and its velocity before the
collision is 25 m/sec. After the collision, the bat has no horizontal
motion. What is the velocity of the ball after it hits the bat? (Ignore
vertical motions, as is done in the top view in Figure 6-6.)
　　Initially, the momentum of the system is

$$[(0.15 \text{ kg})(25 \text{ m/sec})] + [(1.0 \text{ kg})(-6 \text{ m/sec})] = -2.25 \text{ kg·m/sec}$$

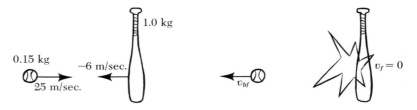

1.0 kg

0.15 kg

-6 m/sec.

25 m/sec.

v_{bf}

$v_f = 0$

Figure 6-6
*The momentum of the bat-ball system remains the same only if the batter is
not holding onto the bat.*

After the collision, the total momentum is the same

$$(0.15 \text{ kg})(v_{bf}) + 0 = -2.25 \text{ kg·m/sec}$$

or,

$$v_{bf} = \frac{-2.25 \text{ kg·m/sec}}{0.15 \text{ kg}} = -15 \text{ m/sec.}$$

As a check, see if the ratio of the masses of bat to ball is equal to the inverse ratio of their velocity changes. According to Equation (4-2), pages 70 and 138, they should be. (Note that the batter must not be holding onto the bat while the collision is occurring. If this happens, the system of ball plus bat is no longer isolated. Forces could be exerted on the system which would change its total momentum.)

Another Look at Newton's Second Law

On page 139 it was stated that momentum conservation is really the application of Newton's first law to an isolated system of objects: $\Delta(\text{momentum}) = 0$, if $F_{net} = 0$. The system can have one or many objects in it. If a system consisting of one or many objects is *not* isolated; if it does experience forces from the outside, then Newton's second law tells us the degree to which momentum is not conserved:

$$\frac{\Delta(\text{momentum})}{\Delta t} = \Delta p/\Delta t = \mathbf{F}_{net}.$$

(From here on, momentum will usually be represented by the symbol p.) This equation states that the rate of change of momentum of a system is equal to the net applied force, and is in the direction of that force.

Up to now, we have assumed that the momentum of an object can only change in one way; that is, by changing the velocity of the components. We assumed the mass of the system to be constant ($\Delta m = 0$). If this is true, then, since $p = mv$:

$$\Delta p = m\Delta v,$$

and

$$\frac{\Delta p}{\Delta t} = m \frac{\Delta v}{\Delta t}$$

$$= ma,$$

and so

$$F_{net} = ma,$$

from Newton's second law.

More generally, though, an object or a system of objects can change their momentum in two ways: either by changing velocity *or* mass. It can then be shown that $\Delta p = m\Delta v + v\Delta m$, so that Newton's second law becomes:

$$F_{net} = \frac{\Delta p}{\Delta t}$$

$$= \frac{m\Delta v}{\Delta t} + \frac{v\Delta m}{\Delta t}$$

$$F_{net} = ma + \frac{v\Delta m}{\Delta t} \qquad \text{Equation (6-3)}$$

Let's consider a few situations in which the mass of a system changes. (The rocket ship on page 140 is an example of this. As fuel is burned to form exhaust gases, the rocket gets lighter and its velocity, per unit mass of exhaust gases, increases.) Let's suppose you are turning a crank to keep a frictionless conveyor belt going. If nothing were dumped on the belt you would not have to exert any force to keep it going at constant speed. But if someone starts dropping sand onto the belt, you will have to exert a force $F_{net} = v(\Delta m/\Delta t)$ just to keep the system moving at the same constant speed. In this equation, $\Delta m/\Delta t$ is the rate of addition of sand in kg/sec. If the sand is thrown on the belt so that it has exactly the same speed as the belt just as it lands, the crank-turner's job is easy. That extra sand doesn't have to be accelerated, so the net force is zero.

In Chapter 9 it is shown that when any object attains a velocity approaching the speed of light, its mass increases greatly. It is more difficult to accelerate. This constitutes a second and more general physical situation in which $F_{net} = ma$ is not quite true. However, the statement $F_{net} = \Delta p/\Delta t$ does seem to be true throughout physics.

Now consider a different sort of problem in which momentum changes. A fire hose with cross-sectional area of 25 cm² (equal to

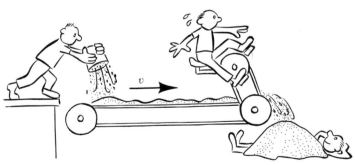

Figure 6-7
The person pedaling must exert a force just to keep the sand on the belt moving at constant speed.

25×10^{-4} m^2) shoots out water at 10 m/sec. What force is required to maintain a 90 degree bend in the hose?

First, the amount of water passing any point in the hose, per second, is that contained in a cylinder whose length is equal to the distance travelled by the water in one second. Thus, the volume of water going through the bend, per second, is simply the area of the hose times the speed of the water, or

$$\text{volume/sec} = (25 \times 10^{-4} \text{ m}^2)(10 \text{ m/sec})$$

$$= 25 \times 10^{-3} \text{ m}^3/\text{sec}.$$

We can obtain the *mass* of the water going through the bend each second by multiplying the volume by the *density* or the mass per unit volume of water, which equals 10^3 kg/m^3;

$$\text{mass/sec} = (25 \times 10^{-3} \text{ m}^3/\text{sec})(10^3 \text{ kg/m}^3)$$

$$= 25 \text{ kg/sec}$$

The momentum change per second, or average force is:

$$\frac{\text{mass}}{\text{sec}}(\Delta v) = \frac{\text{mass}}{\text{sec}}(v_f - v_i).$$

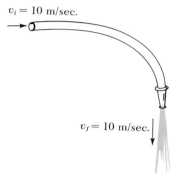

Figure 6-8
*The water entering the hose
maintains the same speed but is
deflected sideways.*

Considering, component by component, the velocity change, Δv, of the water in the hose, let v_x represent the horizontal component, and v_y the vertical component, of the velocity, as shown in Figure 6-9. The initial velocity, v_i, is equal to 10 m/sec to the right: it has the components v_{xi} equal to 10 m/sec and v_{yi} equal to zero. Coming out of the hose, the water has a final velocity, v_f, of 10 m/sec downward, so v_{xf} equals 0 and v_{yf} equals −10 m/sec.

Thus,

$$\Delta v_x = (v_{xf} - v_{xi})$$

$$= (0 - 10 \text{ m/sec}) = -10 \text{ m/sec}$$

$$\Delta v_y = (v_{yf} - v_{yi})$$

$$= (-10 \text{ m/sec} - 0) = -10 \text{ m/sec}$$

Figure 6-9
The slanting arrow represents the change in velocity.

The change from an initial velocity $\overset{v_i}{\longrightarrow}$ to a final velocity $v_f\!\downarrow$ requires a vector $\Delta v \swarrow$.

By vector addition it is found that the *change* in the velocity is slanted downward to the left at 45 degrees. By Pythagoras' theorem the *magnitude* of

$$\Delta v = \sqrt{(\Delta v_x)^2 + (\Delta v_y)^2}$$

$$= \sqrt{200} = 14.14 \text{ m/sec.}$$

The force which must be applied to the hose in order to produce this change in momentum is

$$F_{net} = \frac{\Delta p}{\Delta t}$$

$$= \left(\frac{\text{mass}}{\text{sec}}\right)(\Delta v)$$

$$= (25 \text{ kg/sec})(14.14 \text{ m/sec})$$

$$= 354 \text{ kg·m/sec}^2 = 354 \text{ N.}$$

The direction of application of the force is in the direction of the change of velocity and momentum. Although it might seem puzzling that the total velocity change of the water slanted in the direction shown above, it does make sense that, on the average, the force you apply would have that direction.

Newton's Law of Interaction

Consider again the equation we used for defining and measuring the relative mass of two objects:

$$\frac{m_a}{m_b} = -\frac{\Delta v_b}{\Delta v_a}$$

or

$$m_a \Delta v_a = - m_b \Delta v_b.$$

Divide both sides of this latter equation by the time interval, Δt, in which the changes of velocity occur:

$$m_a \frac{\Delta v_a}{\Delta t} = - m_b \frac{\Delta v_b}{\Delta t}$$

or

$$F_{ab} = - F_{ba} \qquad\qquad \text{Equation (6-4)}$$

F_{ab} is the force exerted on object a by object b and F_{ba} is the force exerted on b by a. This tells us that when two objects interact,

Figure 6-10
Both magnets move toward each other.

one of them exerts a force on the other which is exactly equal and opposite to the force that the other exerts back on it. This is a statement of what is usually called *Newton's third law.*

The forces generated in such mutual interactions *always occur in pairs.* This law deals with interactions between pairs of objects, *not* with the response of a single object to a pair of forces acting on it. If this distinction between Newton's second and third law is kept in mind, certain puzzling phenomena can be understood better.

Two objects need not touch in order to experience the interaction described by Newton's third law. Newton argued that when one object influences another at a distance, then the other object must exert an equal force in the opposite direction back on the first object. This can be illustrated by arranging two magnets to attract one another and pushing one toward the other on a smooth surface. At a certain distance apart a *mutual* jumping toward each other will occur.

Consider a ball of mass m falling toward the earth with acceleration g as shown in Figure 6-11. Without air drag the only force acting on the ball is its weight, mg, which can be represented by F_{be}, (force exerted *on* the ball *by* the earth.) By Newton's third law, an equal but opposite force F_{eb} is exerted on the earth by the ball. Although the earth's acceleration is too small to measure, it actually falls up toward the ball! Now suppose the ball rests on the earth's surface. A new set of forces, H_{be} and H_{eb} arise when the ball and earth touch. The ball and the surface on which it rests will both flatten or indent slightly, which produces this new "elastic" set of forces. For simplicity, the F_{be}, F_{eb} force pair has not been drawn in Figure 6-12, but they are still acting. By Newton's *third* law, the gravitational pair of forces, F_{be} and F_{eb} are always equal and oppositely directed, and the elastic pair of forces, H_{be} and H_{eb} are *always* equal (and opposite) no matter what the ball is doing. If we now look at just the ball and the forces

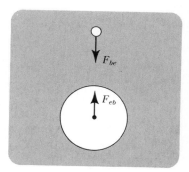

Figure 6-11
The earth attracts the falling ball with the gravitational force F_{be} but the ball also attracts the earth with a force F_{eb}.

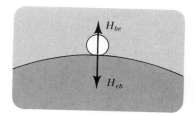

Figure 6-12
Here, the earth pushes up on the ball with a contact force H_{be} but the ball also pushes down on the earth with an equal force H_{eb}.

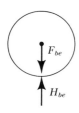

acting on it, we find F_{be} and H_{be}. The two might or might *not* be equal and opposite. By Newton's second law,

$$F_{net} = H_{be} - F_{be} = ma$$

F_{be} and H_{be} are equal and opposite only in the special case where the ball is not accelerating. In which case, they can be shown as being equal and opposite by applying Newton's *second* law.

As a second example, let us consider again the moon-earth rotations mentioned on page 115. From Newton's third law, the force exerted on and by the moon and the earth are equal and opposite: $F_{em} = -F_{me}$. Then the centripetal force, F_c, on each is equal and opposite:

$$F_c = \frac{m_e v_e^2}{r_e} = -\frac{m_m v_m^2}{r_m} \qquad \text{Equation (6-5)}$$

where r_e and r_m are the radii of the circles in which the earth and moon travel around one another.

In addition, the period of their rotation around each other, τ, must be the same. Therefore,

$$\tau = \frac{2\pi r_e}{v_e} = \frac{2\pi r_m}{v_m}$$

Solving these expressions for the velocities of the earth and moon respectively yields:

$$v_e = \frac{2\pi r_e}{\tau}, \text{ and } v_m = \frac{2\pi r_m}{\tau}.$$

Substituting these values for velocities into Equation (6-5) gives:

$$\frac{m_e \left(\dfrac{2\pi r_e}{\tau} \right)^2}{r_e} = -\frac{m_m \left(\dfrac{2\pi r_m}{\tau} \right)^2}{r_m}$$

Squaring the quantities in brackets gives:

$$\frac{\dfrac{m_e (4\pi^2)(r_e^2)}{(\tau^2)}}{r_e} = -\frac{\dfrac{m_m (4\pi^2)(r_m^2)}{(\tau^2)}}{r_m}$$

which is equivalent to:

$$\frac{m_e (4\pi^2)(r_e^2)}{r_e(\tau^2)} = -\frac{m_m (4\pi^2)(r_m^2)}{r_m(\tau^2)}$$

Performing the indicated divisions gives:

$$\frac{m_e (4\pi^2)(r_e)}{\tau^2} = -\frac{m_m (4\pi^2)(r_m)}{\tau^2}$$

Dividing both sides by the common factor $(4\pi^2)/\tau^2$ yields:

$$m_e r_e = - m_m r_m \qquad \text{Equation (6-6)}$$

The mass of the earth, m_e, is 81.3 times the mass of the moon, m_m. Thus, by substitution in Equation (6-6)

$$(81.3 \; m_m) r_e = m_m r_m$$

or

$$81.3 \; r_e = r_m.$$

The distance from the center of the earth to the center of the moon, $r_e + r_m = 240{,}000$ miles. Substituting $(81.3 \; r_e)$ for r_m,

$$r_e + (81.3 \; r_e) = 240{,}000 \text{ miles};$$

$$r_e = \frac{240{,}000}{82.3} = 2.91 \times 10^3 \text{ miles} = 4.7 \times 10^3 \text{ km}$$

This is the distance, from the center of the earth, to the axis of the earth's wobble which was previously stated to exist (pg. 115), without any explanation for the effect being given. Figure 6-13 represents (not to scale) the extent of the wobble of the center of the earth around this point, known as the *center of mass* of the earth-moon system.

Even though an object reacts to an applied force in an inert fashion, Newton's third law tells us that, when two objects interact, there is no separation into an active and a passive agent.

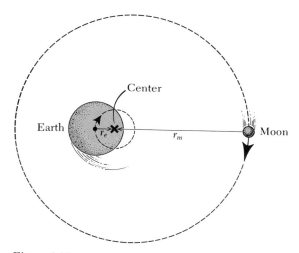

Figure 6-13
The earth and the moon rotate about a common center, X, and that center travels around the sun in an elliptical orbit.

Figure 6-14
*When the rock is accelerated, it pulls back on the string with
a force equal but opposite to ma, where m is the mass of the
stone and a is its acceleration. The string is stretched taut.*

Both objects always pull back or push back with an equal and
opposite force on the other one, whether or not they remain in
one position. If a string is tied to a rock and a tension is applied
to the string, the rock will accelerate immediately in the direction of the tension if it is on a frictionless surface. It *appears* as if
the force exerted on the rock by the string is the only force acting
in the system. In other words, in order to accelerate the rock, the
string must exert a force equal to the mass of the rock times its
acceleration. This force is obvious, and is applied to the rock at
the point where the string is attached, *but* a not so obvious equal
force is exerted backward by the rock on the string. Its presence
is indicated by noting that the initially limp string has become
taut, and remains so, *because of the tension exerted by the inertial
lag of the rock.*

When you lean against a wall, it pushes back on you just as
hard as you push on it. If it didn't, you would accelerate into the
wall as a result of unbalanced forces. However, every wall has
its strength limit. If you exceed this, it no longer pushes back
with sufficient force to keep you from accelerating, but the wall
(or at least the pieces) still pushes back on you just as hard as you
push on it, by being an inert object and requiring force for its acceleration. This is illustrated in Figure 6-15. As can be seen, it's
pretty hard to keep pushing very hard on a wall after it has
broken through.

How are you able to push on a wall in the first place? By pushing back on the floor with your feet! The floor in turn pushes
forward on your feet, enabling you to exert forward forces on
other objects.

If the floor were frictionless, then you would have no base
from which to push on *anything*. Persons abandoned in the
middle of a frictionless ice pond with no velocity in any direction
cannot get off the pond by walking or even crawling. They are
without momentum initially, and given no means for exerting
forces on their surroundings, or for receiving influences back

Figure 6-15

*The pieces of broken wall still push back on the person's hand with a force
equal but opposite to the force exerted on the pieces: that is, with a force
equal to −ma.*

from them, they cannot change their momentum in order to begin to move. Nevertheless, there are a couple of ways they could
get off. Try to think of some, using some of the ideas mentioned
earlier in this chapter.

Many of these ideas can be illustrated with an example of two
boys engaged in a tug-of-war. The boy on the left (Rap) pulls on
the rope with a force of 80 N. The boy on the right (Hud) pulls
with a force of 78 N.

The rope has a mass of 2 kg and acts as a third object in the
system, separating the two boys. The problem: find the acceleration of the system and the forces acting on each boy. The mass of
each boy is 40 kg.

If the two boys are holding rigidly on to the rope, then the
acceleration of the system will be the same as the rope's:

$$a = \frac{F_{net}}{m} = \frac{-80 \text{ N} + 78 \text{ N}}{2 \text{ kg}} = -1.0 \text{ N/kg} = -1.0 \text{ m/sec}^2.$$

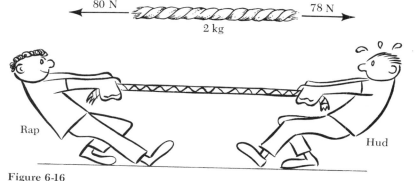

Figure 6-16

*Even if Hud is losing the tug-of-war, the only difference between the pull
he exerts and the pull Rap exerts is the mass of the rope between them
times its acceleration. If the mass of the rope were zero, the force exerted
by Hud is equal to the force exerted by Rap, no matter who's winning.*

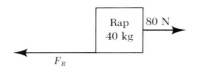

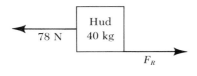

The rope pulls to the right on Rap with a force of 80 N. The ground pushes to the left on the bottom of his feet with a force F_R. This is a third law reaction to the force he exerts on the ground to the right. Now, applying Newton's second law to Rap:

$$F_{net} = -F_R + 80 \text{ N} = ma$$
$$= (40 \text{ kg})(-1.0 \text{ m/sec}^2)$$
$$= -40 \text{ kg·m/sec}^2 = -40 \text{ N}.$$

Then

$$-F_R = ma - 80 \text{ N} = -40 \text{ N} - 80 \text{ N} = -120 \text{ N}$$
$$F_R = 120 \text{ N, to the left.}$$

Now, let's investigate the forces acting on Hud:

$$F_{net} = F_R - 78 \text{ N} = ma$$
$$= (40 \text{ kg})(-1.0 \text{ m/sec}^2)$$
$$= -40 \text{ N}.$$
$$F_R = -40 \text{ N} + 78 \text{ N} = +38 \text{ N, to the right.}$$

From all of this, we can see that Rap pushes on the ground to the right with a force of 120 newtons and, by Newton's third law, the ground pushes back on Rap with a force of 120 newtons to the left. In the same manner, the ground pushes on Hud with a force of 38 newtons to the right.

To check our acceleration calculation, we can look again at the system as a whole. It receives, from the rest of the world, a force of 120 newtons to the left, applied to the system at Rap's feet, plus a force of 38 newtons to the right, applied at Hud's feet. The total mass of the system (both boys and the rope) is 82 kg. Thus, for the whole system:

$$a = \frac{F_{net}}{m} = \frac{-120 \text{ N} + 38 \text{ N}}{82 \text{ kg}}$$
$$= -\frac{82 \text{ N}}{82 \text{ kg}}$$
$$= -1.0 \text{ m/sec}^2.$$

Figure 6-17
Forces acting during the tug-of-war.

Summary

When an object possesses symmetry it is left unchanged by such operations as reflection, rotation, or translation. In the same way, a physical law is symmetrical if it is invariant after changes in frames of reference.

The momentum of an object is defined as its mass times its velocity. Newton's first law states that when an object is isolated, that is, when the net force, F_{net}, applied to it is zero

$$F_{net} = 0 = \frac{m\Delta v}{\Delta t} = \frac{\Delta p}{\Delta t}$$

The same may be said for a system of objects which is isolated. The momentum is conserved even though collisions might change the momenta, p, of individual objects in the system. For a system of two objects a and b,

$$[(m_a v_a) + (m_b v_b)]_{final} = [(m_a v_a) + (m_b v_b)]_{initial} \quad \text{Equation (6-1)}$$

If the mass of a system changes, then Newton's second law must be stated in a more general way:

$$F_{net} = \frac{\Delta p}{\Delta t} = \frac{\Delta(mv)}{\Delta t} = \left(\frac{m\Delta v}{\Delta t} \right) + \left(\frac{v\Delta m}{\Delta t} \right)$$

$$= ma + \left(\frac{v\Delta m}{\Delta t} \right) \qquad \text{Equation (6-3)}$$

Newton's third law says that when two objects interact, the forces always occur in pairs:

$$F_{ab} = -F_{ba} \qquad \text{Equation (6-4)}$$

F_{ab} equals the force exerted on a by b, and F_{ba} equals the force exerted on b by a.

You cannot exert a force on an object without its exerting an equal force back against you.

Further Reading

Feinberg, Gerald, and Goldhaber, Maurice, "The Conservation Laws of Physics," *Scientific American*, 209, Oct. 1963, pg. 36. (Modern physics rests largely on classical conservation laws. The extension of physics into new realms, however, required that these laws be re-examined.)

Feynman, Richard P., "Symmetry in Physical Law," chapter 4, *The Character of Physical Law*, British Broadcasting Company, Cox and Wyman, 1965. (Written in Feynman's chatty but thought-provoking style.)

Gardner, Martin, *The Ambidextrous Universe*, Mentor Books, 1964, 1969. (A popular account of symmetries and asymmetries in nature.)

Gerholm, Tor Ragner, Part V, "Symmetry Principles: Rules For the Laws of Nature," in *Physics and Man*, Bedminster Press, 1967.

Kaempffer, F. A., *The Elements of Physics*, Blaisdell, 1967, chapters 6, and 16. (Discusses symmetry in physical laws.)

Wigner, Eugene P., "Symmetry and Conservation Laws, *Physics Today*, *17*, March 1964, pg. 34. (Although symmetry was inherent in early physics, it took on real importance only after Einstein's special relativity. This article, the material referred to in footnote 1, Chapter 6, and 22 other Wigner essays are collected in *Symmetries and Reflections*, M.I.T. Press, Cambridge, 1967.)

Problems

6-1 Can you think of any biological reasons for human beings to be symmetrical in appearance?

6-2 Draw a pattern which is invariant under a translation *and* a 180-degree rotation.

6-3 Why should physical laws be invariant but perceived events not?

6-4 Is the momentum of a system always conserved during a collision? What is "a system?" How does this relate to Newton's first law?

6-5 A railroad car of mass m moving at 10 m/sec catches and couples onto two other cars, also of mass m each, which were moving at 5 m/sec. What is the speed of the three cars after they couple together? What would happen if they had the same original speeds, but were traveling toward each other before coupling?

6-6 Suppose that two cars on an air track collide. Car a has a mass of 2 kg and is at rest before the collision. Car b strikes it with a velocity of 30 cm/sec toward the right. After the collision, car b has a velocity of 10 cm/sec to the left and car a has a velocity of 20 cm/sec to the right. What is the mass of car b? If the bumpers on the cars are in contact for 0.01 sec during the collision, what is the average acceleration of each car during impact?

6-7 A projectile whose mass is 50 kg is fired directly forward by an airplane whose mass without the projectile is 5000 kg. If the plane's original velocity is 200 m/sec and the projectile's velocity is 300 m/sec, both relative to the earth, what is the plane's speed just after firing the projectile?

6-8 A grocery cart having a mass of 8 kg rolls along the floor of a store at a speed of 50 cm/sec. A shopper drops a five kg sack of potatoes straight down into the cart. Now what is its velocity? (Assume that no one is pushing on the handle and that the wheels of the cart are frictionless.) Suppose now that the same thing is done again but someone else is pushing on the cart, trying to keep it moving at 50 cm/sec. What impulse must be exerted to keep the speed the same? What kind of assumptions do you have to make to figure out what force is exerted, on the average?

6-9 A steam locomotive pulling a train having a mass of 100,000 kg is traveling at 20 m/sec. It comes to a 100 m long water trough which is imbedded between the rails. The locomotive lowers a scoop into the trough and draws in 1000 kg of water. What extra force must be exerted by the locomotive in order to keep the train moving at a constant speed?

6-10 A horse has a mass of 200 kg. It is pulling a cart which has a mass of 100 kg. The wheels of the cart exert backward friction amounting to 400 N. The horse digs in its feet and accelerates the cart at 0.2 m/sec². Assuming that the horse has two feet on the ground at a time, find out the force which the earth exerts on each foot of the horse.

6-11 What is the net force exerted on the horse-cart system? What goes into making up the net force?

6-12 A 70 kg man stands in one end of a canoe which is four meters long. The mass of the canoe is 40 kg. He walks to the other end of the canoe in five seconds, then stops. What is the speed of the man relative to the canoe while he is walking? What is the speed of the canoe relative to him? What is his speed relative to the water (or shore)? What is the speed of the canoe relative to the shore? Where is the canoe the instant after he quits walking? That is, how far does it move during his walk?

7

PHYSICAL WORK AND ENERGY

One of the major changes between medieval and Newtonian physics was that acceleration, rather than velocity, was seen as requiring an external force. Rest, and motion at a constant velocity, came to be "equivalent" states, insofar as objects experience them when no net force acts on them. Nevertheless, there *is* something real about motion itself. Experience shows that a car in motion is *not* the same as a car at rest relative to us. How do modern physicists account for this? They do so, not by attributing an innate spirit or volition to the moving object but by simply saying that, "sometime in the past a net force was applied to the object, which changed its state from one of rest to one of motion. If the net force was then removed from the object, it just continued on in a straight line at a constant speed, carrying with it the effect produced by the net force." In the same manner, a heavy box that is tipping and ready to fall from the top shelf in a closet, is different from a box sitting on the floor of the closet, due to the effects of the forces used to lift it up to the shelf. In this chapter we will examine how physics describes and measures the various sorts of changes which the application of a force can produce in a system.

The Effects of Forces

First, let's examine the effect of a 10 N force when it acts on a 10 kg mass *for one second.* If it is the *only* force acting on the object, then the object simply accelerates from rest to a final velocity

$$v_f = a\Delta t = \frac{F_{net}}{m} \Delta t$$

$$= \left(\frac{10 \text{ N}}{10 \text{ kg}}\right) (1 \text{ sec})$$

$$= 1 \text{ m/sec}.$$

(The result is given with the dimensions m/sec since newtons have the dimensions $m \cdot kg \ sec^{-2}$ which in this example is divided by kg and multiplied by sec.) If the object was moving initially, its velocity will increase by one m/sec. Now suppose the box sits on a rough surface having a coefficient of friction $\mu = 0.2$. The force required to overcome friction would be

$$\mu mg = (0.2)(10)(9.8) = 19.6 \text{ N},$$

so you could push with a 10 N force for one second and nothing would happen.

Thus, we cannot be sure what the effect of the force would be until we know about all the other forces acting on the object. The only case in which the effect of a force acting over a certain period of time is unambiguous is when the force is the *net* force. Then we call its effect *impulse*, as in Chapter 4.

Next, let us examine the effect of a 10 N force which acts on an object through *one meter of displacement.* If the force acts alone, it is the net force, and it will produce an acceleration of 1 m/sec² on a 10-kg mass. From Equation (3-9), the change in velocity:

$$\Delta v = \sqrt{2a\Delta x}$$

$$= \sqrt{2(1 \text{ m}/(\text{sec})^2)(1 \text{ m})}$$

$$= 1.414 \text{ m/sec}.$$

Figure 7-1
The person may push all day, but unless a force greater than the friction force acting on the box is exerted, no work will be done on it.

If, along with other forces, the 10 N force is applied to the 10 kg object sitting on the rough floor, a definite change still occurs: the object will move over the rough floor a distance of one meter, even though the change in velocity will be less than that calculated above. The same is true if an upward force were exerted on an object which is in a downward-acting gravitational field. The velocity of the object might not change at all. We also might not be overcoming any frictional forces. But, if the object is lifted 1 meter in height, its state has been changed to a definite extent.

Work

The effect of a force acting over a certain distance represents a definite physical change. The effort exerted by the force is called *work*, and it is measured either in newton·meters or *joules* (J) (in the MKSA system of units).[1] In the CGS (Centimeter-gram-second) system of units, the unit of work is the dyne·cm or *erg*. (1 joule = 10^7 erg.)

An expression or formula for work can be derived if several points are kept in mind. First, the displacement, Δx, need not be in the direction of the force that is applied and whose work we would like to express. (It also need not be horizontal.) The amount of work done by a force depends on the angle between the direction of the force and the direction in which the object actually moves, as illustrated in Figure 7-2.

Remember: acceleration is in the direction of *net* force. The force, F, we are concerned with might be acting simultaneously with other forces, or it could be the net force, acting alone.

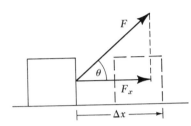

The effectiveness of a force depends on the component of the force in the direction of the displacement, F_x, or $F \cos \theta$, where θ is the angle between the applied force and the actual displacement, Δx. So far, then, we have

$$\text{Work} = F_x \Delta x.$$

If F and Δx are parallel, then $F_x = F$. ($\theta = 0$, $\cos \theta = 1$, (see Appendix B) so $F_x = F \cos \theta = F$.) If θ is more than 90 degrees, $F_x = F \cos \theta$ is negative, and negative work is done by the force! If you push sideways on a moving object, such as a railroad car, then θ is 90 degrees, F_x is zero, and you are doing no physical work!

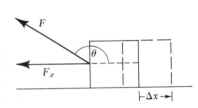

So far, it has been assumed that the force doing the work is constant. If it varies, then the *average* value of the force must be used to calculate the work it does.

$$\text{Work} = (F_x)_{avg} \Delta x \qquad \text{Equation (7-1)}$$

Going back to the examples in the previous paragraph, an alternative expression for the work done by three particular

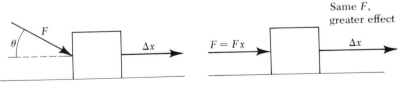

Figure 7-2
A given force F does more work on an object if it acts in the same direction as the movement of the object.

(1) Meter-Kilogram-Second-Ampere. (See page 18.)

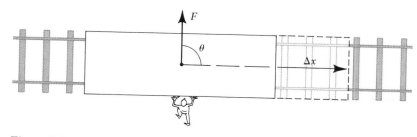

Figure 7-3
If a force is exerted in the wrong direction on a moving object, it may do zero (or even negative) work on the object.

kinds of force can be derived as follows: the work done by a net force equals $F_{net}\Delta x$. However, from Equation (3-8), $\Delta x = (v^2 - v_0^2)/2a$. Thus

$$F_{net}\Delta x = ma \frac{(v^2 - v_0^2)}{2a},$$

$$= \frac{m(v^2 - v_0^2)}{2}.$$

Thus, work done by a *net force* is:

$$F_{net}\Delta x = \frac{m(v^2 - v_0^2)}{2}. \qquad \text{Equation (7-2)}$$

If the object is accelerated from rest, then the work done equals $\frac{1}{2}mv^2$.

In a like manner, work done *against friction* is:

$$F_{fric}\Delta x = \mu mg\Delta x \qquad \text{Equation (7-3)}$$

In Equation (7-3), F_{fric} represents the force *you* would have to exert in order to overcome the opposing friction force.

At this point a comment should be made upon the strict way in which work is defined in physics.

A waiter carrying a heavy tray of dishes back to the kitchen of a restaurant can be exerting large forces to keep the tray above his head, but as long as he just moves the tray horizontally in the vertical gravitational field of the earth, he does no physical work on the tray. He exerts a force upward to overcome gravity but Δx is horizontal, at right angles to the applied force. Thus F_x is zero. The waiter does no physical work, so the physical state of the tray does not change. (Of course, if the waiter starts and stops he is doing positive and negative work during the periods of positive and negative acceleration.) In this case, what counts is how far the tray moves *up*. This is because the force exerted by the waiter's hand must just oppose the weight of the tray.

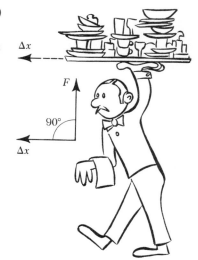

Figure 7-4
The waiter exerts an upward force on the tray. Horizontal movements of the tray thus do not constitute physical work done on the tray. During such movements, Δx and F are perpendicular, so F_x is equal to zero.

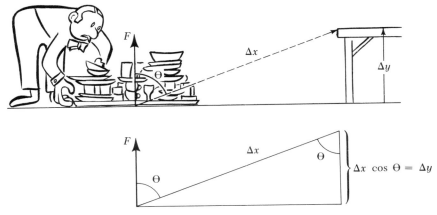

Figure 7-5
*All that counts is the distance moved in the direction of the applied force F:
that is, upward.*

Since he applies his force upward, only the component of displacement or motion which is upward counts. Suppose the waiter lifts the tray diagonally from the floor up to a table. The work he does is

$$F_x \Delta x = F \cos \theta \, \Delta x$$

$$= mg \, \Delta x \cos \theta.$$

But note that $\Delta x \cos \theta$ is the vertical distance, Δy, through which he lifts the tray. The work done in *lifting an object* is:

$$mg\Delta y \qquad\qquad \text{Equation (7-4)}$$

The waiter gets no physical "credit" for carrying the tray sideways. In the gravitational field that the waiter is in, Δy is taken as the vertical rise. When other fields of force, such as electrical or nuclear are considered, Δy will represent (in more general terms) the distance moved in opposition to the field force, and it need not be vertical.

The waiter who lugs the tray around without changing y might object and point out that he really sweated in order to get that tray to the kitchen. If we *were* to examine his muscles microscopically while he carries the tray along, we might see enough motion of muscle fiber (Δx or Δy) to retain our idea of work as forces acting over distances.

Where Work Goes

Having developed a scheme for numerically measuring the work done on an object, we might well ask how an object that has been speeded up, or lifted, is different after work has been done on it. We can answer some of these questions with an example.

An elevator whose mass is 400 kg is raised three stories (10 m) by a force of 4000 N. The work done by the force F is

$$F\Delta y = (4000 \text{ N})(10 \text{ m}) = 40,000 \text{ N·m}$$

$$= 40,000 \text{ J}$$

Part of this work is required just to lift the elevator against the earth's gravitational pull. This force must be just equal and opposite to the weight of the elevator,

$$mg = (400 \text{ kg})(9.8 \text{ N/kg})$$

$$= 3920 \text{ N}.$$

The work, W, necessary just to *lift* the elevator the 10 m can now be calculated, and is:

$$W = mg\Delta y$$

$$= (3920 \text{ N})(10 \text{ m})$$

$$= 39,200 \text{ J}.$$

When an object has been lifted in a gravitational field, it clearly has some differences from an object below it. Work has been done on it and, in turn, the object has a capacity for doing things such as physical work. It can be used to lift another elevator. Or, its cable can be cut and it will fall, picking up velocity and eventually doing more destructive sorts of work on things below. Such a capacity for doing work, by virtue of position, is called *potential energy*, where *energy* is the capacity for doing work. The expression ($mg\Delta y$) represents the *change* in potential energy, $\Delta P.E.$, when an object is moved in opposition to a force field a distance Δy:

$$\Delta P.E. = mg\Delta y \qquad \text{Equation (7-5)}$$

By combining Equations (7-4) and (7-5), the following relations between work done and energy change is obtained:

$$\text{Work done in lifting} = \Delta P.E. \qquad \text{Equation (7-6)}$$

What *is* the potential energy of the elevator when it reaches the third floor? All that has been said thus far is that the potential energy has *increased* by 39,200 joules. At this point, an arbitrary choice has to be made. If the potential energy is taken to be zero at the ground floor, then the potential energy three stories above is 39,200 J. However, if there are one or more basement levels in the building the ground floor might be assigned a potential energy larger than zero. The potential energy

$F = 4000$ N

$W = mg$
$= 3920$ N

Figure 7-6
Even though the upward force F is not the only force acting on the elevator, the work done by F is still equal to F times the distance traveled upward.

of the elevator at the third floor would correspond to a value which is 39,200 J above that potential energy.

If the pull of the earth were suddenly "turned off", the energy distinction between floors in a building would disappear. So would "up" and "down". One cannot overestimate the importance of gravity in our activities, thoughts, and metaphors. (For instance, we speak of people "on the way up", and envy birds, and pay high rent to live in penthouses.)

Since a force of 4000 N acts upon the elevator (which, with a 400 kg mass, weighs only 3920 N) it experiences a net force of 80 N *up*. The *net force* will accelerate the elevator upward. If it starts from rest, its final velocity after rising 10 meters will be

$$v = \sqrt{2a\Delta y} = \sqrt{2\left(\frac{F_{net}}{m}\right)\Delta y}$$

$$= \sqrt{2\left(\frac{F_{net}}{m}\right)\Delta y} = \sqrt{2\left(\frac{80 \text{ N}}{400 \text{ kg}}\right)(10 \text{ m})}$$

$$= \sqrt{2(0.2 \text{ m/sec})^2(10 \text{ m})} = \sqrt{4m^2/(\text{sec})^2}$$

$$v = 2 \text{ m/sec}$$

From Equation (7-2), the work done by the net force equals $\frac{1}{2}mv^2$.

$$\frac{1}{2}mv^2 = (400 \text{ kg})(2 \text{ m/sec})^2$$

$$= 800 \text{ kg·m/sec}^2$$

$$= 800 \text{ J}$$

The name *kinetic energy (K.E.)* has been given to the quantity $\frac{1}{2}mv^2$. It represents the capacity for doing work that results solely from *being in motion*. Then, in general, the

Work done by $F_{net} = \Delta(K.E.)$ Equation (7-7)

If the potential energy increase of 39,200 J and the kinetic energy increase of 800 J are added together the 40,000 J of work done by the 4000 N applied force acting through a distance of 10 m is completely accounted for.

That kinetic energy really does represent the capacity for doing work can be seen from the following: suppose the elevator arrives at the third floor with a velocity of 2 m/sec, as we calculated. Now let us cut the cable which is pulling up on the elevator. The only force still acting on the elevator is its weight, and it will begin to decelerate at 9.8 m/sec². The elevator will rise until its velocity has dropped to zero and then it will begin to drop. How high will the elevator go? From Equation (3-8),

$$y = y_o + \frac{(v^2 - v_o^2)}{2a}$$

Since $y - y_o = \Delta y$, this equation reduces to:

$$\Delta y = \frac{v^2 - v_o^2}{2a}.$$

At the point of highest ascent, $v = 0$, so:

$$\Delta y = -\frac{v_o^2}{2a}$$

$$= -\frac{(2 \text{ m/sec})^2}{-9.8 \text{ m/sec}^2}$$

$$= +.204 \text{ m.}$$

By virtue of its upward velocity, the elevator was able to rise an extra .204 meters. This corresponds to an increase in potential energy, $mg\Delta y$, equal to 3920 N times .204 m or 800 J. This is exactly the amount of kinetic energy the elevator had when the cable was cut!

Conservation of Energy

Figure 7-7 shows that at the top of its path, the elevator has zero kinetic energy, but its potential energy (relative to the ground) is:

$$mg\Delta y = (3920 \text{ N})(10.204 \text{ m}) = 40,000 \text{ J}$$

Now suppose the elevator falls freely from 10.204 m down to a height of 5.0 m. What will the kinetic energy and potential energy now be?

$$K.E. = F_{net}\Delta y = (-3920 \text{ N})(-5.204 \text{ m})$$

$$= 20,400 \text{ J}$$

The minus signs show that F_{net} and Δy are both directed down.

$P.E.$ at $y = 5$ m will be

$$mgy = (3920 \text{ N})(5 \text{ m})$$

$$= 19,600 \text{ J}$$

As the elevator reaches the ground floor, the potential energy equals zero. But, the kinetic energy,

$$K.E. = F_{net}\Delta y = (-3920 \text{ N})(-10.204 \text{ m})$$

$$= 40,000 \text{ joules.}$$

By now, you might have noticed something about the energies of the elevator once the cable was cut. At 10.204 m, 5 m, and 0

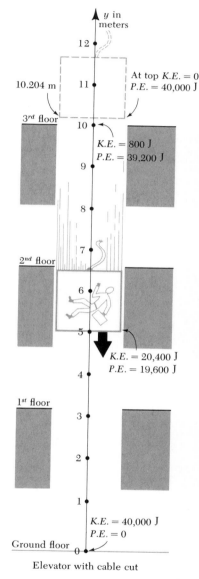

y in meters

12

At top $K.E. = 0$
$P.E. = 40,000$ J

10.204 m   11

3rd floor

10

$K.E. = 800$ J
$P.E. = 39,200$ J

9

8

7

2nd floor

6

5

$K.E. = 20,400$ J
$P.E. = 19,600$ J

4

1st floor

3

2

1

$K.E. = 40,000$ J
$P.E. = 0$

Ground floor   0

Elevator with cable cut

Figure 7-7
Diagram of an elevator with the cable cut.

*Physical Work
and Energy*

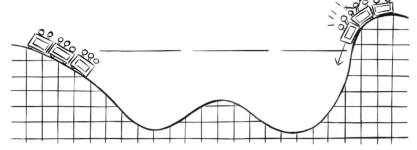

Figure 7-8
The roller coaster car cannot climb back up quite as high as the level from which it started.

meters above the ground, the sums of the kinetic and potential energies are the same: 40,000 joules. Once you quit doing external work on a system (in this case, the elevator plus the earth's gravitational field) the total *mechanical energy*—that is, the kinetic plus potential energy—is *conserved*.

The trouble with energy conservation is that in real physical systems the sum of the kinetic and potential energy does *not* stay the same: Some energy is lost. A roller coaster released from the top of a hill on the track will still have greater kinetic energy and less potential energy at the bottom of dips, and more potential energy and less kinetic energy at the top of a rise. However, the total of these energies is never quite as high as it was originally, unless a chain pulls the car up.

After bouncing, a tennis ball does not return up to the level from which it was dropped. As a person goes back and forth on a swing there is a continuous exchange between kinetic energy and potential energy, but the oscillation eventually dies down as both the kinetic and potential energies disappear.

In each of these cases the initial mechanical energy of the system (kinetic plus potential) is consumed as work done against friction. The sum of kinetic and potential energy would thus seem to be an arbitrary, uninteresting quantity for a given system. It was not until late in the eighteenth century that Benjamin Thompson (1753–1814), better known as Count Rumford, suggested that friction does not destroy energy, but converts it from mechanical energy to a new form: *thermal* energy, U.

When this was realized and confirmed by careful experiments, the concept of energy conservation (and of energy itself) took on wide ranging significance in physics; energy became a far more useful concept in the study of physical systems.

After Rumford's work, the conservation of energy of a system could then be stated in this way: the total energy E of an isolated system remains constant:

$$E = K.E. + P.E. + U = constant. \qquad \text{Equation (7-8)}$$

Zero *K.E.*   Minimum *P.E.*
Maximum *P.E.*   Maximum *K.E.*

Figure 7-9
The swing and rider have maximum P.E. and zero K.E. at the highest point of swing, but maximum K.E. and minimum P.E. (could be called zero) at the bottom.

If energy exchanges occur within an isolated system, then

$$\Delta K.E. + \Delta P.E. + \Delta U = 0 \qquad \text{Equation (7-9)}$$

Whenever the sum of the kinetic and potential energy decreases, the thermal energy increases. The system warms up. The total energy remains the same.

Let us consider the elevator again, this time making it more real by guiding it up and down with rails which exert a frictional drag of 20 N.

If the energy changes which occur as the elevator falls from rest at a height of 10.204 m to the ground floor are reexamined:

$$\Delta P.E. = mg\Delta y = (3920\ \text{N})(-10.204\ \text{m})$$

$$= -40{,}000\ \text{joules}$$

$$\Delta K.E. = F_{net}\Delta y = (-3920 + 20\ \text{N})(-10.204\ \text{m})$$

$$= (-3900\ \text{N})(-10.204\ \text{m})$$

$$= +39{,}796\ \text{joules.}$$

The net loss in energy appears to be 204 joules. However, if we account for the work done against frictional forces:

$$\Delta U = F_{fric}\Delta y = (-20\ \text{N})(-10.204\ \text{m})$$

$$= +204\ \text{joules.}$$

The principle of energy conservation is confirmed. The increase in kinetic energy plus the work done against friction is just equal to the loss of potential energy.

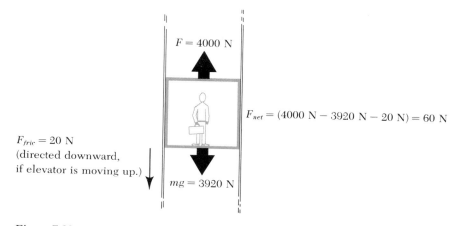

Figure 7-10

Three forces act on the elevator as it moves upward: F, equal to 4000 N and acting upward; the frictional force, F_{fric} equal to 20 N and acting downward; and the weight, 3920 N, and acting downward.

In general, it can be said that whenever an object is moved a distance, Δx, against opposing friction forces, the work going into thermal energy, U, is

Work done against friction $= \Delta U$ Equation (7–10)

This change in thermal energy can show up as a slight warming of the rails which guide the elevator, as well as of the elevator itself. Thermal energy, heat exchanges, and the warming of objects are further investigated in Chapters 15 and 16.

To further illustrate and summarize some of the concepts of work and energy that have been discussed, consider the following example.

A 20-kg sled sits at rest at the bottom of a frictionless 30° slope 20 m long and 10 m high. A person exerts a steady force, F, equal to 110 N on the sled in a direction which is parallel to the slope. How much work is done by the person exerting the force, as the sled moves to the top of the hill? Into what kinds of energy does this work go?

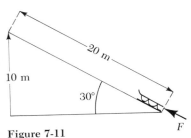

First, the work done is

$$F_x \Delta x = (110 \text{ N})(20 \text{ m}) = 2200 \text{ J}.$$

Figure 7-11

(Notice that, because F is parallel to Δx, $F = F_x$; and that Δx is not horizontal.) Of this work, a portion goes into lifting the sled a vertical distance, Δy, equal to 10 m:

$$\Delta P.E. = mg\Delta y = (20 \text{ kg})(9.8 \text{ N/kg})(10 \text{ m})$$

$$= 1960 \text{ J}.$$

Where did the rest of the work go? The slope is frictionless, so the remaining 240 joules of work probably went into kinetic energy. To check this, we need first to find the net force acting on the sled.

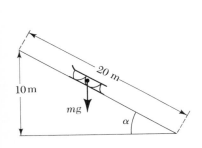

The weight of the sled can be represented by a vector which points straight down: In so doing, it must have a component acting downward, but parallel to the slope of the ramp. By examining the angle involved, you can see that the component of the weight force down along the ramp is $mg \sin \alpha$. (Sin α increases with α, as does the component of force urging the sled down the hill.) For our hill, $\alpha = 30°$ and $\sin \alpha$ equals 0.5 (see Appendix F) so,

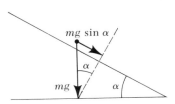

$$mg \sin \alpha = (20 \text{ kg})(9.8 \text{ m/sec}^2)(0.5)$$

$$= 98 \text{ N}.$$

The net force acting along the ramp can now be calculated, and is:

$$F_{net} = (110 \text{ N}) - (98 \text{ N}) = 12 \text{ N},$$

Figure 7-12
*mg sin α is the force acting
downward on the sled, parallel
to the hill.*

$$\Delta K.E. = F_{net}\Delta x$$

$$= (12\text{N})(20 \text{ m}) = 240 \text{ J}.$$

It's important to note that the energy of the sled is *not* conserved while it is being pushed up the hill. It increases by 2200 joules, thanks to an outside agent which reaches in and does work on the sled. As a system, the sled is not isolated. If we had to, we could include the person who is pushing as part of the system. The system would then be closer to being isolated, but it also would be vastly more complicated. The increase of kinetic energy and potential energy of the sled plus pusher as they move up the hill (a neat trick if the slope is frictionless) must be matched by the disappearance of some energy which had been stored in the body of the person who is pushing. Along with this come questions of what the person ate for breakfast, metabolism, perspiration, loss of weight, etc. It's easier to leave the agent which exerts the force outside the system and simply recognize the work that it does to increase or decrease the energy of the (non-isolated) system. We don't care whether the agent pushes the sled with a pole; pulls it up with a rope; or, pushes on the sled from behind while walking up the slope in hobnailed boots. When gravity acts on an object it is usually included as part of the system because it is simple and its effects can be completely accounted for by means of the potential energy gained from the work done in opposition to it. If gravity were perceived as being external to the system, then all the kinetic and thermal energy of objects would seldom be conserved. The energy of the system could change as a result of work done by the "external" gravitational field.

There is thus a close connection between the degree to which energy (and momentum as well) is conserved and one's choice of boundary for a system.

Suppose that someone now sprinkles sand on the slope, setting up a frictional resistance of 10 N to the motion of the sled. The work done by the force F is still the same, as long as F is held at 110 N. The change in potential energy is still 1960 joules if the change in height is 10 m. However, the gain in kinetic energy is less:

$$F_{net} = (110 \text{ N}) - (98 \text{ N}) - (10 \text{ N}) = 2 \text{ N}.$$

$$\Delta K.E. = F_{net}\Delta x = (2 \text{ N})(20 \text{ m}) = 40 \text{ J}.$$

And there is an increase in thermal energy, equal to:

$$\Delta U = F_{fric}\Delta x = (10 \text{ N})(20 \text{ m}) = 200 \text{ J}.$$

$$\Delta P.E. + \Delta K.E. + \Delta U = \text{work done}$$

$$(1960 + 40 + 200) \text{ J} = 2200 \text{ J}.$$

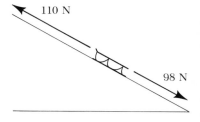

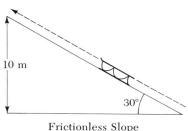

Frictionless Slope

Work = 2200 joules
$\Delta P.E.$ = 1960 joules
$\Delta K.E.$ = 240 joules

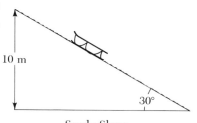

Sandy Slope

Work = 2200 joules
$\Delta P.E.$ = 1960 joules
$\Delta K.E.$ = 40 joules
ΔU = 200 joules

Elastic and Inelastic Collisions

Although momentum is always conserved during collisions in isolated systems, the same is not true for the sum of the kinetic and potential energy of the system. If no mechanical energy is lost in a collision, it is said to be *elastic*. In *inelastic* collisions, a portion of the initial mechanical energy is converted to thermal energy.

Consider a collision between a 10-kg object moving at 10 m/sec, and a stationary 5-kg object, as shown in Figure 7-13. Before collision the momentum of the system is

$$p = (10 \text{ kg})(10 \text{ m/sec})$$
$$= 100 \text{ kg·m/sec.}$$

The kinetic energy of the system is

$$K.E. = \tfrac{1}{2}mv^2$$
$$= \tfrac{1}{2}(10 \text{ kg})(10 \text{ m/sec})^2$$
$$= 500 \text{ J.}$$

A number of "after" situations are physically possible, depending upon the sort of bumpers that are between the objects as shown in Fig. 7-14. But in every situation, the momentum of the system after the collision must be the same as it was before the collision. Let us consider three of the possible final conditions of the system and calculate the total energy still possessed by the system under each condition.

(a) The objects couple together during the collision. Therefore, they must have the same final velocity, v_f. For momentum to be conserved during the collision, $(10 + 5)v_f$ must equal $(10 \text{ kg})(10 \text{ m/sec})$, or 100 kg·m/sec. Then:

$$v_f = \frac{100 \text{ kg·m/sec}}{15 \text{ kg}} = 6.67 \text{ m/sec.}$$

The kinetic energy of the system after collision will be

$$K.E. = \tfrac{1}{2}m(v_f)^2$$
$$= \tfrac{1}{2}(10 + 5)\text{kg } (6.67 \text{ m/sec})^2$$
$$= 333 \text{ J.}$$

Figure 7-13
Initial situation before the collision.

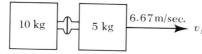

(a) Objects couple together.

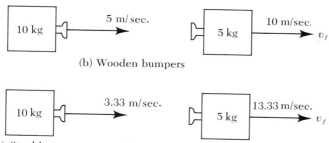

(b) Wooden bumpers

(c) Steel bumpers cause rebound and are not dented.

Figure 7-14
Three of many possible final situations after the collision.

One-third of the original mechanical energy disappeared! For this collision, this value represents the limit of inelasticity. Having faith in the conservation of energy, we guess that the 167 lost joules have gone into heating the bumpers between the objects. The "guess" is experimentally testable using sensitive thermometers, and (in similar instances) has been confirmed.

(b) The objects are equipped with wooden bumpers. After rebound the 10-kg object has a velocity of 10 m/sec and the 5-kg object some other velocity, v_f. Again, this velocity can be found from momentum conservations conditions:

$$(10 \text{ kg})(5 \text{ m/sec}) + 5 \text{ kg } v_f = 100 \text{ kg·m/sec}$$
$$5 \text{ kg } v_f = 50 \text{ kg}$$
$$v_f = 10 \text{ m/sec.}$$

The kinetic energy of the system after the collision will be

$$K.E. = \tfrac{1}{2}(10 \text{ kg})(5 \text{ m/sec})^2 + \tfrac{1}{2}(5 \text{ kg})(10 \text{ m/sec})^2$$
$$= 125 + 250 = 375 \text{ J.}$$

125 J of the original energy of the system have been converted to heat during the collision.

(c) Good spring steel bumpers are now put on both objects and they collide as before. The 10 kg object comes out of the collision with a measured velocity of 3.333 m/sec. To find v_f the principle of momentum conservation is again applied:

$$(10 \text{ kg})(3.333 \text{ m/sec}) + 5 \text{ kg } v_f = 100 \text{ kg·m/sec}$$
$$5 \text{ kg } v_f = 66.666 \text{ kg}$$
$$v_f = 13.333 \text{ m/sec.}$$

The kinetic energy of the system will be:

$$K.E. = \tfrac{1}{2}(10 \text{ kg})(3.333 \text{ m/sec})^2 + \tfrac{1}{2}(5 \text{ kg})(13.333 \text{ m/sec})^2$$
$$= 55.5 + 444.5 = 500 \text{ joules!}$$

Mechanical energy was conserved: the collision was elastic. Note that the relative velocity of the objects after the collision is: $13.333 - 3.333 = 10$ m/sec. The relative speed of departure after the collision is just equal to the speed of approach before the collision. This is found to be true of all elastic collisions between two objects. All real collisions involving human-sized objects are inelastic to some extent. Some thermal energy is always produced. It appears that genuinely elastic collisions occur only on an atomic scale.

Historically speaking, it took European scientists more than a century to develop the ideas that have just been described. In 1695 Gottfried W. H. Leibnitz (1646–1716) coined the name *vis viva* for the quantity mv^2. He and his followers said that this was the proper expression for the effect of force on an object. In opposition to them were the followers of Descartes, who insisted that momentum, mv, was more appropriate. Huygens was able to show a few years later that the *vis viva* was conserved in special kinds of collisions, whereas momentum was conserved in *all* collisions, within isolated systems. Momentum and *vis viva* were thus independent. The arguments dwindled when both were finally seen as deductions from Newton's second law: a net force acting for a certain time produces momentum, and a net force acting over a certain distance produces *vis viva*. Rumford's work in the 1790's finally showed that the loss of *vis viva* in collisions can be accounted for by noting that thermal energy is produced.

During the late eighteenth and early nineteenth centuries *vis viva* or energy became a powerful tool for studying motions in systems far too complex to handle in terms of force, acceleration and momentum. However, it was not until 1829 that Gaspard G. de Coriolis, a French mathematical physicist, gave definite meaning to the word "work". In a paper entitled "Calculations of the Effect of Machines or Considerations of the Use of Motors and of Their Evolution" he defined work as we do now: force times distance. He also divided *vis viva* by 2, giving us the modern expression for kinetic energy, $\tfrac{1}{2}mv^2$. When the principle of energy conservation was at last presented formally in the 1840's, four different scientists had "discovered" it independently.[2] This is perhaps not surprising in view of the long incubation period it had gone through.

(2) Kuhn, T. S., "Energy Conservation as an Example of Simultaneous Discovery", in *Critical Problems in the History of Science*, Marshall Clagett, ed. University of Wisconsin Press, Madison, 1959.

Use of Conservation Principles
in Physics

169

*Use of
Conservation
Principles
in Physics*

Once the idea of energy conservation is accepted, it can some-times be used to solve physics problems more easily. For exam-ple, consider the textbook throwing problem on page 44. We could find how high the book would go by first finding the time, Δt, required for its velocity, v_o initially, to decrease to zero. $v = v_o - g\Delta t$, so that when $v = 0$, at the top of the trajectory, $\Delta t = v_o/g$. In the time Δt, the book changes its vertical position, Δy, by

$$\Delta y = v_o \Delta t - \tfrac{1}{2}g(\Delta t)^2,$$

and if v_o/g is substituted for Δt, then:

$$\Delta y = v_o \left(\frac{v_o}{g} \right) - \tfrac{1}{2}g \left(\frac{v_o}{g} \right)^2$$

$$= \frac{v_o^2}{g} - \tfrac{1}{2}g \left(\frac{v_o^2}{g^2} \right)$$

$$= \frac{v_o^2}{g} - \frac{g v_o^2}{2g^2}$$

$$= \frac{v_o^2}{g} - \frac{v_o^2}{2g}$$

$$= \frac{v_o^2}{2g}.$$

Needless to say, that took quite a bit of algebraic manipulation.

On the other hand, if we used energy conservation, we could simply say the initial kinetic energy, $\tfrac{1}{2}mv_o^2$, is converted to po-tential energy as the book rises and loses speed. At the very top of its trajectory the book has zero kinetic energy, but its potential energy is numerically equal to the initial kinetic energy:

$$P.E. = mg\Delta y = \tfrac{1}{2}mv_o^2.$$

$$\Delta y = \frac{1}{2} \frac{mv_o^2}{mg}$$

$$= \frac{1}{2} \frac{v_o^2}{g} = \frac{v_o^2}{2g}$$

This is the same result as above, but it is more easily obtained.

If, instead of throwing the book straight up in the air, we had slid it up a complicated but frictionless ramp the advantage of the energy approach would be even more striking. Figuring out how high the book would rise on the basis of forces and accelerations

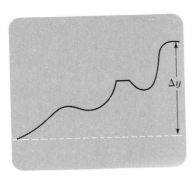

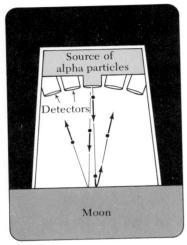

Figure 7-15
Instrument used by the Surveyors, bouncing alpha particles off the lunar surface to determine what the moon is made of.

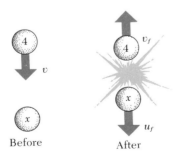

Before   After

Figure 7-16
Representation of the collision of an alpha particle with an unknown atom on the surface of the moon.

would be a gigantic problem requiring the use of a computer for its solution. But by equating initial kinetic energy and the final potential energy, we know that a book thrown onto a ramp with velocity v_o (not necessarily in the vertical direction) will rise to a height, Δy equal to $v_o^2/2g$, no matter what shape the ramp has.

Momentum and energy principles gave us our first hints about the composition of the moon! Prior to the Apollo series of manned moon flights, a series of machines called Surveyors were landed on the moon. One goal of the Surveyors was to identify moon atoms by taking advantage of knowledge of energy and momentum conservation in atomic collisions. The apparatus which was landed included a source of alpha particles (helium nuclei). The source was held in a container pointing down, and was surrounded by detectors which measured the energy of the alphas after they bounced off the moon. Various groups of rebound velocities were observed. In *one* group the maximum rebound energy, corresponding to a single, elastic collision between the alpha and some unknown atom on the moon was found to be 37 percent of the initial energy of the alpha particle.

Let's see if we can find what the mass (x) of the unknown atoms on the moon were. If we take the mass of the bombarding alpha particles to be 4 units (Chapter 9 explains why the number is 4) the collision would look like this:

The initial energy, E, of the alpha particles equals $\frac{1}{2}mv^2$ or, more specifically: $\frac{1}{2}(4)v^2$. From the data given, after the elastic collision, their final energy is $.37(\frac{1}{2}(4)v^2)$ which, in turn, equals $\frac{1}{2}(4)(v_f)^2$. Thus $(v_f)^2 = .37(v)^2$; taking square roots, $v_f = \pm .608v$. We know that the alphas change direction during the collision, so the proper root to use for v_f is the negative one:

$$v_f = -.608v.$$ Equation (7-11)

Now for momentum to be conserved, we require:

Before   After

$$4v = 4v_f + xu_f$$
$$4v = 4(-.608v) + xu_f$$
$$4v = -2.43v + xu_f$$
$$6.43v = xu_f:$$
$$u_f = \frac{6.43v}{x}.$$ Equation (7-12)

For energy to be conserved

Before   After

$$\tfrac{1}{2}(4)v^2 = \tfrac{1}{2}(4)(v_f)^2 + \tfrac{1}{2}(x)(u_f)^2$$

or, using Equations (7-11) and (7-12)

$$2v^2 = \tfrac{1}{2}(4)(-.608v)^2 + \tfrac{1}{2}(x)\left(6.43\,\frac{v}{x}\right)^2$$

$$2v^2 = .74v^2 + \left(\frac{20.65}{x}\right)v^2$$

$$1.26v^2 = \left(\frac{20.65}{x}\right)v^2$$

$$1.26 = \frac{20.65}{x}$$

$$x = \frac{20.65}{1.26} = 16.4 \text{ mass units}$$

Oxygen atoms have a mass number of 16.

This and similar analyses provided the first experimental evidence that the moon was not made of green cheese, but of rocks much like the ones on earth. Since the manned landing in 1969, moon rocks *have* been found to differ from earth rocks in interesting ways, but for the most part they have the same proportions of the same elements as the rocks found on earth.

The physicist's faith in the conservation of energy within isolated systems has been confirmed on many occasions. In the early 1930's detailed studies were being made of a particular sort of radioactivity, in which fast-moving particles (now known as electrons, but commonly called "beta rays" then) were emitted by bismuth atoms (really, from the nuclei of bismuth atoms), after which the bismuth was transformed into lead. The energy of the outcoming electrons ought to have been about 2×10^{-13} joules, (according to principles that will be better understood after the next chapter). Instead, a whole spectrum of electron energies were measured, from zero up to 2×10^{-13} joules. Either energy was not conserved in the transformations, or else energy was going off in a form which could not be detected. Enrico Fermi, Wolfgang Pauli, Niels Bohr, and other leading physicists of the time pondered the phenomenon carefully because the principle of energy conservation was being called into question. Finally, in 1931 Wolfgang Pauli suggested the existence of an invisible particle: the "neutrino", or "little neutral one", as Fermi named it. Its existence was postulated in order to account for the lost energy and momentum. More than 25 years later the neutrino was actually detected experimentally (that is, certain particles and radiation were observed which could only have resulted from the reaction between a neutrino and another particle) in a complicated apparatus set up near a nuclear reactor in the Atomic Energy Commission's Savannah River Plant in South Carolina. The same confidence

that energy, momentum, and other physical quantities are conserved in isolated systems is still evident in contemporary
physics: especially in high energy physics, or the physics of
fundamental particles. Workers in this field have scanty knowledge of the forces or principles involved in the fundamental
processes determining the basic organization of matter. What
they *do* have is the conviction that when new particles appear
or disappear, the overall system of which they are a part remains
unchanged according to some principle of symmetry or invariance. Such aesthetic principles have grown to be important conceptual tools in modern physics.

Early in the previous chapter we introduced conservation
principles as a special sort of symmetry principle. In what sense
are the conservation of momentum and energy symmetry principles? Simply this: if in the macroscopic world the passage of
time could be reflected in a mirror, that is, if time could be made
to go backwards, an examination of momentum and energy conservation in interactions would not tell us the difference. Since
the total momentum and energy are the same before and after
an elastic collision, a motion picture of the collision could be
shown backwards and no one would know the difference. Like
other symmetry principles, it both simplifies and hides things
from us. (In the case of inelastic collisions, where some thermal
energy is produced, we could begin to tell the difference between "before" and "after". The symmetry would have been
upset. This possibility will be further discussed in Chapter 16.)

Power

The words force, energy, and power are used interchangeably
in common parlance, but have distinct meanings in physics.
Power is defined as the rate of doing work. It is measured in
joules/sec or *watts*.[3] How effective is a human being with
respect to doing physical work at a specified rate? Consider a
50-kg woman running to the top of a long set of stairs which
rise 8 meters in 10 seconds. From Equation (7-4) she does work,
$mg\Delta y$, that equals (50 kg) (9.8 m/sec²) (8 m), or 3920 joules. If
she does this work in 10 seconds, her power output is 392 watts,
or $392/746 = .525$ horsepower. It is possible for a person to produce greater power than this, but only for short periods of time.
Consider the shotputter on page 50. Suppose that he accelerated a 16 lb shot (7.27 kg) from 0 to its final velocity of 12 m/sec
while pushing on it over a distance of 1.414 meters. Suppose
also that the shot is raised 1 meter during the push. (1.414 meter
is the length of the hypotenuse of a triangle whose base and

(3) The "horsepower" is equal to 746 watts.

height are both 1 meter.) The athlete gives the shot a kinetic energy equal to $\frac{1}{2}(7.27 \text{ kg})(12 \text{ m/sec})^2$, or 524 joules. He also gives it an extra potential energy:

$$mg\Delta y = (7.27 \text{ kg})(9.8 \text{ m/sec}^2)(1 \text{ m}) = 71.3 \text{ J}.$$

He thus gives the shot energy amounting to 595.3 joules. Assume that he accelerates the shot uniformly over the 1.414 meter pushing distance. The average velocity will be 6 m/sec, since it starts at 0 and ends up traveling at 12 m/sec. He must then push on the shot for 1.414 m/(6 m/sec), or 0.236 seconds. His power output is then 595.3 J/0.236 sec, or 2530 watts. This equals 3.39 horsepower. Add to this the even larger power required for the athlete to accelerate his own body across the ring, and you get a value of between 10 and 15 horsepower. Very few people are strong enough to produce this much power even for such a short period.

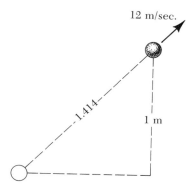

Figure 7-17
The shot is both accelerated and lifted by the athlete as he pushes on it.

Types and Sources of Energy

Most of the energy which ends up doing work, or keeping things warm, originates in radiation from the sun. The side of the earth facing the sun receives energy at the rate of 2000 watts per square meter. About half of this is absorbed or reflected by the earth's atmosphere, so that one kilowatt strikes one square meter of the earth's surface on a sunny day. The portion of energy absorbed is not totally lost because it warms the atmosphere, causing circulation and winds. Sailing ships and windmills are no longer a significant means for tapping the sun's energy. Nevertheless, winds represent a large source of energy, potentially useful to humanity, if it could only be concentrated. Heat from the sun evaporates water from oceans and lakes. When this water condenses and falls as rain on higher portions of the earth's surface, it becomes a source of energy (potential energy of the water). It can be converted into easily transported electrical energy through generators turned by waterfalls or water

flowing through dams. To a lesser extent, solar radiation has been used directly, in solar furnaces and solar batteries, to produce pollution-free thermal energy or electrical energy. The problem up to now has been to concentrate solar energy sufficiently so that it could be utilized on the large scale of a modern power plant. Recent developments make it appear this can be achieved.[4] Still another potential method for using solar energy lies in the world's warm ocean waters. Some advocates[5] have made serious suggestions that by circulating a low-boiling point heat-transfer liquid between the surface waters of the ocean and the cooler water several hundred feet below, some of the thermal energy could be made to run generators, and thus be converted to electric energy. (The conversion of thermal to other forms of energy is discussed further in Chapter 16.)

However, the primary means of utilizing and storing the energy radiated to us by the sun is based on photosynthesis. This is a biochemical process in which sunlight, in the presence of chlorophyll, enables carbon dioxide and water to produce plant tissue and oxygen. The plant tissue can be eaten directly by people to obtain energy, or it can be eaten by domesticated animals which are then eaten. Uneaten tissue in the form of trees might be cut up immediately for fuel, or it might decay over millions of years to produce fossil fuels such as coal, oil, and natural gas.

To get some idea of the possibilities for solar energy storage by photosynthesis, let's calculate the equivalent in horsepower of the energy stored by a one acre field of corn in one second. On the basis of the total mass of the corn plants produced in a 100-day growing season, an acre of corn can store energy from the sun at the rate of 2×10^9 joules per day.[6] Most, but not all, of the growth occurs during daylight hours. If this figure is divided by 86,400, the number of seconds in one day, an average figure of 23,100 joules per second is obtained. This is equivalent to 23,100 watts, or 31 horsepower!

All of the energy stored in corn plants cannot be used as food. The actual food production in various countries around the earth varies from low values up to 50×10^6 joules per acre per day in Japan. A human being requires about 12×10^6 joules per day to live a normal life: 8×10^6 joules for maintaining body temperature and the remainder for walking and doing a little lifting. Thus, an acre of food-growing land can support up to four people if they are content to eat grain. If they want to eat meat, then the efficiency of the energy conversion process is lowered by

(4) Glaser, Peter E., "Solar Energy — An Option For Future Energy Production", *The Physics Teacher 10*, Nov, 1972, pg. 443.

(5) Zener, Clarence, "Solar Sea Power" *Physics Today 26*, Jan, 1973, pg. 48.

(6) Chalmers, Bruce, *Energy*, Academic Press, 1963, chapters 9, 12.

a factor of ten, increasing the acreage required to support humans. Animals store only 10 percent of the energy in the food they eat.

The earth has some energy sources of its own, independent of the sun, and these are just in the initial stages of being tapped. The earth's rotation plus the pull of the moon produces tides, which have been used locally as a means of lifting water to a higher level, after which it is trapped and let run back down through electric generators. Proposals to tap tidal energy on a larger scale have been made.

As a result of its past history, and the release of energy by the decay of radioactive minerals, the core of the earth is much hotter than its surface. In some places the earth's subsurface waters absorb great amounts of energy from hot rocks below them. In Iceland, and in parts of Italy, New Zealand, and California the hot water and steam that are being generated by this means are used to produce electric power and to heat buildings. Given the earth's limited energy resources, and the large amount of energy potentially available[7] from geothermal sources, proposals are now being made to utilize it further. One method involves drilling two holes side by side deep into the earth's crust at a point where the rock is porous. Cool water is pumped down one hole and hot water comes up the other. One problem which arises is that withdrawal of the energy cools the local rock, so the longevity of such power sources is questionable.

Table 7-1 (below) compares the non-polluting power possibilities of tides and geothermal sources on the earth with two indirect solar sources, and sunlight itself.

Nuclear energy is a growing source of terrestrial energy. In most parts of the world, fossil fuels are still cheaper to utilize,

Table 7-1 *Environmental Energy Sources*[8]

Source	Power, in Watts
Sunshine	1730×10^{14}
Wind	3.7×10^{14}
Water (falls & dams)	$.03 \times 10^{14}$
Tides	$.03 \times 10^{14}$
Geothermal	3.23×10^{14}

(7) Barnea, Joseph, "Geothermal Power", *Scientific American 226*, Jan, 1972, pg. 70.

(8) Roberts, Ralph, "Energy Sources and Conversion Techniques", *American Scientist 61*, Jan-Feb, 1973, pg. 59.

Table 7-2 *Practical Sources of Energy in the United States*[10]

	1850	1900	1950	1970	2000?
wood	91%	21%	4%	—	—
gas	—	2.6	18	37%	29%
oil	—	2.4	37	38	34
coal	9	71	36.5	21	18.1
water	—	2.6	3.8	4	1.8
nuclear	—	—	—	0.2	17.1

but there is little question that nuclear energy will grow in importance in the future. Projections indicate that by the late 1970's nuclear energy will constitute 50 percent of all the *new* power sources utilized in the United States.[9] Nuclear energy is discussed further in Chapter 9.

For more than a century, our main source of energy, other than food, has been *fuel*, not the other sources just mentioned. We have burned it for heat, or to boil water and make it expand, push a piston, or turn a turbine in order to run a machine, or a generator of electrical energy. As Table 7-2 shows, there have been drastic shifts in American fuel consumption since 1850. Water power is also included for comparison purposes.

At present, the three fossil fuels are furnishing almost all the energy that Americans consume. They have been accumulating over millions of years but we are consuming them so rapidly that the world's reserves can last only 50–200 more years. (This is at our present rate of consumption. If we recognize that world population *and* per capita energy consumption are both increasing, then the picture is even darker.) Changes in the costs of mining low-grade coal and separating oil from shales can make calculations of fuel reserves uncertain. However, the unavoidable fact is that we are very close to the end of our fuel stockpile. What is perhaps even more crucial is that we need coal and oil for the chemicals they contain. Using them for heat is like burning bread to keep your house warm.

As we tried to point out previously, using up the last of our fuel reserves does not mean we are running out of energy. However, it does mean we are going to have to employ more "environmental" energy sources. They represent a tremendous amount of energy, but they are not as neatly concentrated as the fossil fuels. Using them may cost far more than we are presently accustomed to paying for our energy.

(9) Hogerton, J. F., "The Arrival of Nuclear Power", *Scientific American 218*, Feb, 1968, pg. 21.

(10) Starr, Chauncey, "Energy and Power", *Scientific American 225*, Sept., 1971, pg. 37.

In his book, *The Science of Culture*, Leslie White,[11] an anthropologist, argues that the evolution of culture is intimately linked to the harnessing and control of energy for use by people. A rapid evolution in culture followed the first cultivation of crops. With food energy more readily available *and storable*, people had leisure, out of which came art, science (eventually), government, and the other mental structures we associate with advanced culture. Fire was used, but largely for keeping warm or making food taste better and easier to chew. (Not to make it more energetic.) The machines that people had were powered by themselves or by using animals, moving water, or wind. White calls the revolution which occurred in the late eighteenth and early nineteenth century not the Industrial Revolution or the beginning of the Machine Age, but the *Coal Age*. For the first time, people were able to use heat from fuel *to do work*; to run their machines. After this, both the energy available to people, and the efficiency with which it was used increased sharply. Energy came to be recognized as a basic natural resource. At the same time, claims White, population soared, innovation and capital investment mushroomed, and a wave of social and governmental revolutions, still in progress, began.

In surveying the economic health of a nation, economists like to use measures of productivity such as *per capita income*, or the *Gross National Product* (called GNP by all, and translated as "Gross National Pollution" by recent critics who are oriented more toward environmental quality rather than to ever increasing productivity). Attempts are often made to correlate these measures with agricultural productivity, railroad mileage, literacy rate, etc. But what seems to correlate best with per capita income is *per capita energy consumption*. Figure 7-18 shows those data for a number of countries.[12]

Earlier it was stated that a human being requires about 12×10^6 joules per day to live. That amounts to about 4.4×10^9 joules per year. The chart indicates that in several nations, the energy consumption to keep *everything* going is less than the average amount of energy needed for humans to live and move about a bit. (Doubtless many people in these nations grow their own food so that it never gets counted in the national energy economy.) Nevertheless, the statistics provide evidence that some nations are literally starving. In contrast to this, the average American consumes 50–100 times as much energy as is needed to stay alive.

(11) Chapter 13, "Energy and the Evolution of Culture" in *The Science of Culture*, Farrar, Straus & Cudahy, N.Y., 1949.
(12) United Nations Data, reprinted in Bach, George L., *Economics: An Introduction to Analysis and Policy.* Prentice-Hall, 1963.

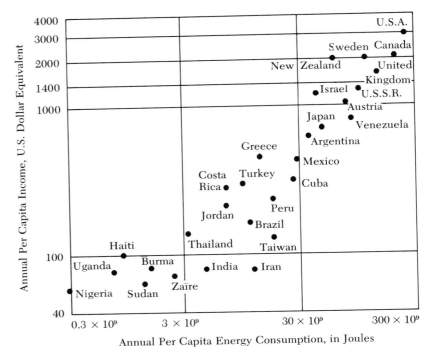

Figure 7-18

*Annual Per Capita Energy Consumption, in joules, plotted vs per capita
income, in U.S. Dollars.*

Transportation and Energy

About one-fourth of America's stupendous energy budget goes
into transportation. People are therefore beginning to examine
our various modes of transportation in terms of the energy that
they consume.[13] Such analyses may help us to make greater use
of transportation systems which consume less energy.

To begin at the bottom, consider walking. This consumes 1.2
to 2.0×10^4 joules of energy per minute, in addition to the basic
5 to 6×10^3 joules per minute required for body warmth. In an
hour of walking, a healthy person consumes (2×10^4 joules/
minute) (60 minute/hour), or 1.2×10^6 joules, and covers about
4 km (2.5 miles). The distance per joule of extra energy consumed in walking is thus:

$$\frac{(4 \text{ km/hour})}{(1.2 \times 10^6 \text{ J/hour})} = 3.33 \times 10^{-6} \text{ km/J,}$$

approximately.

(13) Rice, Richard A., "System Energy and Future Transportation", *Technology
Review 73*, January, 1972, pg. 31.

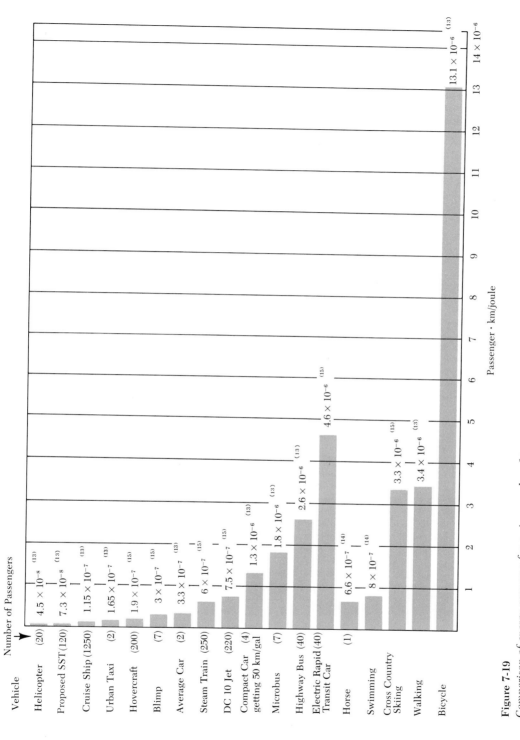

Figure 7-19
Comparison of energy economy for various modes of transportation.

(14) Tucker, V. A., "Energetic Cost of Transportation", *Comparative Biochemistry and Physiology 34*, p. 841, 1970.
(15) Frisch, Klaus, Notes: *Physics in Transportation and Environmental Problems*, Copy Center, John Carroll University, Cleveland, Ohio.

Riding a bike requires less energy per kilometer of travel even though more energy per minute is expended. A 20-km bike ride lasting one hour consumes 1.5×10^6 joules, or about 2.3×10^4 joules/minute. The distance per joule when bike riding is therefore

$$\frac{(20 \text{ km/hour})}{(1.5 \times 10^6 \text{ J/hour})} = 13 \times 10^{-6} \text{ km/J},$$

approximately.

It is striking to compare these means of self-transportation with the motor driven vehicles which the word transportation usually signifies. To make a comparison possible, the energy content of a gallon of fuel is taken to be 1.42×10^8 joules. By multiplying the number of passengers in the vehicle by the distance the vehicle can move per joule of energy, we get a net "propulsion efficiency" in passenger-kilometers/joule, as shown in Figure 7-19.

The vast energetic advantage of walking and cycling for urban trips is clearly evident. Rice points out that if private automobiles were used only for trips longer than three miles, gasoline consumption by automobiles in the United States would be cut in half. When the fact that cars are least efficient (and pollute the most) when they are used for stop and go city driving, and the high cost of parking space are considered, the overall cost of operating private automobiles in cities becomes very high indeed.

The Harmless Dissipation of Energy

Since energy represents the capacity for doing work and is generally something that is paid for, people like to stretch it as far as possible. A great deal of research goes into reducing the friction between sliding parts in machines so that less energy is converted to heat. Nevertheless, if you are traveling fast in a car, you might have to get rid of kinetic energy quickly in an emergency. This is the function of brakes. A brake must exert decelerating forces on the moving car, but more than this, it must be able to absorb the great amount of heat which appears as the kinetic energy disappears. Sometimes truck drivers are seen sitting next to their heavily loaded trucks at the bottom of a long descent waiting for a set of smoking brakes to cool off. The potential energy which the truck and contents had at the top of the hill either appears as kinetic energy as the truck moves down the hill — an undesirable state — or else the potential energy is absorbed by downshifting gears, and using the brakes. (In terms of Equation (7-9), $\Delta(P.E.)$ is negative, ΔU is positive

Figure 7-20
Einstein on Wheels. (Courtesy of the California Institute of Technology Archives.)

and $\Delta(K.E.)$ is zero, if all goes well.) The electric trains which traverse the mountainous regions in Europe are designed so that some of the electrical energy used to pull the train up a hill is reclaimed on the downgrade by running the motors in the locomotives as generators of electrical energy which can be used elsewhere. This increases the efficiency of the overall system. Rather than converting the potential energy to heat by mechanically braking the wheels as the trains go downhill, the potential energy is converted to electrical energy which is sent out over the wires to help other trains go uphill. (This will be better understood after Chapter 11 is studied.)

To be of use, energy must be concentrated. But concentrations of energy are also dangerous. In his article, "On the Escape of Tigers: an Ecologic Note"[16] William Haddon, Jr. suggests a number of measures to prevent energy from having harmful affects. He suggests first the prevention of the marshalling of energy before it happens: preventing the manufacture of gunpowder and nuclear weapons; preventing the buildup of hurricanes and stresses in the earth's crust, and accumulations of snow which could lead to avalanches. Second, he suggests reducing the amount of energy marshalled: reducing the amounts and concentrations of high school chemical reagents, the size of bombs and firecrackers, the speed of cars, and the height from which swimmers dive into pools. Third, he suggests preventing the release of the energy: preventing the discharge of nuclear devices, gunpowder, or electricity; the jumping of would-be suicides; or the escape of tigers. Other strategies involve slowing the rate at which the energy is released; by making beginners' ski slopes gradual, or by putting a collapsible bumper around automobiles (as was suggested on page 79); by separating the energy and the people who might be injured; by using protective barriers, masks or guards; by making things rounded or flexible; by making the object or person who may be

(16) Haddon, William, Jr., *Technology Review* 72, May, 1970.

victim of energy release stronger, or more supple; and finally, by improving alarm systems to allow quicker reactions to unwanted release of energy.

Summary

The work done by an external force on an object is defined as

$$\text{Work} = (F_{avg})_x \Delta x \qquad\qquad \text{Equation (7-1)}$$

where Δx is the change of position of the object and $(F_{avg})_x$ is the component of the average force in the direction of the change of position. If F is constant, then $F_{avg} = F$. This definition applies whether F acts alone, or in concert with other forces.

$$\text{Work done by } F_{net} = \frac{m(v^2 - v_o^2)}{2} \qquad\qquad \text{Equation (7-2)}$$

$$\text{Work done against friction} = \mu mg \Delta x \qquad\qquad \text{Equation (7-3)}$$

$$\text{Work done in lifting} = mg \Delta y \qquad\qquad \text{Equation (7-4)}$$

The work done by forces increases the energy of the object on which the work is done. Energy is in turn defined as the capacity for doing work. Both work and energy are measured in joules.

Work done by F_{net} equals $\Delta(K.E.)$, the change in kinetic energy (the capacity for doing work due to being in motion).

Work done against friction equals ΔU, the change in thermal energy.

Work done in lifting equals $\Delta(P.E.)$, the change in potential energy (the capacity for doing work due to position).

Energy has become a powerful concept in physics because it has been found that, in isolated systems, the total energy, E, is conserved:

$$\Delta E = \Delta(K.E.) + \Delta(P.E.) + \Delta U = 0 \qquad\qquad \text{Equation (7-9)}$$

An increase of thermal energy requires a decrease in the mechanical energy ($K.E. + P.E.$) of the system.

The energy of a system is not conserved while work is being done on it by external sources. Energy conservation depends upon where one puts the boundary of the system.

During collisions between objects in an isolated system the total momentum is always conserved, but the total mechanical energy is conserved only in "elastic" collisions.

Energy conservation permits the prediction of changes to be made which would be prohibitively difficult calculations using only the forces and accelerations involved.

Conservation of energy, momentum, and other physical quantities has come to play an important role in modern physics. Unknown objects have been identified and entirely new particles "discovered" by application of these and other conservation principles.

Insofar as mechanical conservation principles prevent us from distinguishing "before" from "after" they are also symmetry principles. If time could be reflected, as in a mirror, and made to run backwards we would not be able to tell the difference, mechanically.

Power is the rate of doing work, or of consuming energy. It is measured in joules per second or watts. The horsepower, 746 watts, is also used as a unit of power. Few non-athletes can work at the rate of one horsepower for more than a fraction of a second.

The sun furnishes most, but not all, of our energy. It produces winds, water power, direct solar energy, food, and wood, and its energy has also been stored as fossil fuel: gas, coal, and oil.

Besides the sun, tides (from the moon's gravity), geothermal energy, and nuclear energy are also available.

There are close ties between energy consumption and such societal factors as social change and per capita income. However, we are near the end of the fossil fuel on which our technology is based. We must curtail the rate of expansion of our fuel consumption and find ways to use more environmental, less pollutive energy sources.

To be useful, energy must be concentrated, but concentrated energy is also a source of danger.

Further Reading

Brown, Sanborn C., *Count Rumford—Physicist Extraordinary*, Doubleday Anchor Book, Garden City, N.Y., 1962. (Paperback account of the life and loves of an Eighteenth Century Yankee inventor who became a British spy, Minister of Defense to the King of Bavaria, and laid the foundations for the science of thermodynamics.)

Bowen, Richard G., "Geothermal—Earth's Primordial Energy", *Technology Review* 73, Oct/Nov, 1971, pg. 42. (Geothermal sources are badly underdeveloped at present. They bear promise of cheap energy.)

Cook, Earl, "Energy For Millenium Three", *Technology Review* 75, Dec., 1972, pg. 16. (Gives evidence that population and cultural achievement both depend on energy supply.)

Dyson, Geoffrey, *The Mechanics of Athletics*, University of London Press, 1962. (Investigates the physics of track and field.)

Hiebert, Erwin N., *The Historical Roots of the Principle of Conservation of Energy*. Dep't. of History, University of Wisconsin, Madison, 1962. (Traces energy conservation from the Greeks to the 18th century.)

Jammer, Max, *Concepts of Force*, Harvard University Press, Cambridge, 1952. (An historical discussion of the developments of concepts of inertia, force, momentum and energy.)

Manners, Gerald, *The Geography of Energy*, Hutchinson University Library, London, 1964. (The spatial characteristics of energy production, transportation, and consumption.)

Meinel, A. B., and Meinel, M. P., "Physics Looks at Solar Energy", *Physics Today 25*, Feb. 1972, pg. 44. (New solid state physics technology is making large-scale solar energy conversion a possibility.)

National Academy of Sciences — National Research Council, Committee on Resources and Man, *Resources and Man: A Study and Recommendations*, W. H. Freeman & Co., San Francisco, 1969. (Possibilities and problems of food, mineral, and energy resources on earth.)

Ross, Charles R., "Electricity as a Social Force", *Annals of the Amer. Acad. of Political & Social Sciences 405*, Jan, 1973, pg. 47. (Consumption of electric power as a measure of economic and social prosperity is being criticized.)

Schmidt-Nielsen, Knut, "Locomotion: Energy Cost of Swimming, Flying, and Running", *Science 177*, July 21, 1972, pg. 222. (For a given size of animal, running requires more energy than flying and flying more energy than swimming!)

Schurr, S. H. and Wetschert, B., *Energy in the American Economy*, John Hopkins Press, Baltimore, 1960. (Contains a table which compares the energy per kilogram available from coal, gasoline, wood, bread, butter, and sugar.)

Scientific American 225, Sept. 1971. (Entire issue devoted to energy and power.)

White, David C., "Energy, The Economy, and the Environment", *Technology Review 73*, Oct/Nov. 1971, pg. 18. (Presents some of the dilemmas in our demands for more and more energy.)

Problems

7-1 Compute the work done against gravity when a 75 kg man climbs a mountain whose summit is 500 m above its base. If he expands energy at the rate of 100 watts on the average and 1/3 of that energy is used to lift himself, how long will the climb take?

7-2 How much work does it take to push a 20 kg packing crate up a flat frictionless slope which is 5 meters long and 1 meter high? Does it make any difference whether you push horizontally, or parallel to the ramp; or pull on a rope upward, parallel to the ramp?

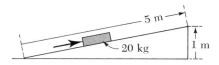

7-3 Suppose you try to stretch a steel spring. The amount of stretch will be very close to proportional to the force you exert on it. (If you double the force, the stretch will double.) Then the relation between the stretch, Δs and the force, F, could be written $F = k\Delta s$. Show that the work you have to do to stretch the spring a distance Δs is $\frac{1}{2}k(\Delta x)^2$.

7-4 Two sleds, starting from rest, slide down frictionless hills 10 m high. Both sleds have masses of 40 kg. One hill slants straight down:

What is the speed of the sled at the bottom of this hill?

The other hill is shaped like this:

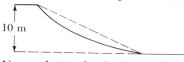

Now, what is the final speed?

What do your calculations tell you about the time required to go down the 2 hills? How would the two final speeds compare if the hills were not frictionless?

7-5 A very elastic ball is dropped from the top of a building and bounces nine-tenths of the way back up. Now the same ball is thrown horizontally from the top of the building. Does it still bounce up to a point nine-tenths as high as the building?

Finally, the same ball is slid down a frictionless ramp from the top of the building to the ground:

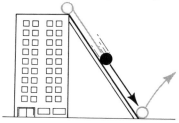

It strikes the ground and rebounds elastically again. Does it still rebound to a point 9/10 as high as the building?

7-6 If you throw a ball straight up in the air, how far will it go in comparison with a ball thrown at half the speed?

7-7 A shopper exerts a horizontal force of 60 N on a grocery cart whose mass is 10 kg. A frictional force of 15 N acts backward on the wheels of the cart. (a) What work is done against friction if the cart is pushed horizontally 10 m? (b) Compute the gain in kinetic energy of the cart.

7-8   An astronaut is sitting on a space platform. The mass of the platform, astronaut, and everything on it totals M. A cartridge having a mass m containing a message is fired toward a co-worker some distance away. Suppose chemical energy amounting to Q joules was expended in shooting the cartridge from the space gun. The platform will recoil a little just as the cartridge is fired. How much of the chemical energy Q is actually carried away in the form of kinetic energy by the cartridge?.

7-9   A surprising amount of fuel carried by jetliners is used simply to get the plane up to a high altitude. Calculate (a) the kinetic energy, and (b) the potential energy, of an airliner having a mass of 10^5 kg, a speed of 600 miles per hour (268 m/sec), and an altitude of 33,000 feet (10,000 m). If the fuel that the plane burns furnishes 4.6×10^7 joules of energy per kg, how many pounds of fuel will be consumed to obtain the requisite kinetic and potential energies?

7-10   Suppose a dam that can generate 100 megawatts of electricity has been built. It is 100 m high. Assuming that the initial potential energy of the water is completely converted into electrical energy as it passes through the dam, how much water, in kilograms, must pass through the dam each second?

7-11   A 1 kg camera is dropped from the side of a cruise ship into the water. The rail of the ship is 10 m above the water, and the water is 10 m deep. If the potential energy of the camera is assumed to be 0 at the surface of the water, account for what happens to the original potential energy of the camera when it is (a) 5 m above the water and falling, (b) 1 meter above the water and falling, and (c) on the bottom of the harbor, at rest. How much thermal energy do the camera and water receive?

7-12   Two old cars engage in a demolition derby in the middle of a frozen lake. They each have a mass of 1500 kg. One comes from the left at a velocity of 10 m/sec and the other from the right with a velocity of 15 m/sec. They crash into each other, hook bumpers, and slither along the ice together after the collision. Calculate their final speed. How much original kinetic energy is converted to thermal energy in their inelastic collision?

7-13   If gasoline provided about 1.4×10^8 joules of energy per gallon and electricity costs three cents per kilowatt-hour, which is the cheaper source of energy?

7-14   Suppose a person is able to convert into useful work one-third of the food energy that he or she consumes. If bread contains 11×10^6 joules of energy per kilogram, how many kilograms of bread will this person need to eat in order to obtain enough energy for a 20 km (12 mile) walk?

7-15   From Table 7-1 and any other factors you can think of, which seems to offer the best prospect for development as a power source: water (falls and dams), wind, tides, or geothermal energy?

8
RELATIVITY

Frames of reference have been mentioned as well as the fact
that some physical quantities are invariant; that they have the
same value when observed from various frames of reference.
However, a lengthy discussion of the effects of the observer's
frame of reference on the measurements made has been delayed
until this chapter. There are good arguments for tackling this
problem at the very start of a textbook on physics: your choice
of frame of reference will affect all subsequent observations
that you make. The branch of physics in which the problem is
handled is called *relativity*. It was not mentioned at the time,
but the switch from the Ptolemaic to the Copernican model of
the solar system which is discussed in Chapter 5 was one of
relativity; of admitting that our frame of reference is moving.
The reason the discussion of relativity has been held off until
now is that it is at once the most simple, and most profound, element in physics. It is also one of the most misunderstood. Some
understanding of physical law and of invariance is needed before
one can appreciate what relativity seeks to do.

First, the ideas of classical relativity held by Galileo and
Newton will be developed. Einstein's special theory of relativity, a set of postulates which put physics on a new basis at

the beginning of this century, will then be discussed. It will seem, as the consequences of Einstein's relativity theory are discussed, that it simply wiped out the permanence—or invariance—of things (such as length and time measurements), that people have always depended upon. In some ways it will seem to add to the absurdity of the world. However, it must be kept in mind that Einstein's purpose in presenting his theory was to find new ways of expressing the laws of physics so that they would have the same form in different frames of reference. He was searching for invariance. He believed that physical reality lay in law, not perception. His theory calls for very puzzling results at the level of measurement, but in return it gives a more comprehensive explanation of phenomena, and has a wider range of validity, than Newton's physics.

Galileo's and Newton's Relativity

Suppose that two observers set out to describe the behavior of a certain object. One of them, S, stands at the side of a highway and the other, S', stands on a railroad flatcar being pulled to the right with a velocity, v, as shown in Figure 8-1. They each have a clock and a meter stick, or steel tape measure, which they hold in their hands, parallel to their relative motion. With these instruments, S is able to determine the location, x, and time, t, of events according to his earth-based frame of reference and S' is able to measure the position, x' and time t', of events in his flatcar-based frame of reference. They both notice a truck moving along a highway. They record the position of the front bumper of the truck and both decide that at the time $t = t' = 0$, the bumper is at position $x = 0$ (according to S) and $x' = 0$ (according to S'). As time passes, their two position observations will disagree. Suppose the truck were at rest relative to the highway and to S. After three seconds pass, S would say x is equal to zero. That is, the front bumper of the truck will still be even with the zero mark on his stick. But if v were 10 m/sec, S' would observe

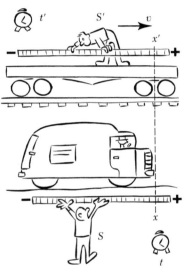

Figure 8-1

Position of a truck as observed in two different frames of reference, S and S'.

$$x' = (-10 \text{ m/sec}) (3 \text{ sec}) = -30 \text{ m}.$$

S' claims the truck is backing up so fast that after three seconds it's 30 meters in back of the zero mark on the flatcar meter stick! More generally, it can be said that the two position measurements, x and x', disagree by an amount equal to vt where v is the relative velocity of the two observers and t is the time elapsed since they matched meter sticks. In our example, the disagreement vt is subtracted from the original observation made by S'. Therefore, x' will read less than x, by the amount vt. Saying it all in one equation;

$$x' = x - vt. \qquad \text{Equation (8-1)}$$

Whose observation is the correct one? The earth based observer's? The wind blowing in the face of S', and the trees and pavement at rest with respect to S, lead us to believe that S is really standing still. But evidence from astronomy indicates that the earth moves rapidly around the sun and also spins. In turn, the sun moves relative to the Milky Way galaxy. All we can say is that the two observers are moving *relative to each other*, and write down Equation (8-1) to express the degree of that movement, and hence of their disagreement.

Now let the truck move relative to the highway. Both observers record the position of the truck in their frame of reference with repeating cameras and clocks. S finds that the velocity of the truck, u, equals $\Delta x/\Delta t$. S', however, finds that the velocity of the truck, u', equals $\Delta x'/\Delta t'$, and that

$$\frac{\Delta x'}{\Delta t'} = \frac{\Delta x'}{\Delta t},$$

since the time intervals should be the same in the two frames of reference ($\Delta t = \Delta t'$). But

$$\frac{\Delta x'}{\Delta t} = \frac{\Delta x - v\Delta t}{\Delta t} = \frac{\Delta x}{\Delta t} - v.$$

Thus:

$$u' = u - v \qquad\qquad \text{Equation (8-2)}$$

If the velocity, u, of the truck relative to S is 20 m/sec, then its velocity, u', in the flatcar frame of reference, will be equal to 20 m/sec − 10 m/sec, or 10 m/sec.

If an observer in a space ship chanced to pass the earth at this moment, traveling at 20,000 miles per hour (8,940 m/sec) toward the right relative to our Figure 8-1, the truck would appear to have a velocity

$$u'' = u - v$$
$$= 20 \text{ m/sec} - 8940 \text{ m/sec}$$
$$= -8920 \text{ m/sec}$$

in the space ship's frame of reference.

To an observer on the sun, the truck would be traveling at 20 m/sec on a platform which is spinning once every 24 hours, and hurtling along at more than 66,000 miles/hour.

The truck now accelerates relative to S, increasing its velocity, u, from 20 to 24 m/sec in 2 seconds. The acceleration

$$a = \frac{(24 - 20) \text{ m/sec}}{2 \text{ sec}} = 2 \text{ m/sec}^2.$$

The flatcar observer, S', says the truck has an acceleration

$$a' = \frac{\Delta u'}{\Delta t'} = \frac{\Delta u'}{\Delta t} = \frac{u_f' - u_i'}{\Delta t},$$

where u_f' and u_i' are the final and initial velocities, respectively, as observed by S'. But, from Equation (8-2):

$$u_f' = u_f - v$$
$$= 24 \text{ m/sec} - 10 \text{ m/sec}$$
$$= 14 \text{ m/sec}$$

and

$$u_i' = (u_i - v)$$
$$= 20 \text{ m/sec} - 10 \text{ m/sec}$$
$$= 10 \text{ m/sec}.$$

Thus,

$$a' = \frac{\Delta u'}{\Delta t} = \frac{u_f' - u_i'}{\Delta t}$$

$$= \frac{14 \text{ m/sec} - 10 \text{ m/sec}}{2 \text{ sec}} = 2 \text{ m/sec}^2.$$

The two observers say the truck has the same acceleration! This invariance of acceleration between the two frames of reference, we can now argue, gives strong support for making acceleration, rather than velocity, the basic quantity in the mechanics we have studied thus far. Physicists like to work in terms of invariant quantities.

Figure 8-2
Frames of reference set up in two racers moving at different speeds.

Some Sample Problems Involving
Relative Velocities

191

*Some Sample
Problems
Involving
Relative
Velocities*

(1) A race car circles the 4.03 km (2.5 mile) Indianapolis race track once per minute and gains five seconds per lap on the car in front of it. How fast are both cars moving, and what is their relative velocity?

First, a car which travels 4.03 km in one minute has an average speed of 4.03 km/1 min, or

$$\left(\frac{4.03 \text{ km}}{1 \text{ min}}\right)\left(\frac{1 \text{ min}}{60 \text{ sec}}\right) = .0672 \text{ km/sec} = 67.2 \text{ m/sec}.$$

If it gains five seconds per lap, then the first car must be taking 65 seconds to circle the track. The first car, then, has an average speed of

$$\frac{4.03 \text{ km}}{65 \text{ sec}} = .062 \text{ km/sec} = 62 \text{ m/sec}.$$

Their relative speed would be, 67.2 m/sec − 62 m/sec, or 5.2 m/sec. Looking at things from the frame of reference of the faster car, we obtain a relative velocity of −5.2 m/sec for the slower car. It appears to back up at 5.2 m/sec relative to the faster car. If we switch our point of view to the slower car, the other car has a positive velocity of 5.2 m/sec.

Now suppose a miniature physics lab is set up in each car and the drivers start making observations of the velocity of the main grandstand. Let S sit in the slower car, as shown in Figure 8-2. He will observe the grandstand to shoot by with a velocity u equal to −62.2 m/sec. (It seems headed in a negative direction.) S′, back in the faster car, measures a grandstand velocity u′ equal to −67.2 m/sec. This information can also be used to find the relative velocity, v. From Equation (8-2), u′ = u − v or v = u − u′. Then, v, the relative velocity equals [−62.2 − (−67.2)] m/sec, or +5.2 m/sec. We get a positive answer here because of our choice of reference frames.

(2) A woman strides up a moving escalator, climbing 2 steps per second. It is 1.5 feet from one step to the next (measured parallel to the motion of the escalator). It takes the woman 20 seconds to ascend the 30 foot escalator. How fast is the escalator moving?

Let's call her speed relative to the escalator u′, her speed relative to the store u, and the speed of the escalator v.

$$u' = \left(\frac{1.5 \text{ ft}}{\text{step}}\right)\left(2 \frac{\text{steps}}{\text{sec}}\right) = 3 \text{ ft/sec}$$

$$u = \frac{30 \text{ ft}}{20 \text{ sec}} = 1.5 \text{ ft/sec}$$

Then since u′ = u − v,

$$v = u - u'$$

$$= 1.5 - 3$$

$$= -1.5 \text{ ft/sec} \qquad v \text{ is negative.}$$

She's on the *down* escalator!

Figure 8-3

Person moving at a speed u′ relative to an escalator which is also moving, at a speed v, relative to the store.

The treatment of relativity which we have discussed thus far is often called Galilean relativity, after Galileo. The equations

$$x' = x - vt, \quad \text{and} \quad t' = t, \quad \text{Equations (8-3)}$$

are often called the *Galilean transformations*. They work well for problems involving classical mechanics; that is, for the motion of massive objects at familiar speeds. However, troubles developed when they were applied to problems involving electricity and magnetism. Further discussion about this follows shortly. But, for now, let us look at the problem of choosing a "good" frame of reference.

Inertial Frames of Reference and Relativity

Newton's second law is a summary of observations, not a scheme for proving that forces are at work on a body. You realize this very soon when you try to decide whether or not an object is accelerating! The word acceleration in Newton's second law implies a measurement *relative to some frame of reference.* If your frame of reference were attached to a dune buggy and you were studying the motion of an object in it being pulled on by a spring, you will have a difficult time separating the "real" accelerations of the object from the apparent changes of motion resulting from lurches in the frame of reference. We can get around the problem of defining the acceleration of an object by using the concept of an *inertial frame of reference.* An inertial frame of reference is one in which a seemingly isolated object—one on which no forces appear to act—remains at rest or travels with a constant velocity. Acceleration is then measured relative to such a frame. Any other frame of reference moving at a constant velocity relative to it will also be an inertial frame of reference.

We saw on page 190 that when the acceleration of an object is measured in two different frames which are moving at a constant velocity relative to each other (two inertial frames) the same result is obtained: acceleration is invariant between them. If acceleration is basic to our study of mechanics then two such frames are *equivalent* as far as a description of mechanical phenomena is concerned. From all of this, a Newtonian "relativity theory" begins to emerge: *laws of motion are invariant between two (inertial) frames of reference moving at a constant velocity relative to each other.* If $F = ma$, then $F' = m'a'$. No mechanical experiment will tell you whether your frame of reference is moving or standing still.[1] Constant velocity itself has no physi-

(1) This does constitute a rather modern interpretation of Newton's laws of motion. Although there is nothing in his second law to suggest that there is a difference between rest and absolute motion, Newton remained convinced that there should be *some* means for defining an absolute space through which all objects move, or in which they remain at rest.

cal effect. And whereas velocities are relative (u' is not equal to u), acceleration takes on the aspects of an "absolute". As far as physics is concerned, the closest you can get to absolutes are invariants.

Late in the nineteenth century things began going wrong with Newtonian relativity. When the common sense Galilean transformations were used to find electric or magnetic effects in a moving inertial frame S', in terms of measured effects in S, forces began showing up that shouldn't have. It seemed untenable that the laws of mechanics (motion) should be invariant between two inertial frames but laws of electricity and magnetism should not. The Dutch physicist, Hendrik A. Lorentz (1853–1928), after great effort, came to the conclusion in the mid 1890's that when electric and magnetic phenomena were transformed from one inertial frame to another, a strange new set of position and time transformations had to be used:

$$x' = \frac{x - vt}{\sqrt{1 - v^2/c^2}} = \gamma(x - vt)$$

and

Equations (8-4)

$$t' = \frac{t - xv/c^2}{\sqrt{1 - v^2/c^2}} = \gamma\left(t - \frac{xv}{c^2}\right)$$

In these equations, γ, pronounced "gamma", equals $1/\sqrt{1 - (v^2/c^2)}$; v is the velocity of the S' frame relative to the S frame; x and t are position and time measurements in S frame; and, x' and t' are the transformed position and time measurements in the S' frame.

These equations are still known as the "Lorentz transformations". Their most startling consequence is that time might not pass with equal rapidity in two inertial frames moving relative to each other. Perhaps saying that $\Delta t = \Delta t'$ prior to deriving Equation (8-2) was not justified.

Until 1905 the Lorentz transformations were seen as a kind of curiosity, necessary for preserving the invariance of electricity and magnetism, but not for the laws of motion of large-scale objects.

Motion and Rest Again

Several pages ago four different observers examined a truck moving along a highway, and they came up with vastly different values for its speed. It was suggested that none of the values was any better than any other. However, there seemed to be a tendency (shared by Newton, among others) to believe that there must be *some* platform, somewhere, that is standing still which would allow you to find out what the absolute velocities of objects in the universe really are. This belief seems to have strengthened among physicists in the 1860's when James Clerk Maxwell

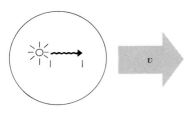

Figure 8-4a
A beam of light traveling parallel to the direction of the earth's movement around its orbit.

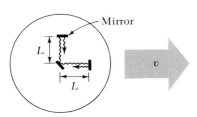

Figure 8-4b
Two beams of light, both traveling out and back along a path of length L. One beam travels parallel to the earth's motion and the other travels perpendicular to it.

(1831–1879), a Scots physicist, was able to show that light has the properties of an electric-magnetic vibration (or wave) which travels with a velocity of 3×10^8 m/sec or 186,000 miles/sec *in empty space.* Until late in the nineteenth century physicists had difficulty conceptualizing empty space as "waving", so a theoretical substance, "aether", was proposed as a medium to carry the electromagnetic vibrations. Right away, scientists thought this might provide a way to settle the question of absolute motion. If an observer were really at rest in space (at rest relative to the aether) then, if the speed of light were carefully measured, the result ought to be 3×10^8 m/sec. Persons on a moving platform ought to obtain a lower or higher result, depending on whether they were traveling with or against the beam. Since light has such a high speed, the experimental detection of these velocity changes would require a highly refined apparatus and procedure.

By 1887, Albert A. Michelson (1852–1931) and Edward Morley (1838–1923) had developed the requisite technique and began a series of delicate measurements at the Case School of Applied Science in Cleveland. Their intent was to measure the speed of the earth through space. They figured that if a beam of light traveled in a forward direction at a fixed speed relative to space, then the motion of the earth would subtract from the measured value of the speed of light. They did not measure the speed directly. Instead, they compared the time required for the light to travel the same distance forward and back parallel to the earth's motion with the time required for the light to travel an equal distance back and forth perpendicular to it (Figure 8-4b). They calculated that the forward-back trip would take longer, according to the following argument.

Suppose the speed of light relative to the aether is c, and that the earth moves through the aether with velocity v. If a beam of light travels forward (toward the right in Figure 8-4b,) its speed relative to the earth should be reduced from c to $(c - v)$. (It is quite analogous to a swimmer in a river trying to make headway against the current.) To travel the distance L should require a time equal to L divided by the effective speed, or $L/(c - v)$. On the return trip (to the left in Figure 8-4b) the light travels "downstream", so its effective speed is increased to $(c + v)$ and the time required to cover the same distance equals $L/(c + v)$. The total time for the forward-back trip over the distance L would thus be

$$\frac{L}{c - v} + \frac{L}{c + v}.$$

Multiplying to obtain the least common denominator, we get

$$\frac{L(c + v) + L(c - v)}{(c - v)(c + v)} = \frac{2Lc}{c^2 - v^2}.$$

When it is simplified by dividing the numerator and denominator by c^2, the expression becomes

$$\frac{2L\not{c}}{\not{c^2}(1 - v^2/c^2)} = \frac{2L}{c(1 - v^2/c^2)}.$$

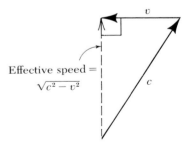

If $v = 0$, the round trip takes the time $2L/c$.

The perpendicular, back-and-forth trip requires some more thought. In order to travel straight across the "stream", the light must travel forward at an angle relative to the direction of motion. It heads in this direction all the time, but is continually pushed back by the aether so that its direction of progress becomes perpendicular to the earth's motion. This reduces the effective speed of the light from c to $\sqrt{c^2 - v^2}$. (The time required for the light to travel a given distance relative to the earth is increased by the factor $c/\sqrt{(c^2 - v^2)}$. Think about it.) The time required for the perpendicular round trip, instead of simply being $2L/c$, becomes $(2L/c)[c/\sqrt{(c^2 - v^2)}]$, or:

$$\frac{2Lc}{c\sqrt{c^2 - v^2}} = \frac{2L}{\sqrt{c^2 - v^2}}.$$

Now let us remove the factor $\sqrt{c^2}$ from the denominator.

$$\frac{2L}{\sqrt{c^2}\sqrt{\frac{c^2}{c^2} - \frac{v^2}{c^2}}} = \frac{2L}{\sqrt{c^2}\sqrt{1 - v^2/c^2}} = \frac{2L}{c\sqrt{1 - v^2/c^2}}.$$

Note that if we multiply top and bottom by $\sqrt{1 - v^2/c^2}$, the result is:

$$\frac{2L\sqrt{1 - v^2/c^2}}{c(1 - v^2/c^2)}.$$

But $2L[c(1 - v^2/c^2)]$ is just the time required for the forward-back trip. The time required for the perpendicular trip is thus shorter by the factor $\sqrt{1 - v^2/c^2}$.

Although the speed of the earth in its orbit, v (equal to about 30 km/sec, as we saw earlier), is much less than c, Michelson and Morley's apparatus was sensitive enough to measure the difference between 1 and $\sqrt{1 - v^2/c^2}$. It should have been able to provide a direct experimental measure of the velocity of the earth through the aether; that is, through absolute space. But, they found *no* difference in the times required for the two trips! They repeated the experiment during various parts of the year, when the earth would be traveling in different directions around its orbit, but there was still no difference. (Recall page 126, where we discussed a sidelight of this.)

How could this be? One explanation was that the earth really is standing still in space, just as Aristotle and Ptolemy had thought.

This way, the speed of light in both beams would be the same ($v = 0$). A second proposal was that as the earth moved it dragged some of the aether with it. Near the surface of the earth, effects would be *as if* the earth were motionless. Finally, Lorentz suggested that meter sticks—that is, length measurements—were contracted in a direction parallel to the motion of the earth and that *this* made it appear that the forward-back length L was the same as the perpendicular length although it was really longer. But what would make a moving meter stick contract?

None of these explanations satisfied many scientists of the day.

Einstein's Special Theory of Relativity

In 1905, Albert Einstein, a 26 year old patent clerk in Berne, Switzerland, published an historic paper.[2] It contained the elements of his special theory of relativity. Physics has not been the same since. It's tempting to describe Einstein's paper as a response to—and inspired by—the Michelson-Morley experiment. However, there's evidence that Einstein wasn't influenced by the Michelson-Morley experiment when he proposed his theory.[3] Another fascinating question is why Lorentz or J. Henri Poincaré (1854–1912), a French mathematician, both of whom came so close, failed to complete the invention of special relativity. Lorentz seems to have been too closely involved in electromagnetic theory to see the wider application of his ideas. Poincaré had all the ingredients for a theory and even lectured on the "Principle of Relativity" in St. Louis during 1904. However, he backed off at a crucial point, saying that special relativity still had to be explained. Some experimental evidence which seemed, at the time, to contradict the predictions of his relativity principle left him doubtful of its validity to the end of his life.[4]

The key to the success of Einstein's special theory of relativity is its profound simplicity. With just two postulates, Einstein made sense out of a mass of contradictory experiments and *ad hoc* theory. The postulates require that you visualize space and time in a new way. But they give "natural" explanations for phenomena which Newtonian physics could not handle.

The two postulates basic to the special theory of relativity can be stated as follows:

> (1) A beam of light passes all observers at the same speed, whether they are moving relative to each other or not. The speed of light is invariant with respect to motion of the observer.[5]

(2) "The Electrodynamics of Moving Bodies", *Annalen der Physik 17*, pg. 891.

(3) Holton, Gerald, "Einstein and the 'Crucial' Experiment", *American Journal of Physics 37*, October, 1969, pg. 968.

(4) Bernstein, Jeremy, *Einstein*, Viking Press, N.Y., 1973, pg. 103.

(5) This is not the same as saying that the velocity of light always has the same value. We shall see in Chapter 13 that light waves travel more slowly in material media than in a vacuum.

(2) All laws of physics (not just laws of motion) are invariant with respect to transformations from one inertial frame of reference to another. All such frames are physically equivalent in every way.

A third idea, more implication than postulate, is that no signal can travel faster than the speed of light. There is no instantaneous communication over any distance.

Together, these ideas are called Einstein's "special" theory of relativity because they are limited to inertial frames—frames of reference moving relative to each other at constant velocity.

Let us concentrate on the first postulate for now. It says, for example, that even if two observers are moving relative to each other at a speed of 2.5×10^8 m/sec, a given beam of light will travel past both of them at the velocity c, equal to 3×10^8 m/sec. It claims that if the highway observer, the flatcar observer, the space ship observer, and the observer on the sun who were looking at the truck on page 189 had been measuring the speed of a beam of light instead, they would all have obtained the same result!

If this strange set of events is true, then we could say that, for a beam of light:

$$\left(\frac{\Delta x}{\Delta t}\right)^2 = c^2 = \left(\frac{\Delta x'}{\Delta t'}\right)^2$$

or

$$\left(\frac{\Delta x}{\Delta t}\right)^2 - c^2 = 0 = \left(\frac{\Delta x'}{\Delta t'}\right)^2 - c^2$$

$$(\Delta x)^2 - c^2\Delta t^2 = 0 = (\Delta x')^2 - c^2(\Delta t')^2 \qquad \text{Equation (8-5)}$$

For a beam of light, both sides of the equation are equal to zero. By inspecting the structure of the equation it can be seen that for other objects moving distances in certain elapsed times, both sides of the equation are negative. (It will be seen later that

Figure 8-5
When observers in two different frames of reference measure the speed of a given beam of light, they both get the same numerical result.

($\Delta x/\Delta t$) for material objects cannot reach or exceed the velocity of light, c.)

For future reference remember that:

(a) Δx equals the distance that an object moves relative to the S frame of reference.

(b) Δt equals the time elapsed in the S frame of reference.

(c) $\Delta x'$ equals the distance that an object moves relative to the S' frame of reference.

(d) $\Delta t'$ equals the time elapsed in the S' frame of reference.

Equation (8-5) can be interpreted as stating that even though Δx and $\Delta x'$ might differ, and Δt and $\Delta t'$ might differ, the numerical value of $(\Delta x)^2 - c^2(\Delta t)^2$, as measured in one frame of reference, will be the same in every other frame of reference even if the other frame is moving at a constant speed relative to the first. It is often called the *space-time interval*, and is invariant.[6]

Equation (8-5) also gives a means for comparing position and time intervals between two frames of reference. We are beginning to develop new rules for *transforming* observations from one frame of reference into another.

Suppose we now consider a single event which occurs at one place and at one time and which is observed by S and S'. From Equation (8-5),

$$x^2 - c^2t^2 = (x')^2 - c^2(t')^2. \qquad \text{Equation (8-5a)}$$

Clearly this equation is incompatible with the Galilean transformations, Equation (8-3), since by using them, the inequality $x^2 - c^2t^2 \neq (x - vt)^2 - c^2t^2$ is derived.

That means we have to come up with some other set of equations relating x to x', and t to t'. Appendix H shows that the new equations are none other than the Lorentz transformations of Equations (8-4). As we saw, they give you the rules for finding the position and time of an event in the S' frame, moving with velocity v, when you know the position and time of the event in the S frame. If, on the other hand, we know the position and time of an event in S' and want to determine the position and time in a frame S moving to the left at velocity $-v$, we use the transformations:

$$x = \gamma(x' + vt'); \; t = \gamma(t' + x'\frac{v}{c^2}). \qquad \text{Equations (8-6)}$$

---

(6) The idea of space and time forming a four-dimensional set comes largely from Herman Minkowsky (1864–1909), a Russian-German mathematician. It was Minkowsky who gave much of the mathematical expression to Einstein's special theory of relativity, which was more philosophical than mathematical when originally published in 1905.

To obtain position *intervals* (lengths) and time *intervals*, the symbol for "a change in", Δ, is placed in front of all the position and time symbols in this equation and Equation (8-4).

When it comes to solving problems in relativity, Equation (8-5) is useful for finding either a distance or a time interval in one frame when you know the other interval in that frame, as well as the time and distance intervals in the other frame. You needn't know the relative velocity of the two systems. Equations (8-4) or (8-6), on the other hand, can be used when you do know the relative velocity between observers and any two distances or time intervals.

New Space-Time Concepts: the Fourth Dimension

The very form of the Lorentz transformations suggests fundamental changes in the way we have to visualize space and time. The *time* of an event in one frame depends on both the time and the position of the event in the other frame of reference. Space and time are mixed together, not separate and independent as we said on page 55. The mixture of position and time in the transformations, plus the possibility that t' is not equal to t (just as x' might not be equal to x) has led to calling time "the fourth dimension." The over-all effect of special relativity was to demote time from a special separate category to the same status as the three length dimensions.

Special relativity tells us that signals can be sent no faster than the speed of light. They cannot pass from one place to another in no time at all. There is thus no longer a simultaneous "now" stretching from one end of the universe to the other because it cannot be established by instantaneous communication. When we look at anything, we are looking back in time. The twinkle from a little star might have left the star millions of years ago.

Einstein did not deny that the simultaneity of two faraway events might be established, but he laid down very strict conditions for its establishment. He said that the observer must be located at a point exactly halfway between the two places where the events are to occur. If the observer then receives a signal, such as a flash of light, from both sources simultaneously, the observer will know that the events *occurred* simultaneously. Or, if the observer is located near one event and knows the distance to another event, their simultaneity can be established by noting the time when the flash of light indicating the occurrence of the faraway event is received. The observer can deduce that the events occurred simultaneously if the time lag between signals from the near and far events is exactly L/c, where L is the distance to the far event, c is the speed of light, and L/c, of course, is the time required for a light signal to travel the distance L.

Figure 8-6
The problem of simultaneity.

By doing one of these things observers can establish the simultaneity of two events in their frames of reference. This does not mean that they will be simultaneous for an observer moving relative to the first observer. To see that this is so, consider two observers. One of them, S', is seated on a high-speed train moving toward the right with velocity, v. The other observer, S, stands on the ground at the side of the tracks.

Just as the two observers are opposite each other, lightning strikes at two places an equal distance, L, to the right, and to the left, of the observers. We'll take the simultaneity to be in the S frame of reference. S checks this out an instant later when the light from the equidistant lightning strokes reaches him at the same time. On the other hand, S' moves a distance to the right during that instant, thereby receiving the light from the right hand flash first. S', of course, concludes that the right-hand flash occurred first and the left-hand flash a little later, because the light gets to S' later. To him, the flashes are *not* simultaneous.

There are many variations of the problem that has just been sketched out, but they all point out that many of the paradoxes in special relativity can be resolved by understanding that the observers do not agree on the simultaneity of measured events.

Let's look again at Einstein's statement that nothing, and in particular no signal, can travel faster than the speed of light.

Consider a lighthouse that sends out a beam of light which rotates very rapidly about a vertical axis. Suppose the beam hits a very long vertical cliff. As the beam becomes almost parallel to the cliff, it will sweep along the cliff at a higher and higher speed, which will eventually exceed the speed of a light beam sent from a to b along the top of the cliff. If a message were to be sent from point a on the cliff to point b, couldn't the lighthouse beam be blinked on and off, so that a message would race from a to b at a speed in excess of the speed of light? The answer is no, because a message would not really be going from a to b. For

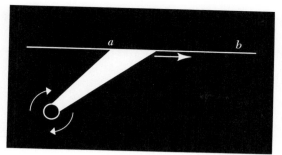

Figure 8-7
Diagram illustrating the problem of whether the spot of light made by a lighthouse beacon can move along a cliff faster than the speed of light.

a to signal *b*, word would have to be sent back to the lighthouse keeper, because the keeper is the one blinking out the message.

The fact that news can be sent no faster than 186,000 miles per second, or 3×10^8 m/sec, does not affect life on earth very much, since the 0.1 second lag in sending a signal halfway around the earth is fairly small and not too noticeable. However, in communicating with the astronauts on the moon, we became conscious of a certain fuzziness in "now." The three-second lag in their replies to questions sent out from the earth was not because of slow thinking on their part, but because it took 1.5 seconds for our questions to reach them and 1.5 seconds for their answers to travel the quarter million miles back.

Some New Rules for Adding Velocities

We have just said that nothing can travel faster than the speed of light. Let's look into this a bit further by considering another example. Observer S is on earth and S′ roars by in a rocket ship traveling at a relative velocity $v = 3/4\ c$. They both look up and see a meteor shooting to the right at an even higher speed. S′ measures its speed as being $u' = 3/4\ c$. Then, if we use Equation (8-2), the Galilean velocity addition formula, we obtain:

$$u = u' + v$$
$$= \tfrac{3}{4}c + \tfrac{3}{4}c$$
$$= \tfrac{3}{2}c$$

This answer is greater than the velocity of light! What's gone wrong?

We need to take a closer look at the relation between u and u'. Remember that $u = \Delta x/\Delta t$ and $u' = \Delta x'/\Delta t'$. Using the Lorentz

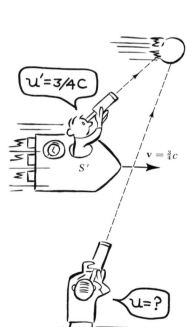

Figure 8-8
Can the meteor be traveling past observer S with a speed greater than c?

transformations [Equations (8-6), since we want to transform from S' to S],

$$x = \gamma(x' + vt'), \qquad \text{so } \Delta x = \gamma(\Delta x' + v\Delta t');$$

and

$$t = \gamma\left(t' + x'\frac{v}{c^2}\right), \qquad \text{so } \Delta t = \gamma\left(\Delta t' + \Delta x'\frac{v}{c^2}\right).$$

To get u, divide one by the other.

$$u = \frac{\Delta x}{\Delta t} = \frac{\gamma(\Delta x' + v\Delta t')}{\gamma\left(\Delta t' + \left(\frac{v}{c^2}\right)\Delta x'\right).}$$

If the numerator and denominator are next divided by Δt, we obtain:

$$u = \frac{\dfrac{\Delta x'}{\Delta t'} + v}{1 + \left(\dfrac{v}{c^2}\right)\dfrac{\Delta x'}{\Delta t'}}$$

But note that $\Delta x'/\Delta t'$ is just u', the speed of the meteor relative to S'. By substitution:

$$u = \frac{u' + v}{1 + \dfrac{u'v}{c^2}} \qquad\qquad \text{Equation (8-7)}$$

Similarly,

$$u' = \frac{u - v}{1 - \dfrac{uv}{c^2}} \qquad\qquad \text{Equation (8-8)}$$

[Compare Equation (8-8) with Equation (8-2), pg. 189.]

Now, according to the new relativistic velocity addition rules, what is the velocity of the meteor relative to the earth? From Equation (8-7)

$$u = \frac{\frac{3}{4}c + \frac{3}{4}c}{1 + \dfrac{\left(\frac{3}{4}c\right)\left(\frac{3}{4}c\right)}{c^2}}$$

$$= \frac{\frac{3}{2}c}{1 + \frac{9}{16}} = \frac{\frac{3}{2}c}{\frac{25}{16}} = \frac{(3)(16)c}{(2)(25)} = \frac{48}{50}c = .96c.$$

This velocity is less than the velocity of light! Our length and time measurements conspire to keep the meteor moving past us at a speed less than that of light. Special relativity says that NO object can move past another at a speed greater than c. In fact, ordinary massive objects cannot be accelerated up to the velocity of light, if Einstein's first postulate is true. This problem is discussed in the next chapter.

Suppose that instead of a meteor, a beam of light moved past the rocket ship so that u' is now equal to c. What is u?

$$u = \frac{c + \frac{3}{4}c}{1 + \frac{(\frac{3}{4}c)(c)}{c^2}} = \frac{1\frac{3}{4}c}{1\frac{3}{4}} = c.$$

Therefore $u = u'$ if $u' = c$! We have completed a logical circle, coming back to Einstein's first postulate: light has the same velocity in all frames of reference.

Moving Sticks Shrink

Let us look into some further deductions which can be drawn from the Lorentz transformations. Suppose that an observer S' carries a meter stick, and is moving to the right with velocity v, relative to observer S. The meter stick will be at rest relative to S', but will have a velocity v in the S frame of reference. To measure the length of the moving meter stick, S holds up a stick too, and matches up both ends of the sticks at a certain instant. Observer S has now measured the length, Δx, of the moving stick as $\Delta x = x_2 - x_1$. Since S made the measurements simultaneously, Δt equals zero.

How long does observer S' think the same meter stick is, if it is at rest relative to him?

$$\Delta x' = x_2' - x_1'$$

$$= \gamma(x_2 - vt_2) - \gamma(x_1 - vt_1)$$

$$= \gamma(x_2 - x_1 - vt_2 + vt_1).$$

Since $t_2 = t_1$, by substitution

$$\Delta x' = \gamma(x_2 - x_1).$$

Thus, $\Delta x' = \gamma \Delta x$. Rearranging the terms to solve for Δx yields:

$$\Delta x = \frac{\Delta x'}{\gamma} = \left(\sqrt{1 - \frac{v^2}{c_2}} \right) \Delta x' \qquad \text{Equation (8-9)}$$

The moving meter stick looks shorter to S than it does to S', who is at rest relative to it. If the relative velocity, v, equals $0.5c$, then, in this case, the value of $\sqrt{1 - v^2/c^2}$ equals:

$$\sqrt{1 - \frac{(0.5c)^2}{c^2}} = \sqrt{1 - \frac{0.25c^2}{c^2}} = \sqrt{.75} = .865$$

Figure 8-9
*Each observer measures the length
of the other's meter stick.*

Substituting this value back into Equation (8-9) yields:

$$\Delta x = \frac{\Delta x'}{\gamma} = (.865)\,\Delta x'$$

The stick appears shorter by almost 14 percent in the direction parallel to which it is traveling.

Conversely, it is not difficult to show that a stick at rest in the S frame of reference appears shorter to S' by the same amount. If this equivalence did not occur then the shortening of the stick might provide a way to show who is moving and who is standing still.

Does the meter stick "really" shrink? There is no way to find out, so the question has no operational answer—or meaning. There are no ways to obtain absolute information about length and time intervals. It follows from this that an object can have many "lengths", depending upon its velocity relative to us. What the theory of special relativity does is to give us a more detailed description of what the word "length" means operationally. The *rest length* of an object is the distance between its ends *when it is at rest relative to us*. In everyday experience, we can neglect the effects of motion on length measurements, but Einstein's analysis has described a realm of experience in which relative motion does alter measurement. Note that the "shrinkage" occurs only in the direction of motion.

The Slowing of Clocks

Now suppose a "clock" is set up in the moving frame of reference. This could be a light which flashes on and off with some regular frequency. S and S' both measure the time interval between flashes. In this case, $\Delta x'$ is equal to zero. (We will assume that the flashing light is at rest relative to S', so that, to this observer, successive flashes will seem to originate from the same place.)

Using Equation (8-6),

$$\Delta t = \gamma \left(\Delta t' + \frac{v \Delta x'}{c^2} \right),$$

which, since $\Delta x' = 0$, becomes:

$$\Delta t = \frac{\Delta t'}{\sqrt{1 - \dfrac{v^2}{c^2}}}.$$

From this, we can see that Δt is larger than $\Delta t'$. Therefore the interval between flashes is larger for S than it is for S'. The frequency of oscillators decreases when they move past us at high speeds. Moving clocks slow down!

This slowing down of time, known as *time dilation*, has been observed experimentally with certain unstable sub-atomic particles.[7] These particles have a specific lifetime when they are at rest or moving slowly, but they exist for longer periods of time when traveling at high speeds relative to the laboratory. The particles are apparently "carrying clocks" with them which are slowed down by the motion!

Two observers moving relative to each other will each see the other's clock as running more slowly than his or her own. Otherwise, one observer would have absolute information that he or she were standing still. Einstein's second postulate would be violated. A lot of debate[8] has centered around the possibility that if one identical twin were to take a long, high-speed round trip, not only would his or her clock tick more slowly but this twin would age more slowly than the twin who stays home. When the twin on the trip returns, both twins would agree that this twin had aged less, and relativity would be "overthrown". The paradox disappears (that is, special relativity is saved) if we look more critically at their process of getting together the second time in order to compare wrinkles. One, but not the other, of the twins actually has to *turn around* and come back. While this twin is undergoing the accelerations involved in changing directions, he or she is not in an inertial frame of reference. All bets on the relative equivalence of the frames of reference are off. If just one of the twins turns around and comes back and the other doesn't, then the genuine stay-at-home knows it. This twin has

(7) Frisch, D. H. and Smith, J. H., "Measurement of the Relativistic Time Dilation Using Mu Mesons", *American J. of Physics 31*, pg. 342, 1963. This work was later made into a film which is available from the Educational Development Center, Newton, Mass.

(8) Sachs, Mendel, "A Resolution of the Clock Paradox," *Physics Today 24*, Sept. 1971, pg. 23. Rebuttals in *Physics Today 25*, Jan. 1972, pg. 9 ff, in Greenberger, D. M., "The Reality of the Twin Paradox Effect," *American J. of Physics 40*, May 1972, pg. 750, and in Muller, R. A., "The Twin Paradox in Spacial Relativity," *American J. of Physics, 40*, July 1972, pg. 966.

felt no temporary forces throwing him or her to one or the other sides of their room, and the visible stars have maintained their steady march across the night sky. The twin who stayed home should expect to be, and will be older. Just how acceleration brings youth with it is beyond the scope of special relativity.

Up to now, no living thing has taken a round trip long or fast enough to settle the twin paradox experimentally.

A Sample Problem in Relativity

Consider two observers who are moving relative to each other at a speed equal to six-tenths the velocity of light. They observe a pair of flashes which seem, for both of them, to be simultaneous. They receive news of the flashes at time $t = t' = 0$, just as they pass each other. (They have no trouble agreeing on the simultaneity of the *arrival* of the signals because that occurs at the same place, namely $x = x' = 0$. Observers only have trouble agreeing on the simultaneity of events that take place far apart.) We will assume S has evidence that the flashes were caused by explosions occurring at the positions x_1 equal to -50×10^8 meters and x_2 equal to 50×10^8 meters. Where and when did the explosions occur in the S′ frame of reference?

Since S is midway between the two events and receives news of both at the same time, this observer reasons that the events must have occurred at the same time, relative to his or her own frame. The times of the events are

$$t_1 = \frac{x_1}{c} = -\frac{50 \times 10^8 \text{ m}}{3 \times 10^8 \text{ m/sec}} = -16.66 \text{ sec.}$$

$$t_2 = \frac{x_2}{-c} = \frac{50 \times 10^8 \text{ m}}{-3 \times 10^8 \text{ m/sec}} = -16.66 \text{ sec.}$$

(The $(-c)$ appears in the latter calculation because the signal travels toward the left.) S′ therefore says that both events oc-

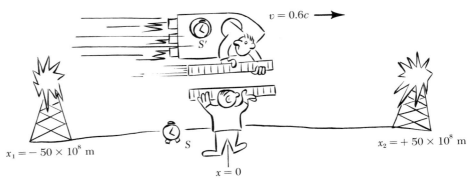

$x_1 = -50 \times 10^8$ m

$v = 0.6c \longrightarrow$

$x_2 = +50 \times 10^8$ m

$x = 0$

Figure 8-10
Observers in two different frames of reference look at two distant flashes.

curred at -16.66 sec, and that it then takes 16.66 sec for both signals to travel the 50×10^8 m distance to his or her location at time $t = 0$. To find the positions and times of the events in S', we need to use Equations (8-4) on page 193. First,

$$\gamma = \frac{1}{\sqrt{1 - \dfrac{v^2}{c^2}}}$$

$$= \frac{1}{\sqrt{1 - \dfrac{(0.6c)^2}{c^2}}} = \frac{1}{\sqrt{1 - .36}} = \frac{1}{\sqrt{.64}} = \frac{1}{.8} = 1.25.$$

Then

$$x_1' = \gamma(x_1 - vt_1)$$
$$= 1.25[-50 \times 10^8 \text{ m} - (0.6c)(-16.66 \text{ sec})]$$
$$= 1.25[-50 \times 10^8 \text{ m} + (0.6)(3 \times 10^8 \text{ m/sec})(16.7 \text{ sec})]$$
$$= 1.25(-50 \times 10^8 \text{ m} + 30 \times 10^8 \text{ m})$$
$$= 1.25(-20 \times 10^8 \text{ m})$$
$$= -25 \times 10^8 \text{ m}.$$

$$x_2' = \gamma(x_2 - vt_2)$$
$$= 1.25[+50 \times 10^8 - 0.6c(-16.66 \text{ sec})]$$
$$= 1.25(+50 \times 10^8 + 30 \times 10^8 \text{ m})$$
$$= 1.25(80 \times 10^8 \text{ m})$$
$$= +100 \times 10^8 \text{ m}.$$

$$t_1' = \gamma\left(t_1 - \frac{x_1 v}{c^2}\right)$$
$$= 1.25\left[-16.66 \text{ sec} - \left(\frac{(-50 \times 10^8)\,\text{m}(0.6 \times 3 \times 10^8)\,\text{m/sec}}{9 \times 10^{16}\,\text{m}^2/\text{sec}^2}\right)\right]$$
$$= 1.25\left[-16.66 \text{ sec} + \left(\frac{(50)(0.6)(3)}{9}\right)\text{sec}\right]$$
$$= 1.25(-16.66 \text{ sec} + 10 \text{ sec})$$
$$= 1.25(-6.66)$$
$$= -8.33 \text{ sec}.$$

$$t_2' = \gamma\left(t_2 - \frac{x_2 v}{c_2}\right)$$
$$= 1.25\left[-16.66 \text{ sec} - \left(\frac{(50 \times 10^8 \text{ m})(0.6 \times 3 \times 10^8 \text{ m/sec})}{9 \times 10^{16}\,\text{m}^2/\text{sec}^2}\right)\right]$$
$$= 1.25(-16.66 \text{ sec} - 10 \text{ sec})$$
$$= 1.25(-26.66 \text{ sec})$$
$$= -33.33 \text{ sec}.$$

S' sees the signals *arrive* simultaneously because one event occurs 25×10^8 m to the left at t_1' equal to -8.33 seconds whereas the other event occurs 100×10^8 m to the right at t_2' equal to -33.33 seconds. Each signal reaches S' at $t' = 0$ because it takes light 8.33 seconds to travel 25×10^8 m and 33.33 seconds to travel 11×10^8 m.

Note that in S,

$$\Delta x = x_2 - x_1$$

$$= 50 \times 10^8 \text{ m} - (-50 \times 10^8 \text{ m})$$

$$= 100 \times 10^8 \text{ m},$$

and

$$\Delta t = t_2 - t_1$$

$$= -16.66 \text{ sec} - (-16.66 \text{ sec})$$

$$= 0.$$

In S'

$$\Delta x' = x_2' - x_1'$$

$$= 100 \times 10^8 \text{ m} - (-25 \times 10^8 \text{ m})$$

$$= 125 \times 10^8 \text{ m},$$

$$\Delta t' = t_2' - t_1'$$

$$= -33.33 \text{ sec} - (-8.33 \text{ sec})$$

$$= -25 \text{ sec}.$$

Note that

$$\frac{\Delta x'}{\Delta x} = \frac{125 \times 10^8 \text{ m}}{100 \times 10^8 \text{ m}}$$

$$= 1.25 = \gamma.$$

If we multiply by Δx on both sides, we obtain $\Delta x' = \gamma \Delta x$, the same sort of contraction as was calculated in the example on page 203. In both cases, the length is seen as shorter by the observer making the "pure" length measurement: i.e.; the observer for whom $\Delta t = 0$.

When Is Relativity Necessary?

The ubiquitous correction factor $\gamma = 1/\sqrt{1 - v^2/c^2}$ is a good measure of the mismatch between classical Newtonian and relativistic descriptions of nature. If you wish, it tells how far wrong you go by using the simpler Galilean transformations. The Lorentz transformations *include* the Galilean transformations, in the limiting case where v is much smaller than c. (Note that when v is small Equations (8-4) and (8-6) reduce to Equation (8-3), and that Equations (8-7) and (8-8) reduce to Equation (8-2).) Suppose you can tolerate or ignore 1 percent relativistic effects. This would mean that $\gamma = 1.01$, or

$$\gamma^2 = \frac{1}{1 - \dfrac{v^2}{c^2}} = 1.02.$$

Inverting, we get

$$1 - \frac{v^2}{c^2} = \frac{1}{1.02.}$$

Multiplying through by (-1) and then subtracting $(+1)$ from both sides, we obtain:

$$\frac{v^2}{c^2} = 1 - \frac{1}{1.02}$$

$$= 1 - .981 = .019.$$

Then, after multiplying through by c^2, we get $v^2 = (.019)c^2$. Taking the square root,

$$v = \sqrt{.019}c = .138c.$$

$$= (.138)(3 \times 10^8 \text{ m/sec})$$

$$= 4.14 \times 10^7 \text{ m/sec}$$

$$= 4.14 \times 10^4 \text{ km/sec}$$

A velocity of 4.14×10^4 kilometers per second equals 25,600 miles per second. This means that even if you were able to accelerate an object up to this velocity, the relativistic corrections in position and time measurements required would still amount to just 1 percent. The highest speeds (relative to the earth, of course) of man-sized objects thus far, those obtained in space travel, are less than 10^{-3} of 4.14×10^6 km/sec. However, the study of atomic-sized particles quite often involves objects which are traveling at speeds close to that of light. Relativistic physics is a necessity for understanding and correctly predicting events in this realm.

Einstein's General Theory of Relativity

The special theory of relativity is concerned with the physical equivalence of inertial frames of reference, all of which are moving at constant velocity relative to each other. Implicit in the arguments has been the idea that we would never have difficulty

"g" = 9.8 m/sec²

(a) Gravity acting down, elevator at rest (or moving at constant velocity).

a = 9.8 m/sec²

(b) Zero gravity, elevator accelerating upwards.

Figure 8-11

separating "real" forces from "fictitious" forces arising because of lurches or turnings of our reference frames. But what of the force of gravity? Can we be sure that it is real?

In 1915, Einstein proposed his *general* theory of relativity.[9] It was concerned with space, time, gravity, and non-inertial, accelerated frames of reference. One consequence of the theory is that fictitious forces arising from the acceleration of one's frame of reference cannot be distinguished from a gravitational force acting in the same direction as the fictitious force. This new *principle of equivalence* states, for example, that a person in an elevator at rest (or moving at a constant velocity) in a gravitational field acting downward experiences precisely the same things as a person in an elevator accelerating upward in zero gravity.

Suppose a physicist sets up a laboratory in a room which sits on a merry-go-round. If a ball is put on a table in such a room, it will undergo funny accelerations without any apparent force being exerted on it. The physicist sitting in the room could claim, according to the general theory of relativity, that he or she is in a perfectly good inertial frame, but that a strange distribution of mass outside the room pulls objects out from a line running up through the middle of the room! This is perhaps one instance where the simpler alternative might be just to admit that the room constitutes a non-inertial frame of reference.

Einstein's theory is a geometrical one, claiming that the presence of mass makes a frame of reference non-inertial (causes fictitious accelerations) by "warping" the space around it. The general theory of relativity will be discussed further in Chapter 12.

Relativity and Human Thought

We shall see in the next chapter the vast implications that special relativity holds for technology and politics. However, the consequences of special relativity which require the greatest concentration to understand are the ones that have just been discussed: the mixing of space and time, the slowing down of the time indicated by moving clocks, and the shrinkage of moving meter sticks. Einstein asked us to give up deeply rooted prejudices about our surroundings and to re-interpret measurements which seem obviously correct. This goes against the grain. Whereas Alexander Pope was able to write, in his "Epitaph Intended for Isaac Newton in Westminster Abbey"

"Nature and Nature's Laws lay hid in night,
God said, 'Let Newton be'! and all was light",

(9) *Annalen der Physik* 49, pg. 769, 1916.

Figure 8-12
M. C. Escher, Relativity. *(Courtesy of Escher Foundation, Haags
Gemeentemuseum, the Hague, and Vorpal Gallery, San Francisco.)*

a latter-day parody reads,

> "Nature and Nature's Laws lay hid in night,
> When God created Newton and there was light.
> But came the Devil and He, with a mighty Ho
> Created Einstein, who restored the status quo."

Are special and general relativity an improvement or a step
toward obfuscation? They constitute a way of describing the
physical world from a broader viewpoint than was possible with
Newtonian mechanics. Einstein broke through the boundaries
of Newton's mechanical understanding of the world to get a more
general physics that gave correct answers whether objects were
moving fast or slow and whether the phenomenon involved just
motion or electricity and magnetism as well.

When you think of it, Einstein's work falls in a natural sequence with the work of Aristotle, Copernicus, and Newton:
with each transition people were asked to shake off more of their
sense impressions and to see a truth which was more basic.

There are definite conflicts between relativity and "common sense." This leads us to take a critical look at common sense—and it is surprising what differences exist between what we really perceive and the "common sense" way we are taught to interpret and understand those perceptions. For example, A. A. Blank[10] has shown that we really see space in a curved fashion, not in the cartesian coordinate-Euclidian manner we are taught. This is not to say that the predictions of special relativity will very soon seem "natural" to us. Its predictions will not be checked out by everyday experience until we start experiencing relative velocities approaching that of light. However, it is healthful for us to realize that some of the "paradoxes" of special relativity are a result of conflicts with a world view *which we have been taught.* (See page 55).

At the start of Chapter 3, the concept of "object" was discussed. With some support from Piaget, it was indicated that an "object" develops in our mind when we receive repeated images from one spot in space. In physical parlance, the perception of invariance is followed by the development of a corresponding mental image.[11] Intelligence itself seems to be based on the ability to find what is invariant in the midst of changes. That which is invariant tends to be seen as real. What Einstein did was to suggest that physical law, rather than perceived objects, or events, should be the most basic invariant. This is probably the most profound idea in this book.

Relativism is much older than Einstein's special theory of relativity. His theory does not just say, "What you see depends on your point of view." In fact, it is possible to think of the theory as the result of a search for an *absolute,* and for an orderly means to make allowance for the differences in the observations of two observers. The absolute he was looking for was physical law which did not change form from one frame of reference to another. It's interesting to realize that Einstein started out calling his theory an "invariance" principle and did not use the word "relativity" in connection with it until 1911, after other people had referred to it in that way.[12]

A number of things have been said about light in this chapter, even though it will not be examined in detail until Chapters 13 and 14. The *finiteness* of the velocity of light was realized as early as the seventeenth century. By then, astronomical observations had made it appear that while light traveled amazingly fast, it did take some time to cross a large distance. What Einstein's special theory of relativity did was to emphasize the *epistemo-*

(10) *Journal of the Optical Society of America* 43, #9, September, 1953, pg. 717.
(11) Bohm, David, "Physics and Perception," pg. 185–230 in *The Special Theory of Relativity,* W. A. Benjamin, N.Y., 1965.
(12) Holton, Gerald, "On Trying to Understand Scientific Genius," *American Scholar* 41, #1, Winter, 1971–72, pg. 95.

Figure 8-13
Pablo Picasso, Les Demoiselles d'Avignon. (Museum of Modern Art, New York. Acquired through the Lillie P. Bliss Bequest.)

logical importance of the limited speed of light. The theory showed that since information could not be sent instantaneously, people had no business conceptualizing in the old Newtonian terms about things happening here and now "at the same time" as they did at some other place. Space and time are mixed together.

An interesting connection between special relativity and the rest of human culture can be postulated if we look at some events that took place at the beginning of this century. At about the same time that Einstein gave us the rules for comparing measurements made in various frames of reference, the Cubist painters were taking apart our traditional absolute, static way of picturing things. For example, Picasso's *Les Demoiselles D'Avignon*, painted in 1907, shows faces as viewed from several frames of reference at once. One of the effects which the painting can have on viewers is to make them aware that they do see objects from a particular frame of reference. Seeing a front view and side view of a face simultaneously can be a jarring experience.

Is it just coincidence that these changes in physics and in painting occurred during the same decade or was the same influence showing up in both of them? This would be a hard question to answer conclusively. Certainly, the case for an over-all zeitgeist (spirit of the times) would be less strong here than for Newton's times (page 89). However, Conrad Waddington[13] has suggested that the cafes of Paris provided at least one place where physicists who knew of Einstein's theory came in direct contact with painters, and the new views of space and time were probably exchanged.

Philipp Frank[14] suggests that one of the main contributions of special relativity to human thought is not that there are no absolutes; but that many things which are considered to be absolutes, ought not to be. It asks for richer, more complete operational definitions of the words we use. It reminds us that unambiguous communication is not easy. The realm of objective, communicable knowledge is narrow. We have seen earlier (pages 17 and 204) the ambiguities possible in the word "length". But consider words such as "free will", "political power", "awareness", "sanity", and "relevance". How many of them mean everything to everybody—or perhaps nothing to anyone? A statement has meaning only if the frame of reference from which it is made is understood. Unfortunately, in our age of action, we usually lack the patience to hear the whole statement.

Summary

The descriptions of position and time as outlined in Chapter 3 lead to "Galilean relativity"; that is, to the rules for the transformation of measurements made in one frame of reference, S, into another frame of reference, S'. The Galilean transformations are:

for position; $x' = x - vt$,   Equations (8-1), (8-3)

for time; $t' = t$,   Equation (8-3)

and for velocity; $u' = u - v$.   Equation (8-2)

In these formulas x', t' and u' are the position, time, and velocity respectively, as measured in a frame S' which is moving to the right with velocity, v, relative to the S frame. These same quantities, as measured in the S frame of reference are x, t, and u, respectively.

An inertial frame of reference is one in which an object acted upon by no forces undergoes no accelerations. According to

(13) Waddington, C., *Behind Appearances: A Study of the Relations Between Painting and the Natural Science in This Century.*, M.I.T. Press, Cambridge, Mass., 1970.

(14) *Relativity: A Richer Truth*, Beacon Press, Boston, 1950.

Newton, all inertial frames are mechanically equivalent. The laws of motion are not affected by the velocity of the frame of reference. Acceleration is invariant between inertial frames.

The Lorentz transformations are a new set of rules for finding x', t' in terms of x and t—or vice versa. They were devised to make the laws of electricity and magnetism invariant between inertial frames. The Lorentz transformations are

$$x' = \gamma(x - vt),$$

$$t' = \gamma\left(t - \frac{xv}{c^2}\right) \qquad \text{Equation (8-4)}$$

$$x = \gamma(x' + vt')$$

$$t = \gamma\left(t' + \frac{x'v}{c^2}\right), \text{ where } \gamma = \frac{1}{\sqrt{1 - \dfrac{v^2}{c^2}}} \qquad \text{Equation (8-6)}$$

x, t, x', t', and v have the same meaning as in Equations (8-1) and (8-3), and c is the speed of light.

The Michelson-Morley experiment failed to detect motion of the earth through an absolute space. With this, the idea of absolute motion lost its main support.

In 1905 Einstein proposed his special theory of relativity, based on two postulates:

(1) The speed of light, $c = 3 \times 10^8$ m/sec is invariant. It has the same value no matter what the velocity of the frame of reference from which it is measured might be.

(2) All the laws of physics are invariant between inertial frames of reference.

He also adds that no signal can travel faster than the speed of light.

From the invariance of the speed of light comes another quantity, the space-time interval, which is also invariant:

$$(\Delta x)^2 - c^2(\Delta t)^2 = (\Delta x')^2 - c^2(\Delta t')^2 \qquad \text{Equation (8-5)}$$

The Lorentz transformations follow as algebraic deductions from Einstein's first postulate, as do new rules for transforming velocities between two inertial frames:

$$u' = \frac{u - v}{1 - u\dfrac{v}{c^2}} \qquad \text{Equation (8-8)}$$

and

$$u = \frac{u' + v}{1 + u'\dfrac{v}{c^2}} \qquad \text{Equation (8-7)}$$

As a result of Einstein's first postulate and the Lorentz transformations, space and time become mixed. Time is a fourth dimension. Observers moving relative to each other cannot agree on the simultaneity of events that occur far from each other.

Moving meter sticks shrink and moving clocks run more slowly.

Relativistic effects are negligible for slow-moving objects but build up to one percent when the speed of the object relative to us is 4.14×10^4 km/sec.

Einstein's general theory of relativity is a geometrical theory connecting space, time, and gravity. It explains phenomena occurring in non-inertial frames of reference.

The most valid basis for understanding the physical world seems to reside at the level of physical law rather than directly in data, or perceptions, which change from one frame of reference to another. Special relativity is the result of searching for ways of expressing the laws of physics so that they are invariant between inertial frames. The one physical invariant—the closest physics comes to an absolute—is the speed of light.

Special relativity underlines the importance of also giving one's background—one's frame of reference—when statements are made.

Further Reading

Baker, Adolph, pp. 42–92 in *Modern Physics and Anti-Physics*, Addison-Wesley, Reading, Mass., 1970. (Contains hip dialogues on relativity and physical reality.)

Barnett, Lincoln, *The Universe and Dr. Einstein*, Harper and Bros., N.Y., 1948. Republished as Signet Science Library Book, William Sloane Assoc., Inc., 1952. (Still one of the best popular discussions of Einstein's theories.)

Bork, Alfred M., "Durrell and Relativity" *Centennial Review* 7, Spring, 1963, pg. 196. (Tells how Lawrence Durrell tried to encompass the four dimensions in his *Alexandrian Quartet*.)

Cassirer, Ernst, *Einstein's Theory of Relativity Considered From the Epistemological Standpoint*, originally published in 1923, republished along with *Substance and Function* by Dover Publications, New York, 1953. (A probing discussion of the impact of special relativity on philosophical concepts of truth and reality.)

Clark, Ronald W., *Einstein: The Life and Times*, World Publishing Co., N.Y. and Cleveland, 1971. (A biography.)

Gamow, George, *Mr. Tompkins in Wonderland*, Cambridge University Press, 1939. (A witty account of some problems in relativity.)

Gardner, Martin, *Relativity For the Million*, Pocket Books, N.Y., 1965. (A clearly written, well-illustrated discussion of both special and general theories of relativity.)

Gerholm, Tor Ragnar, pp. 47–72, 81–111, in *Physics and Man*. The Bedminster Press, 1967. (A different approach to physics. Emphasizes cultural and philosophical background.)

Giedeon, Sigfried, *Space, Time, and Architecture*, Harvard U. Press, Cambridge, Mass., 4th Edition, 1962, pg. 426. (A history of architecture, but it discusses new views of space and time which emerged in physics, architecture, and painting at the beginning of the 20th century.)

Leiber, Lillian R., and Leiber, Hugh G., *The Einstein Theory of Relativity*, Holt, Rinehart and Winston, 1945. (A highly original treatment, starting with the easiest concepts and building quite a comprehensive description of the theory. Written in near-poetic style, with many amusing drawings.)

Russell, Bertrand, "The World of Physics and the World of Sense," Chapter IV in *Our Knowledge of the External World*, W. W. Norton, 1926. Reprinted as a Mentor Book, The New American Library of World Literature, N.Y., 1960. (Says the discrepancy between the world of physics and the world of sense is more apparent than real.)

Schlipp, Paul A., (Ed.), *Albert Einstein: Philosopher-Scientist*, Open Court Press, Lasalle, Ill., 1949, pp. 51, 70. (A series of essays covering the impact of Einstein's work on various fields of thought.)

Problems

8-1   A jet airliner flies from Los Angeles to New York, an air distance of about 2600 miles, in four hours. The plane is aided by a tail wind (a wind blowing from West to East) of 100 miles per hour. (a) What is the average ground speed (speed relative to the earth) of the plane? (b) What is the average speed of the plane relative to the air? (c) How long will the plane take to fly back to Los Angeles if the same wind is blowing from West to East?

8-2   A swimmer jumps into a rapidly moving river at a point 100 meters above a waterfall. If this person swims upstream at 0.8 m/sec and the speed of the water is 2.0 m/sec downstream, when will the swimmer be swept over the waterfall?

8-3   A tortoise and a hare engage in another classic footrace. But this time the rabbit is so confident of winning that he carries with him the end of a very elastic belt. The other end is tied to a tree near the starting line. The tortoise, smart as ever, decides to run his race along the elastic belt instead of on the ground. Will he ever catch up with the rabbit? For a given speed of the rabbit, W, is there a speed above which the tortoise must trot in order to catch the rabbit?

8-4   Two trains of length 75 meters and 100 meters are running past each other on parallel tracks. The short train has twice the speed of the other. A man sitting in the shorter train observes that it takes the longer train 3 seconds to pass his window. Find the velocities of the two trains if the trains are going in opposite directions.

8-5 People get up to the second floor of a building faster and with more energy if they walk up an escalator than if they walk up a stationary flight of stairs. Where and by whom or what is the extra work done? How does the invariance of physical laws between inertial frames of reference enter in?

8-6 How would you answer the following statement? Relativity is concerned only with how we see things, not with what is really happening. Since science deals with reality, relativity must not be science.

8-7 What does it mean when it is said that the speed of light is finite? Invariant? Does this mean that the speed of light is constant?

8-8 Using the Lorentz transformations, show that the observer on the train on page 200 does conclude that the lightning stroke on the right occurred before the one on the left.

8-9 Suppose that you are located near a clock which is at the position $x = 8 \times 10^8$ m. Describe two methods that you could use to synchronize the clock with one located at the origin.

8-10 Although the speed of light is invariant, its vibrational frequency is subject to relativistic dilation. If the vibrational frequency of green light is 5×10^{14} hertz (cycles per second) and the vibrational frequency of red light is 6×10^{14} hertz, how fast would a motorist have to travel past a red light in order to turn it into a green light?

8-11 Relative to some parts of our galaxy, the earth is moving at a velocity which is a significant fraction of the speed of light. Does this mean we should lengthen our international standard meter?

8-12 A space capsule traveling with a speed of $(3/4)c$, relative to the earth, shoots out a projectile in a forward direction at a speed equal to $(1/4)c$. What is the speed of the projectile relative to the earth? What is the speed of the projectile, relative to the earth, if the capsule shoots it backwards?

8-13 A weight on a spring bounces up and down two times per second. If the same weight and spring are then moved past you at half the speed of light, how many times per second does the weight appear to bounce up and down?

8-14 The positive K meson, one of the so-called "fundamental particles" of matter, has a life expectancy of about 10^{-8} seconds. How fast would it have to move in order to exist long enough to travel 15 meters?

8-15 Describe a set of events whose sequence is reversed in two different frames of reference.

8-16 Two events occur at the same place in my frame of reference and are separated by a time interval of three seconds. Find the distance between the events in a frame of reference for which the time interval between the two events is five seconds.

8-17 If a rocket ship were to travel at a speed of a million miles per hour, it would take it almost 3000 years to reach Alpha Centauri, the nearest star to us (other than the sun). Now suppose a space ship were able to accelerate until its speed relative to the earth was 0.9c. How long would it take a ship traveling at this speed to reach Alpha Centauri, 4.3 light years away? (A light year is the distance traveled by light in one year.) Earth observers would say it took the ship 4.78 years, but how long would the crew of the space ship say that it took?

9

AN EXPANSION OF
ENERGY CONSERVATION:
NUCLEAR ENERGY

The evidence thus far indicates that if invariance in physical laws is desired, we have to be ready to give up something in exchange. For example, if it is demanded that all the laws of physics be invariant with respect to velocity of the observer, the Lorentz transformations must be used to compare distance and time intervals, as measured in the observer's frame of reference, with those measured in other frames. As a consequence, moving meter sticks shrink and moving clocks slow down. In the next few pages it is shown that these unfamiliar measurement effects also require that the concept of *mass* be viewed in a new way. The end result is an enlarged definition of the energy possessed by an object, and along with it, a more general statement of the principle of energy conservation. We will find that mass and energy are equivalent and that nuclear reactions which "destroy" mass can release tremendous amounts of energy.

The Variation of Mass with Velocity

We've been saying that lengths, time intervals, and velocities — things at the level of data — are not invariant, but change, or transform, from one frame of reference to another. However, the physicist expects concepts at the level of laws or principles to remain invariant. For example, the numerical value of the total momentum of a system will depend on the frame of reference from which it is viewed, but the momentum as observed from a given frame will be conserved during a collision no matter what the frame.

Consider a very simple and symmetrical collision between two similar balls a and b, as observed by the usual stalwarts S and S'. Observer S holds ball a. S', carrying ball b, roars by to the right at a high speed, v, relative to S. S' travels along a line which is at a distance, y, from the point where S is resting. As the two observers approach each other, they throw balls a and b toward each other with velocities that are the same in their respective frames. We'll assume that these velocities are small in comparison to the relative velocity of the two observers. The symmetry of the situation requires that the two balls meet at a point halfway between the observers. Each observer thinks his or her own ball goes directly out and back.

Observer S now decides to apply the principle of momentum conservation to the collision, concentrating on the back and forth y components of momentum. S calls the mass of his own ball, a, (as measured by himself), M_{aS}, and calls the y component of its velocity (as measured by himself), v_{yaS}. The mass and velocity of ball b as measured by S will be M_{bS} and v_{ybS}. Therefore, conservation of the y momentum of the system as observed in the S frame of reference can be written as

$$M_{aS}v_{yaS} - M_{bS}v_{ybS} = -M_{aS}v_{yaS} + M_{bS}v_{ybS}$$
$$\text{Before} \qquad\qquad \text{After} \qquad\qquad \text{Equation (9-1)}$$

The collision is elastic, so each ball rebounds, its speed staying the same but with its velocity changing sign. Equation (9-1) can be rewritten as;

$$2M_{aS}v_{yaS} = 2M_{bS}v_{ybS} \qquad\qquad \text{Equation (9-2)}$$

If v_y for the two balls were the same, as observed by S, we would find that the observed mass of both balls was the same, and there would be no point in going through all of this. However, v_{yaS} is *not* equal to v_{ybS}. Let's see why.

According to special relativity, distances measured perpendicular to the relative velocity of two observers (in this case, y distances) are unaffected by their relative motion. If, by symmetry, the collision occurs at the midway point, and $(y/2)_{aS}$ is the transverse distance traveled by ball a, as measured by observer S; $(y/2)_{aS'}$ is the transverse distance traveled by ball a as measured by observer S', etc., then:

$$(y/2)_{aS} = (y/2)_{bS'} = (y/2)_{aS'}$$
$$= (y/2)_{bS} = y/2. \qquad\qquad \text{Equation (9-3)}$$

They're all the same.

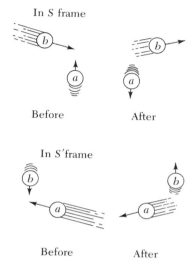

In S frame

Before · · · · · · · · · · · · · · · · After

In S' frame

Before · · · · · · · · · · · · · · · · After

Figure 9-1
Collision as observed from two different frames of reference.

The transverse *velocities* of the two balls, as measured by S, can be obtained by dividing the distance traveled by the times elapsed. Let's call the time intervals $(\Delta t)_{aS}$ and $(\Delta t)_{bS}$. Then

$$v_{yaS} = \frac{y/2}{(\Delta t)_{aS}} \quad \text{and,} \quad v_{ybS} = \frac{y/2}{(\Delta t)_{bS}} \qquad \text{Equation (9-4)}$$

At this point we need to think carefully about the various measured time intervals:

(1) From the symmetry of the interaction, both observers agree that their balls take equal times to cross the distance, $y/2$, as measured in their own frames of reference. Therefore:

$$(\Delta t)_{aS} = (\Delta t)_{bS'} \qquad \text{Equation (9-5)}$$

(2) When the two observers make measurements on the same ball, they disagree. Observer S sees time as dilated because the clock "held by" S' is in motion. Observer S says it takes ball b longer to cover the distance $y/2$ than observer S' thinks it does.

$$(\Delta t)_{bS} = \frac{(\Delta t)_{bS'}}{\sqrt{1 - \dfrac{v^2}{c^2}}} = \gamma(\Delta t)_{bS'} \qquad \text{Equation (9-6)}$$

Now let's put all this together to obtain a relation between v_{yaS} and v_{ybS} which we can put back into Equation (9-2). From the second Equation (9-4),

$$v_{ybS} = \frac{y/2}{(\Delta t)_{bS}}.$$

But, from Equation (9-6), $(\Delta t)_{bS} = \gamma(\Delta t)_{bS'}$, so

$$v_{ybS} = \frac{y/2}{\gamma(\Delta t)_{bS'}}.$$

From Equation (9-5), $(\Delta t)_{aS} = (\Delta t)_{bS'}$, so

$$v_{ybS} = \frac{y/2}{\gamma(\Delta t)_{aS}}.$$

But $y/2/(\Delta t_{aS})$ equals v_{yaS}, the transverse velocity of ball a as measured by observer S. Therefore,

$$v_{ybS} = \frac{v_{yaS}}{\gamma}.$$

Equation (9-2) can then be re-written as

$$m_{aS}v_{yaS} = \frac{m_{bS}v_{yaS}}{\gamma} \qquad \text{Equation (9-7)}$$

or, as

$$m_{bS} = \gamma m_{aS}$$

The ball *b* passing observer S at high speed, appears to have a larger mass than a similar ball *a* which is traveling slowly relative to S.

In this example the masses and velocities were chosen in order to make the symmetries in the collision as obvious as possible. A more involved argument will show that what has been found here holds true in other collisions, inelastic as well as elastic. Equation (9-7) can be rewritten more generally as:

$$m = \frac{m_o}{\sqrt{1 - \dfrac{v^2}{c^2}}}$$

$$= \left(\frac{1}{\sqrt{1 - \dfrac{v^2}{c^2}}} \right) m_o \qquad \text{Equation (9-8)}$$

$$= \gamma \, m_o$$

where *m* is called the *relativistic mass* of an object which has a *rest mass* m_o when it is at rest relative to the observer. An object which moves past you with velocity v appears more massive by the factor γ which is equal to $(1/\sqrt{1 - v^2/c^2})$. Remember that this result was attained simply by allowing for the relativistic slowing down of fast-moving clocks and requiring that momentum be conserved during a collision.

The argument that has just been presented is similar to one used by Einstein in a classic three-page paper,[1] the second one on special relativity that he published in 1905. In it he went on to establish the famous mass-energy connection which will be presented in the next section, and Appendix I. It was one of four epochal papers that Einstein published in the space of one year. Several months before Einstein's papers, Lorentz published a mathematical treatment showing that the mass of an electron increases as $1/\sqrt{1 - v^2/c^2}$. But, as was the case for his space-time transformations, the ideas seemed to apply only to electromagnetic phenomena. Poincaré, in the same 1904 talk which was mentioned in the last chapter, also saw that the inertia of an object should increase with speed, reaching an infinite value at the speed of light. However, as we saw, he was not totally convinced of the validity of his theoretical results.

Approaching the Speed of Light

Does the observed mass of objects really increase with velocity, or is special relativity just a difficult exercise in seeing physical events more elegantly? The experimental evidence, some of

(1) "Does the Inertia of a Body Depend Upon its Energy Content?", *Annalen der Physik 18*, pp. 639–641, 1905.

which actually pre-dates the theoretical work up to 1905, indicated that observed mass does increase with velocity. In 1901, Walter Kaufman and, later, Alfred H. Bucherer found that electrons which had been accelerated to high speeds behaved as if their mass were greater. Equation (9-8) has been confirmed repeatedly in thousands of experiments involving electrons, as well as with other particles. In fact, modern particle accelerators are designed to make allowance for effects due to the increase in mass of particles with very high velocities.

Does this strange increase of mass mean that there is more material in an object when it's moving past us at high speeds? No! It simply means that the object is more difficult to accelerate. The old definition of mass as being "the amount of material in an object" is now outdated. Two bodies moving at different speeds can have the same inertia, or mass, even though they contain different amounts of "material". It can be shown, in a manner similar to the discussion on page 209, that a body must move past you at 4.14×10^4 km/sec for its relativistic mass to be one percent larger than its rest mass. Taking this discussion a step further, reference to Equation (9-8) will show that the observed mass of an object approaches infinity as the velocity of the object relative to the observer, approaches c. It is impossible, according to special relativity, to accelerate an object which has any rest mass up to the speed of light.

Light itself can display particle-like characteristics. It has momentum and can exert pressures, so it must have something equivalent to mass. However, since a particle of light cannot be made to stand still, it has zero rest mass, and Equation (9-8) for the relativistic mass increase does not apply.

There seem to be two kinds of particles: those which have rest mass and cannot be accelerated to the speed of light; and those, such as "particles" of light, which in vacuum, can travel *only* at 3×10^8 m/sec and which have no rest mass. A third class of particles, called "tachyons"[2,3] seems possible theoretically, but they have not yet been detected. In theory these objects can travel *only* at speeds *greater* than 3×10^8 m/sec and can not be decelerated to the speed of light.

Relativistic Expression of Kinetic Energy

On page 142 it was noted that in cases where the mass of a system is subject to change, the response of an object to a force is more complex than $F_{net} = ma$. This also is true with regard to

(2) Bilanuik, Olexa-Myron et al, "More about Tachyons," *Physics Today 22*, Dec., 1969, pg. 47.
(3) Kreisler, Michael N., "Are There Faster-Than-Light Particles?", *American Scientist 61*, March-April, 1973, pg. 201.

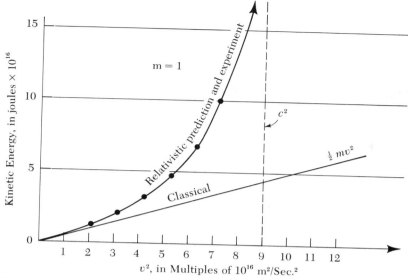

Figure 9-2
Change of K.E. *at very high speed.*

the relation between the work done on a particle by a net force, and the proper expression for kinetic energy. According to the classical expression, a plot of kinetic energy, $(\frac{1}{2})mv^2$, against v^2 should produce a straight line with slope $(\frac{1}{2})m$ as is shown in Figure 9-2. The actual experimental result is shown above this graph, for comparison. As you can see, as v^2 increases, the measured kinetic energy increases at a rate *faster than* a constant times v^2. The best way to interpret this is to say that, sure enough, m increases with velocity. The inertia is not constant. The rate of increase agrees with Equation (9-8).

Since doing work on a particle increases its mass as well as its velocity the possibility arises that mass and energy are connected in some way. Energy, you will recall, has the dimensions of force times distance, or more basically, (mass)(length)^2/(time)^2. If you were to multiply mass by a (velocity)^2, the result would be something that at least has the dimensions of energy. In Appendix I it is shown that if you multiply relativistic mass by the square of a particular velocity, c^2, then subtract the rest mass times c^2, you get an expression which is equal to $\frac{1}{2}mv^2$ when v is small in comparison with c:

$$mc^2 - m_o c^2 \cong \tfrac{1}{2}mv^2, \qquad \text{Equation (9-8)}$$

or by substituting $m = \gamma \, m_o$ from Equation (9-8),

$$(\gamma \, m_o)c^2 - m_o c^2 \cong \tfrac{1}{2}mv^2 \qquad \text{Equation (9-9a)}$$

$$(\gamma - 1) \, m_o c^2 \cong \tfrac{1}{2}mv^2. \qquad \text{Equation (9-9b)}$$

The expression on the right is, of course just the classical expression for kinetic energy. This strongly suggests calling the quantity $mc^2 - m_oc^2$ or $(\gamma - 1)m_oc^2$ the *relativistic kinetic energy* of the object whose rest mass is m_o. For speeds small in comparison with c, it reduces to the classical value, $\frac{1}{2}mv^2$.

This, in turn, suggests calling mc^2 the *total* relativistic energy of the object and m_oc^2 the *rest* energy. If we exclude potential energy and heat;

$$mc^2 = m_oc^2 + mc^2 - m_oc^2$$

$$\frac{\text{total}}{\text{energy}} = \frac{\text{rest}}{\text{energy}} + \frac{\text{kinetic}}{\text{energy}}$$
Equation (9-10)

The Equivalence of Mass and Energy

According to the special theory of relativity a body at rest has energy just because it has mass. If work is done on the object and it is accelerated, kinetic energy is added to its rest energy and its mass is increased from its original value m_o to some larger value, m. This can be shown in the following way:

$$\text{Work done} = \Delta(K.E.) = mc^2 - m_oc^2$$
$$= (m - m_o)c^2$$
$$= \Delta mc^2$$

If an increase in kinetic energy is accompanied by an increase in mass, then perhaps we can say that when the energy, E, of a system is increased, its mass, m, is always increased, as follows:

$$\Delta E = \Delta mc^2.$$

Mass can be thought of as "congealed" energy. Furthermore, we can say that any change which somehow destroys part of the mass of a closed system should release energy. That is, ΔE will be positive if m is negative:

$$\Delta E = -\Delta mc^2$$
Equation (9-11)

How much energy will be released? In other words, how large will ΔE be?

If we could make one gram of mass disappear in a reaction $(\Delta m = -10^{-3}$ kg$)$, the equivalent energy released would be:

$$\Delta E = (-\Delta m)c^2$$
$$= [-(-10^{-3} \text{ kg})] (3 \times 10^8 \text{ m/sec}^2)^2$$
$$= +9 \times 10^{13} \text{ joules.}$$

(Courtesy of Sydney Harris.)

This would run the entire United States for one minute!

We can now re-write the principle of energy conservation for a closed system in its most complete form:

$$E = K.E. + P.E. + \text{Rest energy} + U = \text{Constant}$$

or

$$\Delta(K.E.) + \Delta(P.E.) + \Delta mc^2 + \Delta U = 0 \qquad \text{Equation (9-12)}$$

The Nuclear Atom

The main purpose of this book so far has been to trace some of the history and philosophy of ideas concerning motion, force, and energy. The structure of matter has not been a major concern. However, to really appreciate the practical consequences of special relativity theory—and, in particular, of the equivalence of mass and energy which was just discussed—we have to understand a little more about the way in which material objects seem to be put together.

A lot of physical phenomena can be explained by theorizing that matter is made up of almost countless similar, separate, and invisible units called *atoms*. This is a very old idea, dating back to ancient Greece, when it was believed that if you kept subdividing an object, you'd eventually come to a piece which could not be divided in two (*a-tomos* means uncuttable). The properties of these atoms determined the properties of the objects

Figure 9-3a
*Diagram of the expected small
deflection of alpha particles.*

Figure 9-3b
*Diagram of large deflections
observed.*

that they made up. If everything in the Greek world was made up of mixtures of earth, air, fire, and water, then there must be earth atoms, air atoms, fire atoms, and water atoms. At the time of Galileo and Newton, atomism was not a main concern. Newton saw the universe as made up of multitudes of small interacting corpuscles. However, he made little attempt to speculate as to the differences between corpuscles of wood, and those of iron.

The birth of modern atomic theory (which tries to figure out what the properties of these basic but invisible building blocks must be) took place in late eighteenth century chemistry. The size of the atom—that is, the distance between atoms in a solid, where they are packed as closely as possible—had been narrowed down to a few angstrom units ($1\text{Å} = 10^{-10}$ m). It was also known that in some chemical substances called compounds, the basic building block, called a *molecule*, was made up of several atoms. By the latter half of the nineteenth century, electrical experiments had indicated that small charged particles—electrons—played an important part in the construction of the atom. (Electric charge is discussed in Chapter 11.) At the beginning of the twentieth century the most promising model of the atom was a jelly-like sphere with the electrons embedded in it like raisins in bread. Between 1905 and 1910, two German physicists, Hans Geiger and Ernest Marsden, and Ernest Rutherford, a British scientist born in New Zealand, began studying the "scattering" of high speed alpha particles by very thin gold foil targets. They expected to detect fairly small deflections as the charged alphas plowed through the closely packed atoms of the foil. Great was their astonishment when they observed that a small fraction of the alphas were scattered through very large angles; in some cases, bouncing right back to where they came from. In a classic paper[4] Rutherford was able to show that the only way this could happen was for the positively charged alphas to encounter, and be strongly repelled by, a very small (less than 10^{-14} m) and massive cluster of positive charge. Rutherford called this core the *nucleus* of the atom.

By 1914, the model of the atom as proposed by the Danish physicist, Niels Bohr, looked something like that shown in Figure 9-4. At the center is the nucleus, very small (about 10^{-15} m across), very dense (most of the mass of the entire atom is concentrated here), and positively charged. Much further out are arranged a number of very light, negatively charged electrons. They possess sufficient negative electrical charge amongst them to match the positive charge on the nucleus, making the whole atom electrically neutral. An atom which is missing one or more electrons is called an *ion*.

(4) "The Scattering of Alpha and Beta Particles by Matter and the Structure of the Atom," *Philosophical Mag. 21,* 1911, pg. 669.

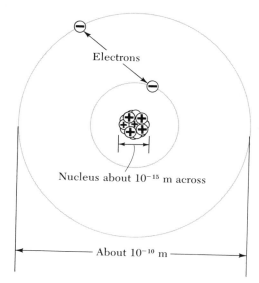

Nucleus about 10^{-15} m across

About 10^{-10} m

Figure 9-4
The nuclear model of the atom.

From its start, the nuclear model of the atom seemed analogous to the solar system: the electrons circulated around the nucleus, drawn inward by electrical forces just as the planets traveling around the sun are drawn in by gravity. One surprising realization was that the atom is more empty than the solar system! The diameter of Pluto's orbit is 7.6 billion miles. The diameter of the sun is about 860,000 miles. If you divide one number by the other, you get about 9000. On the other hand, if you divide the size of the atom; that is, the diameter of the electron orbits by the diameter of the nucleus, you get 100,000.[5] To call the space between the electrons and the nucleus empty is probably misleading. Giant forces act within the atom, and the electrons seem to spread out like miniature clouds rather than being well defined in size, as are the planets. Nevertheless, it was amazing to learn that what seems like a solid, impenetrable piece of material dissolves into a whirl of clouds and force fields when it is examined on the smallest scale.

The nucleus itself seems to be a clump of two kinds of smaller particles called *protons* and *neutrons*. The general term used to include both the proton and neutron is *nucleon*. Protons carry a positive electric charge but neutrons are electrically neutral. The number of protons in an atom's nucleus determines its chemical identity: one for hydrogen; two for helium; three for beryllium, and so up through the periodic chart as shown in

(5) This is true for the "ground state" of the atom. In an "excited" state, where the electrons possess extra energy, they seem to be located even further out.

Table 9-1   *Periodic chart of the elements*

H 1								He 2
Li 3	Be 4	B 5	C 6	N 7	O 8	F 9		Ne 10
Na 11	Mg 12	Ar 13	Si 14	P 15	S 16	Ce 17		A 18
K 19	Ca 20	Sc 21	Ti 22	V 23	Cr 24	Mn 25	Fe Co Ni 26 27 28	
Cu 29	Zn 30	Ga 31	Gp 32	As 33	Se 34	Br 35		Kr 36
Rb 37	Sr 38	Y 39	Zr 40	Nb 41	Mo 42	Tc 43	Ru Rh Pd 44 45 46	
Ag 47	Cd 48	In 49	Sn 50	Sb 51	Te 52	I 53		Xe 54
Cs 55	Ba 56	57-71	Hf 72	Ta 73	W 74	Re 75	Os Ir Pt 76 77 78	
Au 79	Hg 80	Te 81	Pb 82	Bi 83	Po 84	At 85		Rn 86
Fr 87	Ra 88	Ac 89	Th 90	Pa 91	U 92	Np 93	Pu 94	

57 – La	61 – Pm	65 – Tb	69 – Tm
58 – Ce	62 – Sm	66 – Dy	70 – Yb
59 – Pr	63 – Eu	67 – Ho	71 – Lu
60 – Nd	64 – Gd	68 – Er	

Table 9-1. Experiments on single protons showed that the amount of their positive electric charge is equal to the negative charge on single electrons. Therefore in a neutral atom the number of electrons circulating around the nucleus is equal to the number of protons in the nucleus.

For a given number of protons in the nucleus there can be various numbers of neutrons. Corresponding to this variability in neutrons, a given chemical element such as hydrogen, helium, or beryllium, etc., has various *isotopes*. These are symbolized by writing a superscript before the letters representing the symbol for the chemical element. (The atomic number, or position in the periodic chart, is indicated by a subscript before the letter.) The superscript, whose numerical value is equal to the sum of neutrons and protons in the nucleus, is called the *mass number* of the nucleus. For example, carbon exists in seven isotopes, from $^{10}_{6}$C up to $^{16}_{6}$C. In each case the nucleus of the atom contains six protons, but $^{10}_{6}$C nuclei contain only four neutrons whereas $^{16}_{6}$C nuclei contain 10. Isotopes with either a low or a high number of neutrons compared to protons are exceedingly rare. In the case of carbon, only $^{12}_{6}$C (carbon 12), $^{13}_{6}$C (carbon 13), and $^{14}_{7}$C (carbon 14), exist in nature. The rest are man-made. In general, the number of neutrons in the nucleus is equal to, or somewhat larger than, the number of protons.

The krypton 86 isotope ($^{86}_{36}$Kr) which was mentioned on page 13 contains 36 protons and 86 minus 36, or 50 neutrons. The cesium 133 ($^{133}_{55}$Ce) atoms mentioned on page 14 have 55 protons and 78 neutrons in their nuclei.

Chemical reactions, such as the burning of wood or the joining of Hydrogen and Oxygen to form water, have been found to involve just the electrons of an atom (and usually just the farthest from the nucleus). *Nuclear* reactions, on the other hand, involve rearrangements of the central nucleus.

The mass of the electron is small: 9×10^{-31} kg. The protons and neutrons in the nucleus each have masses a little more than 1800 times as great. A unit of mass commonly used in describing atoms is the *Atomic Mass Unit* (AMU). It is defined as 1/12 of the mass of the carbon 12 atom, and amounts to 1.66×10^{-27} kg. In terms of energy equivalence, 1 AMU equals:

$$E = mc^2 = (1.66 \times 10^{-27} \text{ kg}) (3 \times 10^8 \text{ m/sec})^2$$

$$= (1.66 \times 10^{-27} \text{ kg}) (9 \times 10^{16} \text{ m}^2/\text{sec}^2)$$

$$= 1.494 \times 10^{-10} \text{ kg·m}^2/\text{sec}^2$$

$$= 1.494 \times 10^{-10} \text{ joules}$$

One atomic mass unit has a rest energy equivalent to 1.49×10^{-10} joules.

1 A.M.U. $= 1.66 \times 10^{-27}$ kg

1 A.M.U. $\equiv 1.494 \times 10^{-10}$ joules

Nuclear Energy

Now comes the crucial point. The relation between mass and energy, as expressed in Equation (9-11), might have remained an exciting but theoretical curiosity except for one fact which came to light when physicists began measuring nuclear masses in the 1920's: the nucleus has a slightly smaller mass than the total mass of the individual protons and neutrons which make up the nucleus. The whole is *less* than the sum of its parts. This fact of nature makes the use of nuclear energy a possibility. As an example consider the nucleus of the Helium 4 atom; the "alpha" particle. It seems to be made up of two neutrons and two protons. Since neutrons have a mass of 1.008665 AMU and protons have a mass of 1.007825 AMU, you would expect alpha particles to have a mass of

$\left.\begin{array}{l}1.008665 \\ 1.008665\end{array}\right\}$ neutron masses

$\left.\begin{array}{l}1.007825 \\ 1.007825\end{array}\right\}$ proton masses

4.032980 AMU

However, the mass of alphas has been determined accurately to be 4.00260 AMU. This mass is 4.032980 minus 4.002600, or .03038 AMU less than expected. This discrepancy is called the *mass defect* of the alpha particle. For reasons that aren't completely understood, two neutrons and two protons attract each other and form a single composite particle, giving off .03038 AMU, or $(1.494 \times 10^{-10} \text{ J/AMU}) (.03038 \text{ AMU})$ which equals 4.54×10^{-12} joules of free energy. Alternatively a total energy of 4.54×10^{-12} joules $(1.13 \times 10^{-12}$ joules per nucleon) must be furnished to tear an alpha particle into four separate nucleons. For this reason the mass defect can also be seen as a *binding energy*.

As a further illustration, consider the first successful artificial nuclear transformation, which was conducted by Rutherford in 1919. He bombarded nitrogen gas with alpha particles and detected oxygen and hydrogen as products. This reaction is shown below, along with the known masses of the nuclei produced:

$$^{14}_{7}\text{N} \quad + \quad ^{4}_{2}\text{He} \quad \longrightarrow \quad ^{17}_{8}\text{O} \quad + \quad ^{1}_{1}\text{H}$$

nuclear
masses $\longrightarrow$ 14.00307   4.00260   16.999130   1.007825
in AMU

Equation (9-13)

Nitrogen 14 ($^{14}_{7}\text{N}$) is an isotope with a nucleus containing seven protons and seven neutrons. Oxygen 17 ($^{17}_{8}\text{O}$) nuclei contain eight protons and nine neutrons. $^{1}_{1}\text{H}$ represents the nucleus of ordinary hydrogen, which is a single proton. The question is: can this reaction take place spontaneously on its own, or must extra energy be provided? To answer this question, the total mass of the nuclei before and after the reaction must be found:

Before: 14.00307   After: 16.999130
           4.00260   1.007825
          ――――――   ――――――――
          18.00567 AMU   18.006955 AMU

The reaction calls for a mass *increase*,

$$\Delta m = 18.006955 - 18.00567 = .001285 \text{ AMU}$$

Extra energy must be furnished in order to make the reaction take place. This can be done by firing the alpha particle into the nitrogen nucleus with some kinetic energy. We can find out what the K.E. has to be:

$$(.001285 \text{ AMU})(1.494 \times 10^{-10} \text{ J/AMU}) = 1.92 \times 10^{-13} \text{ J}.$$

Will the alphas be travelling at relativistic velocities when they possess this much energy? We can again get some idea by comparing the kinetic energy, 1.92×10^{-13} joules, with the rest energy of an alpha particle:

$$m_o c^2 = (4.002600 \text{ AMU})(1.494 \times 10^{-10} \text{ J/AMU})$$

$$= 5.98 \times 10^{-10} \text{ joule.}$$

The ratio, $K.E./m_o c^2$, is only:

$$\frac{1.92 \times 10^{-13}}{5.98 \times 10^{-10}} = 3.21 \times 10^{-4}$$

The kinetic energy is .03 percent of the rest energy. *Not* large enough to be noticeably relativistic. The kinetic energy is still safely expressed as $\frac{1}{2}mv^2$. (The diligent reader will (of course) want to calculate what the minimum *speed* of the alphas must be to make the reaction take place.)

It will be noted here that the alpha particles used in the moon experiments described on page 170 were traveling too slowly to excite nuclear reactions. They were supposed to bounce back elastically, and they did.

Consider now a naturally occurring nuclear reaction: the alpha decay of uranium 238 ($^{238}_{92}U$). In this process the uranium gives off an alpha particle, changes into thorium 234 ($^{234}_{90}Th$), and releases energy, in contrast to the nitrogen reaction just discussed.

$$^{238}_{92}U \longrightarrow {}^{234}_{90}Th + {}^{4}_{2}He + (6.7 \times 10^{-13}) \text{ J/atom of } {}^{238}_{92}U$$

Equation (9-14)

For centuries, people kept warm by using a common chemical reaction: combustion of wood. It is interesting to compare the energy released in this reaction, where electrons are simply re-arranged, with the energy released in the alpha decay of U^{238}:

$$C + O_2 \longrightarrow CO_2 + (6.5 \times 10^{-19}) \text{ J/atom of Carbon}$$

Equation (9-15)

The energy released per atom in the nuclear reaction is about 10^6, or 1,000,000 times greater than that released in the chemical reaction. Uranium is pretty heavy, but even if we examine the energy released per pound of fuel, the alpha decay of uranium wins over burning wood by a factor of 1.8×10^5, or 180,000. The catch is that the energy released in the alpha decay is released too slowly to even boil a cup of water. The half-life of uranium 238 is 4.51 billion years. (This is the time required for half of the uranium atoms to decay into thorium 234 by emitting alpha particles.)

Now suppose we wanted to set up a nuclear reaction which would produce as much energy, per kilogram of fuel, as possible. We would be concerned with how fast the reaction should occur. But, more basically, we would also want to begin with nuclei which were more loosely bound (have a lower mass defect) and end up with tightly-bound nuclei (having a larger mass defect). *All* nuclei have a binding energy. What we want to do is end up with nuclei which have a larger mass defect or binding energy than the nuclei that were started with. This will maximize the mass lost and the energy released.[6] Referring back to Equation (9-11), we want Δm to be as large and negative as possible so that ΔE, will be as large as possible. Figure 9-5 shows the average binding energy per nucleon for a number of nuclei. The vertical axis of the plot is presented upside-down

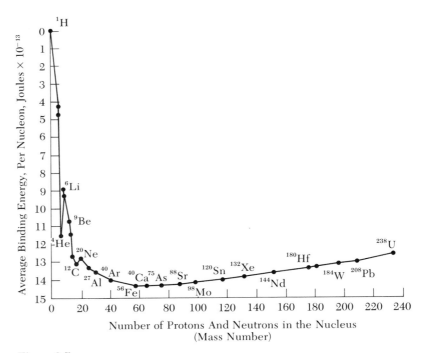

Figure 9-5

A plot of the average binding energy per nucleon: (i.e. the mass defect of the nucleus divided by the number of particles in the nucleus) vs mass number (total number of particles in the nucleus) for various chemical elements.

(6) The energy which is released in a nuclear reaction is carried off by high-speed by-product particles and/or radiation. As we now know, a moving particle has higher mass than a similar one which is at rest. Radiation, by carrying energy, must also be equivalent to mass. Therefore, to be perfectly correct, mass is not destroyed at the moment of a nuclear reaction. It is simply changed from rest mass to kinetic mass. Mass and energy *are* equivalent.

to emphasize those nuclei which have relatively lower mass per nucleon (a larger binding energy per nucleon). This plot is, in effect, a graph of mass per nucleon *vs* number of nucleons in the nucleus. It suggests several possibilities for energy-releasing nuclear reactions. One would be to start with hydrogen nuclei uncombined protons, ($_1^1$H) and synthesize helium nuclei alpha particles, ($_2^4$He). Another would be to start with lithium ($_3^6$Li) and break it down to $_2^4$He. You could also try starting with $_3^6$Li and synthesizing carbon ($_6^{12}$C). A fourth possibility would be to start with the heaviest nuclei and break them down to nuclei having 60 to 120 nucleons. The first and last possibilities are the ones we shall look into.

Fission

Radioactive nuclei emit small pieces of themselves (alphas, positive electrons, negative electrons, etc.). In so doing, they change their chemical identity by one or two places in the periodic chart (as will be indicated by comparing Table 9-1 to the alpha-decay of uranium shown in Equation (9-14). Therefore, when the German physical chemist, Otto Hahn, and Fritz Strassmann bombarded Uranium with neutrons in the late 1930's and obtained evidence that barium ($_{56}$Ba), cerium ($_{58}$Ce), krypton ($_{36}$Kr), lanthanum ($_{57}$La), strontium ($_{38}$Sr), xenon ($_{54}$Xe), and other elements were formed, they were puzzled. These elements lie between 36 and 58 in the periodic chart and have mass numbers between 80 and 150. They were slow to recognize their actual identity, thinking at first they were new elements located above uranium in the periodic chart. Hahn found his own results hard to believe. For some time he felt that some error in the chemical analysis and identification of his products had occurred. This was the first evidence that nuclear species could be produced which corresponded to such large jumps in the periodic chart. A theoretical answer to the problem was soon proposed in Sweden by Lise Meitner, an Austrian physicist and a former associate of Hahn's, and her nephew, Otto Frisch, who later fled from Germany as the Nazi persecutions worsened. They suggested a new nuclear phenomenon, *fission*, whereby a single nucleus breaks into two large pieces. They saw the process as analogous to the splitting of the biological cell, but on a scale billions of times smaller.

When a uranium atom is struck by a neutron ($_0^1$n) moving at the right speed it can temporarily absorb the neutron, and then break into two smaller nuclei. As shown in Figure 9–6, if a uranium 235 atom absorbs a slow-moving neutron, it can split into a barium 141 nucleus plus a krypton 92 nucleus. (Other fragment nuclei are possible with most of them having mass numbers in the neighborhood of 85 to 95 or 135 to 145.)

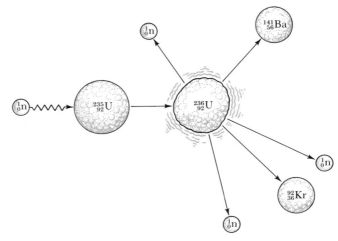

Figure 9-6
One possible scheme for the fission of Uranium 235.

In equation form:

$$^{235}_{92}U + ^{1}_{0}n \longrightarrow ^{141}_{56}Ba + ^{92}_{36}Kr + 3\ ^{1}_{0}n +$$

Equation (9-16)

$$(300 \times 10^{-13})\ \text{J/atom of } ^{239}_{92}U$$

The energy released, 300×10^{-13} joule, is almost 50 times as much per atom of uranium as that released in the alpha decay of uranium 238. [Equation (9-14).] But note that three neutrons are also released during the fission process. (In other fission reactions, only two may be released. The overall average is 2.47 neutrons.) What makes these fission reactions so important for engineering reasons is that more than one neutron is produced in each fission, whereas only one neutron is required to trigger it. If the extra neutrons can be utilized to produce new fissions then the process will keep itself going, and will become a potent source of energy. The name of this self-sustaining process, the nuclear *chain reaction*, is now a common part of the English vocabulary.

Fissioning is only one of several processes which a neutron in the midst of a piece of uranium can take part in. It might escape from the uranium. It might scatter or bounce off uranium nuclei, or other impurities, and lose energy. Or, it might be captured by a uranium 238 nucleus, producing a new nucleus which does not fission. All of these processes, including fission itself, depend for their likelihood on the speed of the neutrons. The work of the Italian physicist, Enrico Fermi in Italy, and in America, at Columbia University and the University of Chicago, was primarily concerned with finding out how these processes depended upon neutron velocity. Non-fission-producing neutron capture, and scattering, were found to increase with neutron velocity. This meant that if the energetic neutrons produced in the process of uranium 235 fission were to be effective in triggering further fissions they had to be slowed down to a speed where the number of neutrons removed in the competing processes of capture and scattering were much smaller. The most effective atoms for slowing down neutrons are those with light nuclei, such as carbon, or the hydrogen atoms in water molecules. (This can be shown by a straightforward application of conservation of momentum.) Materials used for slowing down neutrons came to be called "moderators".

It was soon found, as has been implied previously, that various isotopes of uranium responded differently to neutron bombardment. Uranium 238 seldom fissioned when hit with slow neutrons but uranium 235 did, so uranium 235 became the critical material in the effort to produce a chain reaction in the early 1940's. The difficulty was that uranium in nature contains 99.27 percent uranium 238 and only 0.72 percent uranium 235, along with minor amounts of uranium 234. In order to accomplish a chain reaction a method had to be found to "enrich" natural uranium with uranium 235. This could only be done by separating the desired isotopes from the rest, keeping the uranium 235, and disposing of the unwanted uranium 238; an awesomely difficult task in those days.

The Manhattan Project

During the Summer of 1939, several European-born physicists who had emigrated to America, notably Leo Szilard, met with Einstein at his summer home on Long Island. They decided to call the attention of President Roosevelt to the possibility of using nuclear energy, primarily from fission reactions, as a weapon. Their concern was to beat Hitler to the development of such a weapon. After a long delay, a huge government-sponsored effort, the Manhattan Project, was undertaken to produce an "atomic" bomb. ("Nuclear" bomb would have been a more accurate term.) Much has been written about this most interesting and grim chapter in the history of physics. Scientists were asked to play an exceedingly direct, important, and large-scale part in modern warfare. Their main single task was to find a scheme for separating uranium 235 from ordinary uranium in large amounts. Various scientists and engineers had favorite schemes for performing the separation and did all they could to get their methods adopted.[7]

(7) Davis, Nuell Pharr, *Lawrence and Oppenheimer*, Simon and Schuster, N.Y., 1968.

A huge factory was built at Oak Ridge, Tennessee, to do the separating, first by mass spectrograph, wherein electrically charged ions of uranium 235 and 238 travel in circles of different radii when they are shot into a magnetic field (see page 310). However, the process which eventually predominated at Oak Ridge was one which diffused gaseous compounds of uranium through a long series of permeable membranes. The gas molecules containing the slightly heavier uranium 238 diffused through the membranes more slowly, so that after each diffusion, the resulting gas was enriched with uranium 235. Enough uranium 235-enriched uranium was produced, so that by December 2, 1942, a carefully controlled chain reaction was made to go successfully in a nuclear "pile" at the University of Chicago.[8,9] A new era had begun. On July 16, 1945, an explosively rapid chain reaction was triggered near Alamagordo, New Mexico in the first "atomic" bomb. Several weeks later two bombs were dropped on Japan and the whole world became aware that the Atomic Age had begun. By this time the Allies had defeated Hitler in Europe, and many of the scientists who were responsible for getting the United States government interested in nuclear weapons initially were trying to prevent the dropping of the bombs.[10]

Early in the 1940's it had also been found that some of the neutrons captured by uranium, while not producing fission, are really not wasted. If a neutron is absorbed by a uranium 238 nucleus, an atom of uranium 239 can be formed. It quickly gives off an electron, becoming neptunium 239 ($^{239}_{93}$Np). (Loss of the negative electron from the nucleus *increases the positive charge* in the nucleus by one unit, so that the atom moves up a step in the periodic chart. See Table 9-1.) Then the neptunium 239 loses another electron, producing plutonium 239 ($^{239}_{94}$Pu). We could say that same thing with an equation for the reaction:

$$^{238}_{92}U + ^{1}_{0}n \longrightarrow ^{239}_{92}U \longrightarrow ^{239}_{93}Np + e^- + energy$$
$$\longrightarrow ^{239}_{94}Pu + e^- + energy$$

Equation (9-17)

The plutonium which is produced is dangerous but interesting stuff. Its nucleus has a good probability of undergoing fission when hit by fast *or* slow neutrons, so that moderating their velocity is less of a problem. At the same time, it is a "safer" explosive than uranium 235 in the sense that bombs using it can be de-

(8) Fermi, Laura, "Success", from *Atoms in the Family: My Life With Enrico Fermi*, University of Chicago Press, 1954. Reprinted in *Great Essays in Science*, Martin Gardner, ed., 1961.

(9) Anderson, Herbert L., "Early Days of the Chain Reaction", *Science and Public Affairs XXIX*, April, 1973, pg. 8.

(10) "The Decision to Drop the Bomb", an NBC Television Special. Available from Encyclopaedia Britannica Films, Chicago.

signed so as to begin the explosive chain reaction only when it is under high pressure. Rather than just going through the painstaking process of uranium enrichment, the Manhattan Project also produced plutonium 239 by the reaction shown in Equation (9-17) and used it in a weapon.

Of the two bombs dropped on Japan, the Hiroshima bomb was made from uranium (enriched about 90 percent with uranium 235) and the Nagasaki bomb was made from plutonium.

Peaceful Uses of Fission Energy

Thus far nothing has been said about what the long-range use of nuclear energy should be: the slow, controlled reactions which release heat to generate power. The reaction is carried out in a "pile" or "reactor" which is a carefully designed three-dimensional stack of fuel elements, moderator material, and neutron absorbing dampers, or control elements. The fuel may be uranium enriched one to 20 percent with either uranium 235, or plutonium 239, or in rarer instances, thorium 232 ($^{232}_{90}$Th). The moderator is usually carbon or water. In some reactors, the fuel and moderator are arranged as separate units like pieces in a Chinese puzzle. In others, the fuel is mixed directly with a liquid moderator. The idea is to achieve an arrangement such that when one uranium nucleus fissions, the fast neutrons that are released must travel through enough moderator to be slowed down enough so that when they encounter their first uranium 235 nucleus they will be capable of causing a fission. Since neutrons reaching the surface of the fuel-moderator mass are lost to the fission process, a minimum *critical* mass of uranium is required before the fission process can keep itself going.

(At the end of Chapter 2 it was noted that smaller animals have a relatively larger surface area and have to eat more per unit of volume in order to keep up their body temperature. Somewhat the same principle is at work here, although neutrons, rather than heat, are being lost.) The smallest critical mass assembled thus far, using thin foils of pure uranium 235 separated by thin sheets of polyethylene, amounts to 242 grams of uranium and makes a rectangular solid 15.4 by 15.4 by 12 cm. in size.[11] Ordinary piles are much larger than this.

The control rods perform the important function of making sure that, on the average, only one neutron from each fission goes on to induce another fission. At this point the reaction in the pile will just maintain itself, not accelerating or dying out. To "cool off" a reactor, the control rods are moved in, absorbing more neutrons. To "warm it up", the rods are withdrawn, so

(11) Made by Carroll B. Mills and George A. Jarvis, of Los Alamos Scientific Laboratory.

that more neutrons can find their way to new uranium 235 nuclei, producing more fission and energy release. To use the heat produced in the reactor a liquid heat transfer material (water or sodium) is pumped through the reactor, and a "heat exchanger" as shown. Water is then pumped through the heat exchanger in separate tubes so that it can accept heat from the heat transfer liquid without picking up any of its radioactivity. As the water travels through the heat exchanger it turns to pressurized steam and can be used to drive a turbine connected to an electric generator (The actual electrical energy production is thus done by a converter of mechanical energy to electrical energy).

Besides producing useful heat, nuclear piles are also used to make radioactive isotopes for medical work, to harden certain plastics by irradiating them, to produce beams of neutrons for medical therapy, and for experimental purposes.

In the United States, there are now about 100 nuclear power plants in use or being planned. Thus far they have had difficulties competing economically with coal, oil, and gas as sources of thermal energy. They have also become the center of criticisms by environmentalists because of the radioactive waste

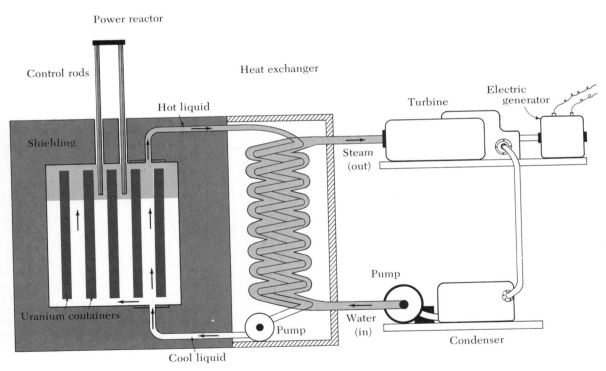

Figure 9-7
Schematic diagram of one type of nuclear reactor.

products they produce.[12] What these critics sometimes over-look is the ecological damage, including radioactive components in smokestack gases, which ordinary fossil fuel power plants produce.[13] For example, a 1000 megawatt coal-fired power plant produces 30,000 tons of air pollutants and about 1000 tons of ash *per day*. Even if the pollutants are trapped, they constitute a huge storage and disposal problem. A similar sized nuclear plant does produce more thermal pollution (discussed further in Chapter 16) but it gives off almost no air pollutants, and only a few tons of radio-active waste *per year*. With fossil fuel running out there is counter pressure to build nuclear plants. As stated in Chapter 7, predictions indicate that very soon half of the new power producing facilities in the United States will be nuclear.

The real heart of the uranium fission reaction is uranium 235. Naturally occurring uranium contains less than one percent of this material. This means that the world's fissionable uranium reserves are also limited. Recognition of this fact has helped generate interest in a different sort of reactor which uses the plutonium-producing reaction shown on page 238. Such reactors are called *breeder reactors*, because they actually produce more fissionable material in the form of plutonium, than is consumed. Since they utilize the relatively plentiful uranium 238 isotope, breeder reactors expand by many factors of ten the world's store of useful uranium fuel. However, their design and use entails some tricky technological problems. The most effective coolant for them seems to be liquid sodium, a metal which is violently reactive, and explodes when mixed with water. Nevertheless, breeder reactors have been on the drawing boards, or under test, since the early 1950's. The next two decades will probably see more of them come into use as producers of energy.[14, 15]

Fusion

In 1939 the German physicist (and refugee) Hans Bethe proposed an explanation for the energy released by the Sun. The process could obviously not be akin to any chemical reaction, such as burning, or else the heat of the Sun would have died out billions of years ago. Bethe suggested instead that the source

(12) Lapp, Ralph E., "The Four Big Fears About Nuclear Power," N.Y. Times Magazine, Feb. 7, 1971.

(13) Foreman, Harry, (Ed.), *Nuclear Power and the Public*, University of Minnesota Press, Minneapolis, 1970.

(14) Seaborg, Glenn T. and Bloom, Justin L., "Fast Breeder Reactors", *Scientific American 223*, November, 1970, pg. 13. Offprint 339.

(15) Culler, Floyd L. and Harms, William O., "Energy From Breeder Reactors", *Physics Today 25*, May, 1972, pg. 28.

of the energy is the joining or *fusion* of hydrogen nuclei (protons) to form helium nuclei (alpha particles) having a large mass defect. Several reaction chains seemed possible, and only in 1967 was enough experimental evidence collected to show which was most likely. It seems to occur in three steps:

(a) $^1_1H + {}^1_1H \rightarrow {}^2_1H +$ positive electron + neutrino + energy

(b) $^2_1H + {}^1_1H \rightarrow {}^3_2He + \gamma$ ray + energy.   Equations (9-18)

(c) $^3_2He + {}^3_2He \rightarrow {}^4_2He + 2 {}^1_1H +$ energy.

In these equations, ^1_1H stands for a proton, the nucleus of ordinary hydrogen. ^2_1H stands for a deuteron, or proton-neutron pair, which is the nucleus of deuterium, an isotope of "heavy" hydrogen. ^3_2He stands for a nucleus of helium 3, made up of two protons plus a neutron. ^4_2He is the alpha particle, the nucleus of helium 4, and is composed of two protons plus two neutrons. In Figure 9-8 e^+ stands for the positive electron, or positron. It is similar to the ordinary electron, but has an opposite electric charge. ν stands for the neutrino, the particle we first encountered on page 171. γ stands for a gamma-ray, a very high-energy x-ray or packet of radiation. (See Table 13-2, page 339.)

Equations (9-18) can be summarized by noting that steps (a) and (b) must occur twice, using a total of six protons, because two ^3_2He nuclei are necessary for step (c). Since two protons are regenerated in step (c), only four new protons are necessary and the total reaction may be written

$$4 {}^1_1H \rightarrow {}^4_2He + 2\gamma + 2 e^+ + 2\nu +$$ energy.

We can estimate the energy released by adding up the mass of the four protons entering the equation and comparing this total with the mass of an alpha particle plus the mass of the two positive electrons, to find how much the mass of the system has been reduced:

Mass of an alpha particle = 4.00260 AMU;
Mass of two positive electrons = 0.0011 AMU.
Total 4.00370 AMU

Mass of four protons = (4)(1.007825 AMU) = 4.0313 AMU;

$\Delta m = 4.0037 - 4.0313 = -0.0276$ AMU $= -2.76 \times 10^{-2}$ AMU

The energy released,

$\Delta E = -\Delta mc^2$

$\quad = -(-2.76 \times 10^{-2}$ AMU$)(1.49 \times 10^{-10})$ J/AMU

$\quad = +4.11 \times 10^{-12}$ J/nucleus of ^4_2He

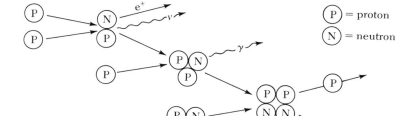

$\boxed{\text{P}}$ = proton

$\boxed{\text{N}}$ = neutron

Figure 9-8
Steps in the Sun's fusion reaction.

This is a gigantic amount of energy per kilogram of fuel: almost ten times as much as in fission, 400 times as much as in the alpha-decay of uranium 238, and 70 million times as much as in the burning of carbon in oxygen. The main technological difficulty in tapping this rich vein of energy is that the reactants must be brought together with great energy—the equivalent of many millions of degrees in temperature—before the fusion process begins. Thus far, scientists have succeeded only in producing rapid destructive fusion processes: the "Hydrogen" bomb. In contrast to the three-step fusion process which the Sun seems to use, the bomb reactions have probably taken place in one step, starting with deuterium and tritium. (In ordinary water, 140 to 150 hydrogen atoms per million are deuterium, ^2_1H, and one atom in 10^{18} is tritium, ^3_1H. The nucleus of ^3_1H, called the triton, is made of one proton and two neutrons.)

$$^2_1\text{H} + ^3_1\text{H} \rightarrow ^4_2\text{He} + ^1_0\text{n}$$
$$+ 2.7 \times 10^{-12} \text{ J/nucleus of } ^4_2\text{He}$$

Equation (9-19)

Since deuterium has an atomic mass of 2.01410 mass units, energy balancing shows that the triton must have a mass of 3.01125 AMU to produce the 2.7×10^{-12} joules of energy. The temperatures required to start the fusion are produced *in weapons* by means of a rapid uranium or plutonium fission process. A fission ("Atomic") bomb is used to trigger the fusion ("Hydrogen") bomb.

Over on the peaceful side, research groups in the U.S.A., U.S.S.R., and, to a lesser extent, England, and France have been edging toward the production of controllable fusion processes in the laboratory. The deuterium-tritium process in Equation (9-19) might be the most promising one to pursue, but the temperature required to start it (upwards of 50 million degrees) will have to be produced by electromagnetic heating and compression, not by a fission bomb. The technical problems are staggering but they are being solved, one by one.[16] Some scientists[17] have suggested that instead of a controlled fusion reaction, a series of mini-Hydrogen bomb explosions is the answer. Pellets of frozen hydrogen isotopes would be set off by energy from intense bursts of laser light. The explosions, like successive detonations of gas in the cylinders of an automobile engine, would still yield a great deal of energy and would be easier to achieve.

The fusion process is a tremendous lure for development because it releases astounding amounts of energy, produces very little radioactive pollution, and uses water for fuel. Perhaps the best reason for being cautious about the installation of fission-powered nuclear plants is that when (and if) fusion processes are finally usable, they will constitute a far superior source of power. The super-high temperatures accompanying the fusion process might even be used to vaporize materials in recycling systems, providing a cheap means for reducing pollution and handling refuse produced by other industrial and urban processes. The energy of the stars is an elusive but tempting thing to reach for and control.

Summary

Momentum conservation and the time dilation called for by Einstein's special theory of relativity lead to the prediction that the mass of an object increases with velocity:

$$m = \frac{m_o}{\sqrt{1 - \dfrac{v^2}{c^2}}} = \gamma m_o \qquad \text{Equation (9-8)}$$

where m_o is the mass of the object when it is at rest relative to the observer. From this equation it appears that no object possessing any rest mass can travel at the speed of light. If it did, its mass would be infinite. Mass measures inertia, not the quantity of matter in an object.

The fact that mass increases with velocity suggests a new expression for kinetic energy. It is

$$mc^2 - m_o c^2$$

or

$$m_o c^2 (\gamma - 1)$$

Equations (9-9)

(16) Post, Richard F., "Prospects for Fusion Power," *Physics Today* 26, April, 1973, pg. 31.
(17) Lubin, Moshe J. and Fraas, Arthur P., "Fusion by Laser," *Scientific American* 224, June, 1971, pg. 21.

For velocities, v, much less than c, the expression reduces to $\frac{1}{2}m_o v^2$, the classical expression for kinetic energy. Then:

$$mc^2 = E = \text{total energy}$$

and

Equations (9-10)

$$m_o c^2 = \text{rest energy}$$

In closed system

$$\Delta E = -\Delta mc^2.$$

Equation (9-11)

A more comprehensive statement of energy conservation in a closed system has now become possible:

$$\Delta E = [\Delta(K.E.) + \Delta(P.E.) + \Delta mc^2 + \Delta U]$$
$$= 0$$

Equation (9-12)

Matter appears to be made up of many tiny (10^{-10} m) atoms. These atoms themselves have a structure. At the center is a tiny nucleus (10^{-15} m), seemingly composed of a clump of protons and neutrons. Most of the mass of the atom is concentrated in this nucleus. The number of protons in the nucleus determines the chemical identity of the atom. Various isotopes of a given element correspond to their nuclei having different numbers of neutrons for a given number of protons. The mass number of a nucleus is equal to the sum of its protons and neutrons.

The mass of a nucleus is less than the sum of the masses of the protons and neutrons of which it is composed. This difference is called the mass defect of the nucleus. In chemical reactions rearrangements of the electrons outside the nucleus occur. In nuclear reactions the nucleus itself is re-arranged; broken apart, or built up. If the sum of all the nuclear masses is less after the reaction, then energy will have been released. To release energy a reaction must produce nuclei having a larger mass defect, or binding energy.

Two types of energy-releasing nuclear reaction have been carried out:

(1) Fission. A heavy nucleus is hit with a neutron of a certain energy. The nucleus splits into two large pieces, (usually radioactive) releasing energy, and several neutrons. These neutrons can then go on to trigger other fissions, starting a chain reaction which can rapidly expand. An explosively fast chain reaction is used in atomic bombs. Chain reactions can also be controlled, and made to go at an even rate, as is done in nuclear power plants.

(2) Fusion. Isotopes of hydrogen are shot together under conditions of extreme pressure and temperature. They "fuse" to form helium

nuclei and release huge amounts of energy per pound of fuel (about 70×10^6 times as much as in burning carbon, and 10 times as much as in fission). Fusion reactions are the source of the energy which the sun radiates. Very little radioactive waste product is formed. The hydrogen bomb is an uncontrolled fusion reaction. Up till the time of writing this, fusion reactions have not been tapped successfully as sources of peacetime power, but the promise is very bright.

Further Reading

Groves, Leslie R., *Now It Can Be Told: The Story of the Manhattan Project*, Harper and Row, N.Y., 1962. (A report by the general who directed the project which produced and detonated the first nuclear bomb.)

Guillemin, Victor, *The Story of Quantum Mechanics*, Charles Scribner's, N.Y. 1968. (Chapters 2 and 3 describe the development of atomic theories.)

Hogerton, John F., *Nuclear Reactors*, United States Atomic Energy Commission Division of Technical Information, 1963. (An informative, easy to understand booklet.)

Holliday, Leslie, "Early Views on Forces Between Atoms", *Scientific American*, May 1970, pg. 116. (Greek philosophers first conceived of atomic theory of matter, and scientists after the Renaissance speculated on inter-atomic forces. More detailed theories awaited 19th century experimental results.)

Jungk, Robert, *Brighter Than a Thousand Suns*, Harcourt, Brace & World, 1956. Translated into English, 1958. (A history of the atomic bomb, beginning with Rutherford in 1917.)

Smith, A. K., "Behind the Decision to Use the Atomic Bomb, Chicago, 1944–45", *Bulletin of the Atomic Scientists 14*, pg. 288, 1958.

Smythe, H. D., *Atomic Energy for Military Purposes: The Official Report on the Development of the Atomic Bomb Under the Auspices of the U.S. Government, 1940–45*, Princeton University Press, 1945.

Weinberg, Alvin M. and Hammond, R. Philip, "Limits to the Use of Power," *American Scientist 58*, pg. 412, July-Aug. 1970. (If breeder reactors or controlled fusion are used, sufficient energy will be available to support a much larger world population. This leaves open the question of the quality of the life available to such a crowded population.)

Whyte, Lancelot L., *Essay on Atomism: From Democritus to 1960*, Wesleyan University Press, Middletown, Conn., 1961.

9-1   An electron has a velocity sufficient to increase its mass by a factor of five. Using Equation (9-8), calculate its velocity. What is its kinetic energy, in joules? (The rest mass of the electron is 9.1×10^{-31} kg.)

9-2   What happens to the density (mass per unit volume) of a rod made of steel when it is made to fly past you at half the speed of light?

9-3   What must be the speed of the alpha particles to trigger the reaction in Equation (9-13)? (Specifying this as a problem does not necessarily mean the author questions your diligence.)

9-4   Solar radiation reaches the earth, which is 93,000,000 miles (1.5×10^{11} meters) from the sun, at the rate of about 2.0×10^3 watts per square meter. At what rate must the mass of the Sun be decreasing due to its radiation in all directions?

9-5   An electron (with rest mass 9.1×10^{-31} kg) is known to be traveling at a speed of $0.9990\,c$. Calculate its total energy on the basis of relativity theory. Find the ratio of the relativistic expression for kinetic energy to the classical expression for kinetic energy.

9-6   The following reaction occurs:

$$^{27}_{13}\text{Al} + {}^{4}_{2}\text{He} \longrightarrow {}^{30}_{14}\text{Si} + {}^{1}_{1}\text{H}$$

Given the following masses:

aluminum 27	26.990109 AMU
alpha particle	4.0026 AMU
silicon 30	29.983307 AMU
proton	1.00783 AMU

will the reaction release energy, or will it require energy to make it go? How much energy is used up, or released?

9-7   Would you say we have no business tampering with the nucleus? Why?

9-8   From Equation (9-19) on page 243 and your knowledge of the masses of the neutron and deuteron (${}^{2}_{1}\text{H}$), show that the mass of the triton is 3.01125 AMU.

9-9   Two objects, each having a rest mass of one kilogram, travel toward each other at a speed equal to $c/2$. They collide, *and* stick together! What is the mass of the composite body?

9-10   When heavy nuclei undergo alpha decay, the product nucleus is often radioactive. It emits beta rays (negatively charged electrons) but never positrons (positively charged electrons). Explain this.

10
PHYSICS, PHYSICISTS, AND SOCIETY

It takes little imagination to realize the impact of nuclear power on our society. There are, however, a number of other issues connected with the growth of science. Some of them are considered in this chapter, as a sort of breather after the section on invariance, prior to plunging into the study of fields and waves.

Physics as a Profession

The professional scientist entered Western society fairly recently. Until the middle of the nineteenth century most scientists had independent means, or else they were professors, doctors, government officials, or wealthy amateurs who did science as a sideline. There were very few openings for "Royal Astronomers". Public support of scientific research began to grow with the founding of the École Polytechnique in Paris in 1794, and the opening of research oriented institutes attached to German Universities in the mid-1800's. European universities, especially those in England, had been surprisingly slow to supplement their instruction in "natural philosophy" with planned scientific research. Societies for the discussion, support, and publication of science had been in existence for over two centuries before universities started switching from preserving ancient knowledge to investigating and gathering new knowledge. But once the switch had been made, late in the 19th century, universities quickly became the place where scientists were trained, and where the vast majority of scientific investigations were carried out.

There are now about 750,000 scientists in the world, studying natural phenomena, or at least talking about them. Of these, about 400,000 live in the United States. Of the American scientists 37 percent have Ph.D.'s, 29 percent have M.S.'s, 30 percent have bachelor's degrees, and 2.5 percent have M.D.'s. About one in ten are women, and one in twelve are physicists. About 53 percent of American physicists are employed by educational institutions, 30 percent by private industry, 11 percent by governments, and 3 percent by non-profit research organizations. The rest are self-employed, or unemployed. However, the salaries of 80 percent of American physicists are government supported.

What kinds of things do physicists do? Some are theoretical or mathematical physicists, some are experimental physicists, some are teachers, and some are research administrators. Their activities are sometimes arbitrarily divided up into pure science, applied science, and development or engineering, though the boundaries are indefinite and debatable. They work on acoustics, astronomy, high-energy particles, atmospheric physics, nuclear physics, optics, plasma physics (including fusion research), properties of solids, and other problems. Some of the most exciting work being done now is in multidisciplinary fields such as biophysics, geophysics, and psychophysics. The growing importance of these areas suggests that the dividing of the study of natural phenomena into biology, chemistry, geology, and physics is artificial. It fits the present organization and funding of laboratories and university departments more than it does the problems to be solved.

Physicists as People

Aside from being slightly younger than average, physicists are not different from any other group of fairly creative people. They display a wide range of personality traits and spare time interests. Among the outstanding twentieth century physicists, you find the garrulous, party-loving Niels Bohr and Enrico Fermi, the reclusive Einstein and Paul A.M. Dirac, the big science executive Ernest O. Lawrence, and the enigmatic, charismatic Robert Oppenheimer. Doing physics does not commit the individual scientist to any particular philosophical, political, or theological point of view. Scientists are about as formally, or informally, religious as any other group of people.

It is important to point out the highly social nature of physics. People are not the physicists' main work, but physicists do bear the burden of ultimately making someone else understand what they are doing. Their search is not so much for truth as for unambiguous communication with other people. Whereas the beauty of art lies in its ambiguity, some of the beauty of science lies in agreed-upon understanding. This calls for nearly continuous human interaction; day to day, in verbalizing one's ideas with colleagues, and at annual gatherings of scientists, where excited talkers jam hotel corridors.

Many of the personality traits which lead to success in business, government, or the older, more established professions also help the scientific career.[1] The same aggressiveness and persuasiveness which "closes the deal" in business gets financial support for research.

Physicists work in groups more than other scientists and mathematicians do.[2] The growth of team research and the use of expensive equipment requires leadership. It also has raised some problems in the scientific career. The experiment which can be designed, instrumented, carried out, and analyzed by one person is rare now-a-days. More typically, one group thinks up experiments to do and designs them, another group might operate the equipment. The data might be collected automatically, so that a computer run by still other people analyzes the evidence. The communication which announced the discovery of the omega minus particle had 33 authors![3] Team esprit may be a good thing, but the personal sense of creating a new understanding can get lost. Then too, young physicists, or mavericks who would like to try something new, have a more difficult time doing this when they are members of a large team.

The Funding of Science

Long before modern science began, rulers realized the military value of understanding nature, especially if the enemy lacked that understanding. For example, by using the magnetic compass, the Vikings were able to navigate in cloudy weather when their competitors, who depended on seeing the sun and stars, had to stay in port. The famous Persian poet of the early twelfth century, Omar Khayyám, was also a skilled astronomer. He is reputed to have gained the favor of Sultan Malik Shah in part by successfully predicting a solar eclipse. The Sultan made plans for battle on the appointed day, and when his superstitious enemies saw the sun disappearing, they took it as an evil omen and fled. Science, or at least technology, has thus been associated with defense and warfare for a long time. The concept of public and private support of science as a national economic or cultural resource is more recent.

(1) The research of Anne Roe on personality traits in scientists has attracted wide attention (*Making of a Scientist*, Dodd-Mead, 1953). She says they were largely introverted, and that they lacked affection as children, but her sample consisted of only 64 persons!

(2) Hagstrom, Warren O., "Traditional and Modern Forms of Scientific Teamwork," *Administrative Sciences Quarterly IX*, #3, pg. 241, December, 1964.

(3) V. E. Barnes *et al.*, "Observations of a Hyperon With Strangeness Minus Three," *Physical Review Letters 12*, #8, pg. 204, 1964.

In America, the idea that the work of the scientists in their laboratories might have an immediate military significance did not develop until the 1920's. During World War I the United States Department of War employed one physicist! Even in the 1930's much of American research in the physical sciences was privately rather than publicly sponsored. Government research funds came first to agriculture and the biological sciences, then finally to the physical sciences. Then the picture changed quickly. By 1940, scores of government-funded scientists were involved in developing a radar system which was powerful and accurate enough to protect England from the German airforce. Within a few years, the Manhattan Project had brought together hundreds of the Western world's best scientific brains in order to design and construct the first nuclear bomb.

Trends in government funding of science and technology can best be shown by the budget figures displayed in Figure 10-10[4] The *federal* funding rose rapidly, then leveled off during the late 1960's. *Industrial* investment in science maintained a more steady 10 percent increase per year until it had reached $10 billion in 1972. (The cost of the research and development actually performed by industry was almost twice this much, but half of it was funded by the government.) The nation's *universities* budgeted another $1.6 billion dollars for research and development in that same year. (They performed an additional $2.6 billion in government-funded research.) The national total thus came to $22 billion dollars for 1971. (About 1/5 of this was

(4) Holloman, J. Herbert, "Technology in the United States: The Options Before Us", *Technology Review 14*, July/Aug, 1972, pg. 32.

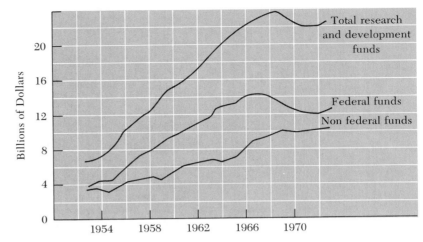

Figure 10-1
Funding of research and development in the United States (1967 dollars).

for basic research; the rest went into applied research and development.)

To summarize, the rapid growth of American science ended with the mid 1960's due in part to economic pressures resulting from the American-Vietnamese conflict. A slower rate of growth seemed to follow. Some felt that this amounted to a national disaster; that the economic and cultural well-being of the United States would suffer drastically if we backed off from research-generated technological growth. Others felt that it was a natural development, coinciding with a need for careful sorting of scientific priorities and a more careful assessment of the human value of the new technologies we undertake.

Priorities and Politics

How does a society decide how much and what kind of science it wants? Until recently, few people have thought about this question. But now it has become a live issue. Alvin Weinberg[5] has suggested two criteria which need to be considered when a proposed science project is under discussion.

(1) How well can this science be done?

(2) Why should society support this activity?

"Can we send a man to the moon", and "Should we send a man to the moon" are two separate questions. The United States government answered the first of these questions in July 1969, but the criteria for answering them were probably not well discussed. As the moon project neared completion in 1973 public

(5) "The Choices of Big Science", Chapter 4 in *Reflections on Big Science*, M.I.T. Press, Cambridge, 1967.

support had dwindled because the reason for going to the moon (i.e., to beat Russia) was no longer there, and the cost of the program loomed larger.

There are at least six ways for a society to answer "yes" to Weinberg's second question above. In a time when we cannot say "yes" to all proposed science projects, we need to be as honest as possible in deciding which of the reasons should have the highest priority. (The question of who decides on the priorities is a thorny one indeed.) Among the reasons for supporting scientific research are these:

(1) cultural—science enlarges the realm of human understanding. We should support scientists for the same reason we support musicians, artists, and city planners.

(2) economic—research leads to new technology and improved productivity.

(3) educational—exposure to science can make people more aware of themselves and their surroundings.

(4) environmental—research is needed to determine the wisest ways in which to exploit our surroundings, without destroying them.

(5) military—defense and weapons are part of being a "world power", and science-based technology leads to new weapons.

(6) social—scientific study of social systems might help us avoid past mistakes in welfare programs.

We have already suggested that, right or wrong, reasons (2) and (5) have predominated in the past. But this is changing.

When science itself becomes large, involving many people, and much money, it becomes (by definition) political. Like any other special interest group, science has to fight for funds in Washington. As Daniel S. Greenberg says,

It is naive to assume that big science can be aloof from the hard-sell tactics that have become standard in every other phase of American life. A quest for millions from the U.S. Treasury—whether for highways, bombers, or chemistry—arouses the question of how to make the most attractive case.[6]

As advisors to government, scientists have a mixed record. Part of this arises from the fact that scientists making trips to Washington or working for agencies there are involved in activities which sometimes conflict. One time they might be advocates, arguing for more money for their own projects. The next they are judges, trimming budgets or deciding who should get the new research grants.

The gulf between controlled experiments and complex social and political systems is a wide one. Conclusions which are definite and valid in research have to be applied very cautiously to the real, uncontrolled world. For this reason, scientists called

(6) "The New Politics of Science", *Technology Review* 73, Feb., 1971, pg. 41.

in as experts are often caught in a bind. Their sense of personal responsibility may lead them to speak out authoritatively when the result of the application of the scientific facts is still uncertain. In the opposite direction, scientists have sometimes avoided commenting on the beneficial or detrimental effects of a particular process until long after the effects were known.

Even when the data are fairly certain, scientists can disagree on their implications. Using the same information about radiation from nuclear fallout (radioactive debris from nuclear weapons being tested in the atmosphere), Edward Teller concluded that the effect on humanity is negligible and Linus Pauling concluded that thousands of babies would be born dead or deformed.

Government Science Policy

In the absence of formal government policy, America has assumed world leadership in science in the last 50 years. It has certainly benefitted from an influx of European scientists, especially in the 1930's, when many well-established German and Austrian scientists came to America as a result of Nazi purges. Still, the growing number of Nobel prizes won by Americans[7] have, in most cases, gone to people trained in the United States. It should also be pointed out that when the refugee scientists arrived in America they found a going scientific community, colleagues to talk to, and, for the most part, were happy to settle in the United States permanently.[8]

The shift in the center of research from Europe to America coincided with the birth of expensive research which the European countries were unable or unwilling to match. But in addition to this the multiplicity of American funding sources — public and private — fostered flexibility. The organization of American research teams gave promise to younger scientists who would have been buried in the hierarchical pyramid which many European professor dominated research centers still are.

With this history, few people in Washington felt that a *centralized* planning agency for science and technology in America was advisable. However, after the fund-cutting of the late 1960's, the tune began to change. Enough projects were undercut by seemingly shortsighted governmental funding changes to make a number of people feel that a Department of Science and Tech-

(7) Up to 1925, German scientists had been awarded 23 Nobel prizes in the sciences; British, 13; French, 11; and American, 3. Between 1926 and 1966, Germans won 22; Britishers, 32; Frenchmen, 9; and Americans, 68! Of the 71 American prizes, 14 were won by scientists who were born or educated in Europe. (Zuckerman, Harriet, "The Sociology of the Nobel Prizes", *Scientific American*, 217, pg. 25, Nov., 1967). The first award to an American was made in 1907. It went to Albert A. Michelson, for his experiments with light, some of which were discussed in Chapter 8.

(8) Fleming, Donald and Bailyn, Bernard, (eds.) *The Intellectual Migration: Europe and America; 1930–1960.* Harvard University Press, Cambridge, 1969.

nology might be a good thing after all. Science policy—a sense of where we want to be headed—is a crucial need.[9]

Whether or not a centralized agency decides on this policy, there are a number of issues or questions which the government must face. Among them, first of all, is the question of what the research and development dollars should be spent for. Early in the 70's, the American government began shifting research funds from the support of military to the support of research into civilian problems, such as environmental quality, and transportation. Even so, Germany and Japan both have four times as many scientists per capita at work on civilian research problems than does the United States.

A second problem facing our policy makers is how to best foster multidisciplinary research on the real problems of society; use of energy, health care, education, etc. The existing government research agencies all have fairly specific objectives, and no single university can afford to attack problems on so large a scale. A partial answer to the question has been the offering by the National Science Foundation of research funds for study of contemporary ecological-urban problems. But research is one thing and reforming our cities is another. The money to provide jobs for the people who might have some answers has been slow to come. According to Jay Forrester[10] this may be just as well. He says the government urban renewal efforts have tended to backfire in the past because important factors were overlooked. We are just now learning that population, pollution, food production, rate of economic development, and rate of consumption of resources *all* have to be controlled simultaneously. If just one is controlled, undesirable results are likely.

A third issue which developed during the 1960's was the *geographical distribution* of governmental research funds. The funds had grown to the point that they could affect the economic climate of the region into which they went. Therefore, members of Congress began fighting for research funds the way they used to fight for roads and dams. Coastal congressmen argued that if you wanted the most for your research dollar, you spent it on a team which had already proved its excellence. Midwestern congressmen on the other hand complained that too much of the money was going to laboratories and universities on the two coasts. The trend *was* toward wider geographical distribution.

An illustration of these factors is the Atomic Energy Commission's giant proton accelerator near Batavia, Illinois. Its location was chosen only after intense public discussion and political maneuvering. The machine was to cost $375 million and would require $60 million per year to operate. 126 bids from 46 states were received. Even after the commission announced the loca-

---

(9) Daddario, Emilio Q., "Needs for a National Policy", *Physics Today 22*, Oct., 1969, pg. 33.
(10) "Counter-intuitive Behavior of Social Systems", *Technology Review 13*, Jan., 1971.

Figure 10-2
*The Geodesic dome atop one of the buildings at the National Accelerator
Laboratory in Batavia. It is constructed of triangular panels of polyester-
reinforced fiberglass. Each triangle is constructed of two panels, that form a
sandwich containing discarded beverage cans, donated by the public. The
dome contains 120,000 cans. (Courtesy of National Accelerator Laboratory.)*

tion late in 1966, the project was almost moved elsewhere because local authorities would not at first guarantee that the
workers in the projected laboratory, of mixed racial backgrounds,
would all be given equal chances to obtain adequate housing.
Congress finally approved the location by a narrow vote, after
further debate. Once the building of the facility began, the
hiring of the construction workers made further history. Many
of them were intentionally recruited from the ghettos of Chicago,
given on-the-job training, and proved to be excellent workmen.
Whenever possible, construction and supply firms run by persons from minority groups were employed[11]

What is interesting is that the magnitude of the undertaking
required that it be funded by the federal government and that
a facility for pure research should become so intimately involved

(11) The Batavia facility is also turning out to be a pace-setter in the ecological
and aesthetic realm. The director, R. R. Wilson, is a skilled sculptor in his spare
time, and has seen to it that color and design are used to make the laboratory
beautiful. The towers bearing the electric power lines were re-designed to make
them more pleasing to look at. The office building rises just 17 stories and no
more because the view from a helicopter 170 feet in the air above the Illinois
plain was found to be just as good as it was higher up. 100,000 trees were planted,
and a buffalo herd was imported and is thriving.

with social movements and economic priorities. At the same time it forced Congress to make some decisions concerning the enlightened use of federal funds to promote social change.

Another question is how the government can best support colleges and universities in science and technology? With the exception of a few industrial laboratories, such as those operated by General Electric and the Bell System; and more recently, the government funded national laboratories at Brookhaven, Oak Ridge, Argonne and elsewhere, most of our scientific discoveries have come from university campuses. When the moon race began, back in the early 1960's, a great deal more government money went to the universities than had gone there previously. It went there in piecemeal fashion, grant by grant. As has already been pointed out, this helped keep our scientific research effort flexible, and was quite successful. What tended to be overlooked was the contribution which the universities added to each government research grant, in the form of buildings, equipment and staff. Seven or eight years later, when fund cutting began, some universities were left holding the bag. At the same time, other government funds were educating young scientists, based on the belief that we were going to need a lot more of them. Few government agencies were looking forward far enough to realize that, for the time being, we already seemed to have as many scientists as our society was prepared to support.

A fifth question asks whether mission oriented agencies such as the Department of Defense and the National Aeronautics and Space Administration (NASA) should conduct basic research. On the one hand, they would have a more difficult time being objective in deciding whether society should support the activity they were about to undertake. (Weinberg's second criterion.) On the other hand, if they are not funding basic research, can we be sure that a sufficiently large part of the government's science and technology money will still go into scientific research and not just into technology? One also wonders if it might not be good for the health of an agency such as the Department of Defense to be concerned not just with weaponry, but with a broad range of scientific activities.

A sixth problem arises from the fact that not only do government agencies instigate and sponsor research and development in science and technology, but these same agencies are also supposed to regulate and control their application. A good example of the conflict of interests resulting from such policies is shown in the setting of safe nuclear radiation standards; that is, in the decision as to how much radiation constitutes a harmful dose for a human being. These standards have been set by the Atomic Energy Commission (AEC). But at the same time, the AEC is our national agency for encouraging the use of nuclear energy.

While the AEC is theoretically under the surveillance of an 18 member joint congressional committee, ever since it was organized in 1946 it has tended to cloak its activities with secrecy, even in areas where national security is not involved. After a number of scientists outside the AEC group complained about loose radiation standards, a separate Federal Radiation Council was set up and now provides an external check on regulations coming out of the AEC. Some safety regulations have been tightened as a result.

As the government moves further into environmental control the problems will multiply in complexity. Should it use a carrot, and reward (subsidize) a company for using presently available, and perhaps inefficient, pollution controls; or, should it use a stick and tax the company so that it will seek better pollution controls?

A seventh, and perhaps the most important question facing the government, is how to obtain an improved input of information and ideas for governmental decisions. The Atomic Energy Commission, The National Aeronautics and Space Administration, The Department of Defense, The National Science Foundation, The National Bureau of Standards (Department of Commerce), The Department of Transportation, and other agencies all receive internal and external advice from scientists. Congress has committees which hear the testimony of scientists, and in 1972, set up the Office of Technological Assessment, to feed it expert advice. For a time, the President had his own Office of Science and Technology. And yet science policy decisions still seem to be made with an eye to special interests. There could be better correlation between the kinds of advice that are utilized in the making of decisions affecting scientific development. There is need for a wider forum in which debate can take place, including informed persons besides scientists and government people. Appointments to science advisory panels should be made public to prevent loading the panels in favor of certain interests. At the same time, scientific advisors should be given a workable assurance that their advice will be made as widely known as possible.

A good example of the need for truly open debate on technological issues was the decision, made in 1971, about the construction of supersonic transports. In spite of adverse scientific advice from various quarters, the project was nearly launched. Only when the bill for appropriations came to Congress did all the issues really get an airing. The project was finally killed; probably a wise decision, in view of the uncertainties in possible environmental effects.

Emilio Daddario, a former Congressman and chairman of the House Subcommittee on Science, has suggested[12] that democracy might not be compatible with science and technology. By the time all sectors in a democracy have been heard from,

(12) "Technology and the Democratic Process", *Technology Review* 73, July-August, 1971, pg. 18.

the technology problem might have become too large to solve. Nevertheless, we can do better in bringing to bear what we do know, scientifically, on problems which have to be solved ethically and politically.

Science and Technology

The growth of science has largely been paid for by technology, whether for military reasons or for economic reasons, in the Baconian (Francis) sense of forcing nature to serve man. However, scientists have traditionally been anxious to keep clear the distinction between science and technology. The scientist studies nature and seeks to describe and understand it. The technologist, or engineer, uses this understanding to build things: roads, radar, fluorescent lamps, and lunar space modules. The line between science and engineering is not clearcut. Derek deSolla Price suggests[13] that the distinction between scientists and engineers can be drawn on the basis of their communications. The scientist's product is a *publication*, whereas the engineer's product is a *device*. Engineers are also more likely to secure a patent protecting their rights to profit from the device, or even to hide information as to just how the device was made. Tradition sees engineering devices as a fruit of scientific research, when actually the devices themselves quite often contribute to science by providing new tools for examining the world. The success Galileo and Tycho Brahe had with the experimental approach was at least partly the result of having better instruments to work with. As was mentioned on page 5 the experimental studies, beginning around 1890, which led to modern atomic physics were made possible by the development of good vacuum pumps.

It is also possible to argue that science and technology lead fairly separate existences. Scientific discovery does not always lead to gadgets. In the course of their experimentation, physicists sometimes develop, for their own purposes, special electronics (for example) which bear little relation to equipment existing anywhere but in research laboratories.

Nevertheless the fact that scientific research does so often — eventually — lead to profitable technology has made it almost *de rigeur* for American corporate executives to include "research" laboratories in their corporate structure, no matter what its size. But research is a very expensive activity. In all but a very few large industrial laboratories, what is called research is really *development*: the inventing, testing, or improving of a device or process which the company can sell. Even in the largest laboratories there is continual tension between the seeking of fundamental knowledge and the development of

(13) "Is Technology Historically Independent of Science? A Study in Statistical Historiography", *Technology and Culture*, Vol. 4, pg. 553, 1965.

something that will make money. An automobile manufacturer may allow a group of physicists the freedom to study the crystalline structure of steel. But, sooner or later, the company will need to be able to make a profit from what is learned by being able to stamp out stronger steel fenders more quickly and cheaply.

As was mentioned in Chapter 1, the scientist's disinterest in engineering and technology can be seen partly as an effort to avoid responsibility for what is finally done with what he or she discovers. Happily, the number of scientists who are socially concerned over the proper directing of our technology is growing rapidly. This concern needs to be followed up with effective communication and action.

Public Attitudes Toward Science

The post-Sputnik adulation of science has faded. Science is under attack from many quarters. Paul Goodman says:

> Given the actual disasters that scientific technology has produced, superstitious respect for the wizards has become tinged with a lust to tear them limb from limb.[14]

The fund cutbacks we discussed earlier had their immediate fiscal causes, but they reflect a more subtle doubt which has been growing in America. The doubt amounts to a disaffection with the science-connected prosperity-and-progress paradigm which has been part of the American Dream for so long. After almost three decades of "progress" and "prosperity" somehow life is not sweeter. (It is easy to forget that life with less technology was, in many other respects, not sweet either.) Don K. Price says

> I suspect the current attacks on science come less from those who have always feared it than from those who were frustrated when they tried to put too much faith in it. To them, it was another god who failed.[15]

Herman Kahn and Anthony Wiener published a book several years ago in which they predicted a number of technological advances which would have been made by the year 2000. Among them, they list[16]

(1) Permanent manned satellite and lunar installations.

(2) Robots and machines "slaved" to humans.

(3) Artificial moons and other means for lighting large areas at night.

(14) *Utopian Essays and Practical Proposals*, Random House, New York, 1962.

(15) *Science 163*, 3 January, 1969, pg. 25. Copyright 1969 by the American Association for the Advancement of Science.

(16) Kahn, Herman and Wiener, Anthony, *The Year 2000: A Framework for Speculation on the Next 33 Years*, Macmillan, New York, 1967.

(4) Non-harmful methods of overindulging.

(5) Human hibernations for months or years.

(6) Programmed dreams.

(7) More reliable educational and propaganda techniques for affecting human behavior—public and private.

(8) New and possibly pervasive techniques for surveillance, monitoring and control of individuals and organizations.

(9) Extensive and permanent cosmetological changes (features, skin coloring, physique).

This list might have excited us once, but we've already had too many such "improvements" to expect them to solve many of the serious human problems existing right now. Educated persons in our society should certainly be aware of these and other technological possibilities in order to take responsible action, both in opposing programs which they feel to be detrimental to society; and, in supporting those which are perceived as being beneficial. There is still a large gap between what scientists and technologists ought to be doing to relieve human misery, and the projects that have actually been worked on. Hopefully, some of the caution and re-examination regarding the support of science and technology that has been taking place will result in a closer match between human needs and scientific investigation.

Some of the more vehement criticisms of technology might arise from the fact that, although it produces more goods and food for us, it also can disrupt and destroy basic cultural patterns. Our culture is in flux. Historically, times of change have coincided with periods of witch hunting.[17] And as Goodman suggested above, wizards can easily become witches. When discussing our "runaway" technology, the thing to remember is that scientists were among the first to call our attention to the dangers of pollution and over-population, and the need for technological assessment. Without the application of science and technology, we're not going to be able to handle these and other national and world problems. It would be extremely foolish to let Neo-romantic longings for a departed non-technological world keep us from employing every scientific tool possible to eliminate some of the evils which technology and human bad taste have produced.

We have already mentioned in Chapters 1 and 4 that the analytical approach which science uses has drawn criticism for years from those who believe in a more organic or holistic view of the world. These critics claim that by chopping the world up into measurable pieces, beauty, emotion, and all that is human is destroyed. Actually, the numerical way of viewing nature is no worse than any other, so long as no one claims that

(17) White, Lynn T., Jr., "The Necessity of Witches", *Chapter 11 in Machine ex Deo: Essays in the Dynamism of Western Culture*, M.I.T. Press, Cambridge, 1968.

it is the *only* way. But there are inherent dangers in the analytical approach. Followed blindly, it leads to fragmentation of knowledge — an over-specialization — in which people ignore the connections between various views of the world. Above all, it leads to a disregard for the societal consequences of technology — the primary problem being discussed in this chapter.

Physics and Human Spirit

The previous section of this chapter might well have been titled "Down With Science". We concentrated on the problems and stresses which have come with science-based technology, and did not bother to enumerate the medical, educational, transportational, and other *benefits* which have accrued. In this section, we hope to pull together thoughts which have been mentioned in various places in the book thus far, along with some new ones, to show that science has, or ought to have, a salutory effect on the human spirit.

Although scientists are just men and women, when they are "sciencing" they need and use personal characteristics such as diligence, discipline, rationality, and honesty. The same could be said for many other human activities, but in the case of science, it is the *main* thing, putting monetary profit and rhetoric in the background. We have commented frequently upon the physicist's search for accuracy and unambiguous communication, the attention to *meaning* which our age needs so badly. Physicists are dedicated to order. They believe the world is *not* absurd.

As a group, physicists and other scientists have played a significant role in fostering world understanding. Science, like music, is a nearly international language. From a time prior to the dropping of the first nuclear bomb, scientists have worked toward limitations in the use of nuclear arms. They were instrumental in establishing the atoms-for-peace program. The annual Pugwash conferences, bringing together American, European, and Asian scientists have provided a prolonged thrust toward arms limitation and mutual trust. Copenhagen, Geneva, and Trieste have become genuinely international research centers where scientists of every political stripe work together. Russia and America are currently competing for the first controlled fusion reaction, but the secrecy on both sides is far less than when fission research was at a similar stage in the 40's.

Moving to the more aesthetic side, we have seen that physics, especially relativity, has drastically altered the way we look at our surroundings. It has been a source of inspiration to authors, painters, and musicians who are excited by new ways of exploring space-time relationships. The basic way in which physics describes the world has, in the last century, become less mechanistic and more aesthetic, using principles of balance and cooperation.

But how many people realize this? Arthur Koestler has said, "Our hypnotic enslavement to the numerical aspects of reality has dulled our perception of non-quantitative values."[18] Is this the fault of science? Who can determine which parts of science or technology John and Jane Doe decide to be affected by? They (we) are bowled over by "fact" and number. We leave a friend and rush to answer the telephone. Technology is a rational scheme for giving us the free time to do irrational (fun) things. So we choose degraded television programming and snowmobiles. It would be difficult to decide whether technology has produced our attitudes or our attitudes have produced our technology. In either case, awareness of what our attitudes really are should help. And it might also be helpful to know the attitudes are not a necessary result of the scientific process.

Summary

Of 400,000 scientists in America about one in twelve are physicists. More than half are employed by educational institutions and about one-third by private industry: 80 percent receive government-supported salaries.

Physics is a very social profession.

Physics, along with the rest of American science, started to grow rapidly in the 1950's when the Federal government began investing in aerospace research and development. At the beginning of the 1970's government expenditures for research and development leveled off at about twelve billion dollars per year. (See Fig. 10-1). Private investment had risen more gradually to about 10 billion, and educational institutions had spent about one billion per year. About one-fifth of the total amount was for basic research.

Science is thus a significant part of American economic life as well as cultural, educational, environmental, military, and social life. Questions of scientific research priorities and who should decide these priorities have become important political issues.

Science and engineering/technology have distinctly different purposes. However, they have led a symbiotic existence for centuries. Public attitudes toward science, especially the technology which it engenders, have cooled since the 1960's. Even the analytical approach which science encourages has received sharp criticism from certain Neo-Romantics. At the same time, science and scientists are becoming more humanistically aware of the dangers of having too narrow a view. Recently, they have been among the foremost to point out the dangers posed by unbridled technological development, and among the first to bridge the world's military and political abysses.

(18) *The Sleepwalkers*, Macmillan, New York, 1959.

Barbour, Ian, *Science and Secularity: The Ethics of Technology,* Harper and Row, New York, 1970.

Bevan, William, "The Welfare of Science in an Era of Change", *Science 176,* 2 June, 1972, pg. 990. (A wide-ranging article, pulling together many public attitudes toward science.)

Bohm, David, "Fragmentation in Science and in Society", *Impact of Science on Society XX,* April-June, 1970, pg. 159. (A physicist claims that science and technology both reflect a strong human tendency: fragmentation. We need to begin thinking in terms of wholes.)

Burke, John G., *The New Technology and Human Values,* Wadsworth Publishing Co., Belmont, California, 1966. (A collection of short pieces on science and society. Note especially part IV, on science, technology and policy making.)

Commoner, Barry, *Science and Survival,* Viking Press, New York, 1967. (An early ecological warning. Chapters six and seven are still one of the best discussions of the dilemma facing the scientist who feels a social concern.)

David, Douglas M., "Art and Technology—The New Combine", *Art in America,* Jan/Feb., 1968, pp. 29–45. (Includes an interview with Gyorgy Kepes, founder of the Center For Advanced Visual Studies at M.I.T.)

Ferry, William H., "Must We Rewrite the Constitution to Control Technology?", *Saturday Review 51,* March 2, 1968. (The list of things we can do but *must not* do is growing.)

Greenberg, Daniel S., *The Politics of Pure Science,* The New American Library, New York, 1967. (Controversial but factual and widely quoted.)

Holton, Gerald, "Modern Science and the Intellectual Tradition", *Science 131,* April 22, 1960, pp. 1187–93. (A masterful and widely quoted discussion of scientific scholarship and its separation from the rest of culture.)

Meadows, Donella H.; Meadows, Dennis L.; Randers, Jørgen; and Behrens, William W., *The Limits of Growth,* New American Library Signet Book, N.Y., 1972. (A widely discussed book which predicts world-wide doom in the twenty-first century unless we control population, food production, air pollution, economic development and consumption of natural resources simultaneously.)

Oppenheimer, J. R., "Physics in the Contemporary World," pp. 188–204 in *Great Essays In Science,* Martin Gardner, Ed., Washington Square Press. (Discusses contributions of physics to human thought.)

Rickover, Hyman G., "A Humanistic Technology", *Nature 208,* pp. 721–726, Nov., 1965. (We have the intelligence and opportunity to use technology to further human rights and meet human needs.)

Rose, Steven and Rose, Hilary, "The Myth of the Neutrality of Science", *Impact of Science on Society XXI,* April/June, 71, pg. 137. (The nature and direction of research is always framed by the social and scientific context of its time.)

Schroeer, Dietrich, *Physics and Its Fifth Dimension: Society,* Addison-Wesley, Reading, Mass., 1972. (A series of short comments on social problems in science, with many suggested readings. See especially Chapter 1, on the two cultures. Chapters 3 and 4 on the growth of science as an estate; Chapters 16, 17, 18, and 21 on science and politics; and Chapters 24 and 25 on the moon race and anti-ballistic missiles also bear on topics discussed in this chapter.)

Skolnikoff, Eugene B., "Technology and the Future Growth of International Organizations", *Technology Review 73,* June, 1971, pg. 39. (A sober examination of the implications of technology for international political action.)

Starr, Chauncey, "Social Benefits vs. Technological Risks", *Science 165,* pg. 1232, Sept. 19, 1969. (How much are we willing to pay to be environmentally safe?)

Ziman, John, "The Impact of Social Responsibility on Science", *Impact of Science on Society XXI,* April/June, 1971, pg. 113. (Uniting passion of spirit with cool rationality, scientists must form the loyal opposition against the irresponsible use of science.)

Problems

10-1 What ingredients do you think are necessary for science to flourish in a society?

10-2 Should a society support a certain number of scientists in comfort? Artists? Physicians? Beauticians? Soldiers? Why?

10-3 What do you think is a reasonable size for science, in terms of budget? What would you see as a minimum role for science?

10-4 Which adds more to human culture; research into particle physics, which studies the basic structure of matter; or flights to the moon (or Mars)?

10-5 Some people have said that the real reason Congress vetoed the supersonic transport was because it feared that if a test model were built, the project would continue, no matter what the results of the test were. What does this suggest to you?

10-6 Do you think that technology increases or decreases your freedom as an individual?

10-7 Which do you think can lead to greater infringements on human freedom; the social sciences, or the physical sciences?

10-8 In an article published in 1963, Bertrand de Jouvenel suggested that the age of science is also an age of great personal ignorance. If there's more to know, then we end up knowing a smaller and smaller fraction of the total knowledge that man possesses. Do you agree? Where do you think knowledge stops and wisdom begins?

10-9 How much technology: communication, transportation, refrigeration, music reproduction, etc. would you be willing to give up in order to protect the environment?

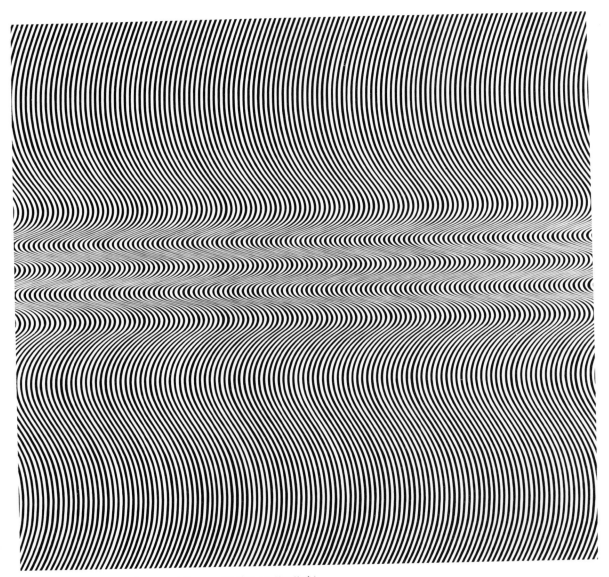

Bridget Riley, *Current*, 1964. *(Courtesy of Museum of Modern Art, New York.)*

FIELDS AND WAVES

11
ELECTRICITY

From at least as early as 500 B.C., people have noticed (and recorded) that certain materials can exert odd forces on other objects around them. After it was rubbed with fur, a piece of amber would attract bits of grass and dust. Pieces of a certain kind of iron ore, called magnets because they were commonly found in Magnesia, attracted other magnets with forces that were strong enough to feel. (In this case, no rubbing had to be done.) The words "electricity"—based upon the Greek word for amber, *elektron*—and "magnetism" came into Western culture quite early. However, the rapid growth of science which put mechanics on solid footing during the seventeenth and eighteenth centuries left electricity amd magnetism lagging, perhaps because they could not be explained very well on a mechanical basis. As the nineteenth century approached, they were still curiosities; material for spectacular lecture demonstrations, but poorly understood.

In this and the next chapter, some electric and magnetic phenomena are discussed, and the development of our present understanding of them is traced from about 1750. In so doing, the concept of *field*, already touched upon with regard to gravitational forces, is discussed in more detail. In turn, this will enable us to do two things. First, we will have an opportunity to examine some of the similarities and differences between the physical forces which hold the world together and make it move. Second, we will continue to move away from (or supplement) the mechanistic point of view which has dominated our discussions thus far. At the end of Chapter 4, we said that the mechanistic world view sought to explain everything in terms of clockwork: hard, unchanging little particles connected by forces, undergoing known pushes and pulls. The discussion of nuclear phenomena in Chapter 9 and of Einstein's special theory of relativity constituted a softening of the mechanistic point of view, insofar as the physical world was no longer seen as being comprised of permanent particles and external energy. The line between matter and energy became fuzzy, so that the little particles were no longer so permanent or well-defined. But, historically speaking, the mechanistic approach was in trouble long before special relativity made the scene. Its troubles began when people tried to explain how it is that two objects *not touching* manage to exert gravitational, electrical or other forces on each other.

Let us begin by looking at some basic electrical phenomena.

Electrification

What is it that happens when you rub amber with a piece of fur — or a hard rubber comb with your hair — or a glass rod with a silk cloth? The amber, comb, and the glass rod all will now pick up bits of paper or cork dust. It is said that they have become "charged" or electrified, which of course just raises other questions. Unlike magnetism, charge (whatever it is) can easily be removed, just by touching the charged object to the ground, or to an object like a metal water pipe, which is in contact with the earth.

There seem to be two classes of charged objects. Those within one class repel each other when charged, but attract objects in the opposite class. This led mid-eighteenth century scientists to believe that there must be two kinds of electric charge at work. However, if you look closely enough into charging phenomena, you notice that objects are always charged *in pairs*, and that after charging, one always lies in one class and the other in the opposite class. Benjamin Franklin (1706–1790), Colonial America's "Renaissance man", noticed these things and in 1747 proposed a simple theory; that charging involves transfer of just one kind of electric charge. When a pair of objects is charged, one loses some charge and the other gains some charge. *Charge cannot be created or destroyed.*[1]

(1) Not only is charge conserved, but it is invariant. The mass of an object depends on the frame of reference from which it is viewed, but experiments indicate that the electric charge on an object measures the same in all frames.

One class of charged object is called "positive" and the other "negative," although the words are arbitrary. Today, by convention, the charge which is transferred is called negative. A positively charged object is simply one which has lost some negative charge. Different materials lose or gain charge in varying degrees. In general, you have best luck in getting things to become charged if you bring together a material which gains charge easily, with one which loses charge easily. You don't have to rub them. For example, a piece of cellophane tape stuck onto and stripped rapidly from a metal surface becomes negatively charged and leaves the metal positively charged. The diagram below shows what is going on, according to the model we've used thus far:[2]

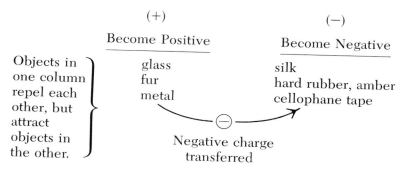

The charging process does not *create* charge. It just *transfers* charge from one object to another.

When two charged objects are touched together they will share their charge, if there is a relative excess or deficiency on one or the other of them. (Without their touching one another, charge can also jump between them by means of a spark.) This means, for example, that an uncharged object touched by a positively charged object tends to become positively charged itself, as shown in Figure 11-1. According to our model, it does this by giving some negative charge to the other object.

If two objects having a fairly equal amount of charge are brought in contact, they will tend to neutralize each other electrically if the charges are of opposite sign (Figure 11-2). The positively charged object, lacking negative charge, picks up the charge it needs from the other object. To be sure of fully discharging an object, it need only be "grounded" by touching it to a water pipe[3] or faucet which connects ultimately with the earth.

(2) Positive charge transferred from right to left would produce the same effect. Historically, Franklin arbitrarily chose moving positive charges. However, experimental evidence on page 273 supports the idea that the charge transferred *is* negative.

(3) Gas pipes should never be used for grounding electrically charged objects because of the possibility that a spark might ignite leaking gas.

The earth acts as a giant sponge for electric charge, capable of accepting or giving large amounts of charge without being much affected itself.

As far as we're concerned, it is the *motion* of charge which we exploit and which we need to be careful about. When cars travel along the highway they tend to become negatively charged. For this reason, tollbooths on turnpikes are fitted with a wire sticking up from the pavement, so that when a car approaches the booth, its charge can escape directly to the earth as it touches the wire. No charge will then jump from the coins you hold to the earth through the bodies of the attendants. This makes them happier to see you coming.

One other electrical characteristic which distinguishes various materials is their ability to *conduct* charge. In some materials, charge placed at one end of the object will stay where it's put, whereas in others the charge seems free to roam over large distances within the object. We call such materials *insulators* and *conductors*, respectively. Since similarly charged objects repel each other, you would expect that charge in a conductor would spread itself as thinly and uniformly as possible, like Britishers in a park. For this reason, when people are given a large electric charge (not harmful if the charge doesn't flow through them to ground), their hair stands on end. Since the hairs are slightly conductive, the charge spreads over them. The re-

Figure 11-1
An excess (or deficiency) of negative charge is shared.

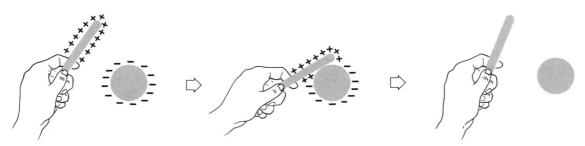

Figure 11-2
Opposite charges tend to neutralize.

pulsive forces within the charge on one hair, as well as between the similarly charged hairs tends to straighten out the hairs.

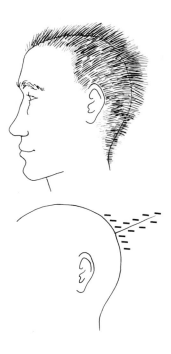

What Is Charge?

Up to now we have avoided visualizing just what the stuff is that is transferred during processes of electrification. Between 1906 and 1911 the American physicist Robert A. Millikan (1868–1953) developed a very delicate means for studying the charging process, and with it, the nature of charge itself. He knew that when oil was sprayed from an atomizer, the droplets became negatively charged. He then contrived to catch some of the droplets between two horizontal parallel plates, the upper one positively charged and the lower one negatively charged. With this arrangement, the negatively charged droplets felt an electrical force F_E upward which balanced the force of weight downward. (They are attracted to the positive (+) plate and repelled by the (−) plate.) As time passed the droplets lost charge to the surrounding air, and would start to fall, since the electrical force up was less. Millikan was able to show that the electrical force, hence the charge on the droplets, did not decrease smoothly, but in jumps! (This had already been shown by the British physicist Joseph J. Thomson and the Irish physicist John S. E. Townsend, but Millikan's set-up permitted a far more accurate measurement.) He found that the jumps always corresponded to a whole number times a basic quantity of charge. Therefore, he concluded that electric charge does not behave like a smooth fluid, but rather, like fine-grained particles. The basic "grain" of charge on these particles amounted to -1.6×10^{-19} Coulomb. (We shall

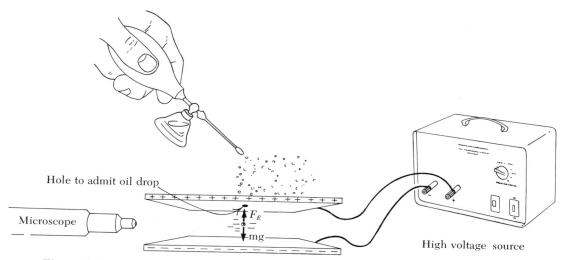

Hole to admit oil drop

Microscope

F_E

mg

High voltage source

Figure 11-3

Diagram of apparatus used in Millikan's oil drop experiment.

see in the following section what a coulomb of charge is.) The "electron" which bears this negative charge is the same electron spoken of on page 228. It possesses a rest mass of 9.1×10^{-31} kg, as well as carrying an electric charge.

Millikan's discovery of the lumpiness of electric charge, along with growing evidence that matter was made up of atoms, was just the beginning of a flood of evidence indicating that the smoothed-out, continuous, large-scale world we perceive is discontinuous on a small scale.

Coulomb's Law

We have said that similarly charged objects repel and that oppositely charged objects attract each other. What is the magnitude of the electrical force between them? How does it depend on their distance apart? In the 1780's, Charles A. Coulomb (1736–1806), a French scientist, began investigating these problems quantitatively, using a balance similar to the one Cavendish used to measure the gravitational constant (page 112). He found that the force, F_E, between two spherical bodies with electrical charges q_a and q_b (in coulombs) on them decreased in proportion to the inverse square of the distance, r, between them; just as in the case of gravitational forces.

$$F_E = \frac{Kq_aq_b}{(r_{ab})^2} \qquad \text{Equation (11-1)}$$

As in the gravitational force equation, r_{ab}^2 is the square of the distance between the *centers* of the charged objects. Coulomb was not able to measure q_a and q_b precisely, so it was not until later that the value of the constant K was determined. K was found to equal 9×10^9 newton meter^2/coulomb^2. The two objects need to be very small, or else spherical, for this relation to be true. The $1/r^2$ dependence is not peculiar to gravity and electricity. It's true of all physical influences which spread out evenly in all directions from a point or ball-shaped source.

From Equation (11-1), we can obtain a definition of the coulomb: If two spherical objects, each bearing a coulomb of charge, are placed with their centers one meter apart, they attract (or repel) each other with an electrical force equal to 9×10^9 newtons. (What will be the magnitude of the force if the centers of the objects are two meters apart?)

When we've calculated the magnitude of the electrical force between two charged objects, how do we know whether it's an attraction or a repulsion? We have said that opposite charges attract. Suppose we assume that q_a equals −3 coulombs and q_b equals +4 coulombs. The product (q_aq_b) equals −12, a negative number. The F_E will be negative. Experiment has shown that

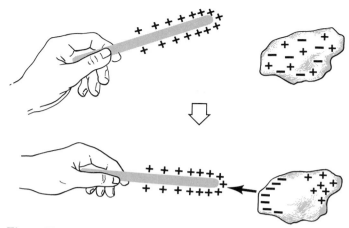

Figure 11-4

*An uncharged object may experience a net electrical force
toward a charged object.*

opposite charges attract. Therefore, a negative force is attractive.
It "tries" to reduce the distance between the objects. (Recall that
on page 191 it was implied that a negative relative velocity corresponded to two things approaching each other, so that the distance between them decreased.) If q_a were to equal +3 and q_b
equaled +4; or if q_a were to equal −3 and q_b equaled −4, the
product $(q_a q_b)$ would equal +12, in either case. Like charges
produce positive forces and therefore, positive forces produce
repulsions.

We began this chapter by saying that amber which has been
rubbed with fur picks up bits of grass. Does this mean that the
grass is charged oppositely to the amber? Not necessarily. Suppose a positively charged rod is placed near a small neutral object. While the rod is nearby, negative charge will tend to be
pulled to the near end of the small object, leaving the far end
positively charged. The rod will then exert an attractive force on
the negatively charged near end of the object and a repulsive
force on the positively charged far end. But from Coulomb's law,
the repulsive force on the far end will be less than the attractive
force on the near end. A net attractive force acts on the object in
spite of its overall electrical neutrality. It follows from this, that
if you wish to show that an object is charged, you must show that
it is *repelled* by another charged body.

Sample Problem

Suppose a charged object, q, is close to *several* other small,
round charged objects. The total force on q will simply be the

sum of the electrical interactions that q would have with the other objects, taken one at a time. To be specific, suppose bodies bearing charges of -3×10^{-6}, $+3 \times 10^{-6}$ and $+5 \times 10^{-6}$ coulombs (abbreviated C) are arranged in a line, as shown.

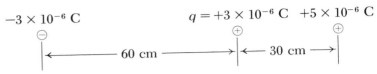

-3×10^{-6} C $\ominus$

$q = +3 \times 10^{-6}$ C $\oplus$ $+5 \times 10^{-6}$ C $\oplus$

$\longleftarrow$ 60 cm $\longrightarrow$ $\longleftarrow$ 30 cm $\longrightarrow$

The problem: find the direction and magnitude of the electric force acting on q.

(a) The force on q due to the left-hand body is

$$F_L = \frac{K(-3 \times 10^{-6}\text{C})(+3 \times 10^{-6}\text{C})}{(.6 \text{ m})^2}$$

$$= \left(9 \times 10^9 \frac{\text{N·m}^2}{(\text{C})^2}\right)\left(\frac{-9 \times 10^{-12}(\text{C})^2}{.36 \text{ m}^2}\right)$$

$$= -\frac{81 \times 10^{-3}}{.36} \text{ N} = -225 \times 10^{-3} \text{ N}$$

$$= -2.25 \times 10^{-1} \text{ N}$$

$\longleftarrow \underset{F_L}{\oplus}$

The negative force, attracting q toward the left-hand body, points to the left.

(b) The force on q due to the right-hand body is

$$F_R = \frac{K(5 \times 10^{-6}\text{C})(3 \times 10^{-6}\text{C})}{(.3 \text{ m})^2}$$

$$= \left(9 \times 10^9 \frac{\text{N·m}^2}{(\text{C})^2}\right)\left(\frac{15 \times 10^{-12}(\text{C})^2}{.09 \text{ m}^2}\right)$$

$$= +\frac{9 \times 15 \times 10^{-3}}{.09} \text{ N} = +1500 \times 10^{-3} \text{ N} = 1.5 \text{ N}$$

$\longleftarrow \underset{F_R}{\oplus}$

This positive force repels q from the right-hand object, so it also points toward the left.

Remember: the positiveness of the force means that it is repulsive, not that it *necessarily* points toward the right, as has been the convention, thus far, in handling forces. The total force acting on q is thus:

$$F_E = F_L + F_R = (-2.25 \times 10^{-1} \text{ N}) + (-1.5 \text{ N})$$

$$= (-.25 \text{ N}) + (-1.5 \text{ N})$$

$$= -1.725 \text{ N}.$$

In *this* instance, the minus signs in the parentheses tell us that the forces both point left.

The Electric Field

If a system contains many charges the vector addition of the electric forces acting on a charged object soon becomes a tedious job, especially if the charges don't lie in a straight line. An alternative approach is to concentrate on and measure the *total* force acting on the charged object right from the beginning. If the 3×10^{-6} coulomb charge in the example above were removed and replaced by an object bearing a charge of 9×10^{-6} coulomb, the electric force on the new object would be three times as great, or 5.175 N. The electric force on a charge q placed at a certain point in space will always be proportional to q. (Assuming, of course, that the charge q is not so large as to distort the arrangement of charges setting up the field.) If we now divide the electric force acting on a charge q at a point in space by the charge q, we get a number which is independent of the magnitude of q. From the problem above, for example,

$$\frac{F_E}{q} = \frac{1.725 \text{ N}}{3 \times 10^{-6} \text{ C}} = 5.75 \times 10^5 \text{ N/C}$$

We could put other charges into the same place, but as long as the rest of the charges in the system don't change, the electric force per unit charge will remain at 5.75×10^5 N/C. We call the ratio of the electric force on an object to the charge on the object the *electric field intensity,* $\mathscr{E}$,[4] at the point in space:

$$\mathscr{E} = \frac{F_E}{q}$$

Equation (11-2)

It is analogous to the gravitational field intensity, g (page 113). The fact that it depends only upon the point in space itself and not on the charged object that we happen to put there draws our attention away from interactions between pairs of charged objects and toward the space around the object itself. Every point in space where an electric field exists has associated with it a vector $\mathscr{E}$ whose direction and magnitude indicates the direction of and the amount of force exerted per unit charge on a charged object placed at that point.

From all of this, we can distill operational definitions for the electric field and its intensity:

(1) If a charge at rest at a certain point in space experiences a force of electrical nature, an electric field, $\mathscr{E}$, is said to exist at that point. By convention, the direction of the electric field is the direction of the force exerted on a *positive* charge at the point (Figure 11-5).

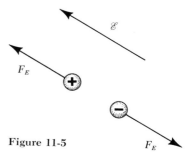

Figure 11-5

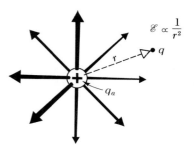

Figure 11-6

The electric field of a positive point (or spherical) charge points outward. Its intensity decreases in proportion to the square of the distance from the center of the charge.

---

(4) We use the script $\mathscr{E}$ to distinguish from the E previously used to symbolize energy.

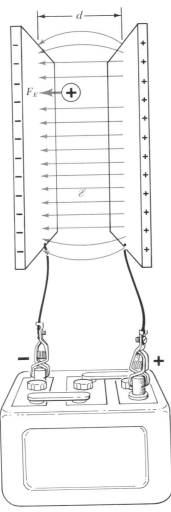

Figure 11-7
The electric field intensity $\mathscr{E}$ is uniform between two large parallel sheets (except near the edges).

A negatively charged object will experience a force in a direction exactly opposite to the electric field.

(2) The *intensity* of the electric field is equal to the force divided by the charge, or the *force per unit charge*.

Electric fields can be uniform, that is, the intensity can remain the same over large distances; or they may be inhomogeneous, in which case the intensity varies with position. It all depends upon the positioning of the charges which produce the electric field. For example, a single positive charge, q_a, produces a field which points radially outward, and which decreases in intensity as the square of the distance from the charge. We know this because the force exerted on q by q_a is $F_E = Kq_aq/r^2$, and the field intensity $\mathscr{E} = F_E/q$ becomes equal to Kq_a/r^2. Fields can be visualized by means of imaginary "lines of electric force." By convention, these lines of force point out from positive charges and in toward negative charges. That is, they show the direction of the force which would be exerted on a positive charge. The distance between lines, or their density, also gives a sense of the magnitude of the field intensity. Note that the lines are closer together nearer to the central point charge, where the field is stronger. Note also that a charged sphere with smaller radius, r, has a higher electric field intensity at its surface. Pointed, charged objects, for which the effective radius is quite small near the point, can produce very strong and concentrated electric fields.

Uniform electric fields can be produced by placing two large conducting sheets parallel to each other and connecting one terminal of a battery to each of the plates. The battery will "pump" charge away from one sheet and onto the other. As was stated previously, the charges will arrange themselves evenly over the sheets, except at the edges, where some crowding and distortion of the electric field occurs. A positive charge put in the space between the sheets will be repelled from the right sheet and attracted toward the left sheet of Figure 11-7. Thus, the $\mathscr{E}$ field between the sheets must point from right to left. It can also be shown that the electric force on the charge is the same anywhere between the sheets, as long as it is not close to the edges of the sheet. The field intensity, the force per unit charge, will be uniform. The lines of the electric field will be parallel and have equal spacing everywhere between the sheets.

If a point charge, or a small charged ball, is placed near a large conducting plate, the electric field shown in Figure 11-8 results: The plot indicates that a positive charge leaving the metal plate may not be forced toward the negatively charged ball along a straight line. Its acceleration would not be constant either, since the field intensity is greater closer to the ball.

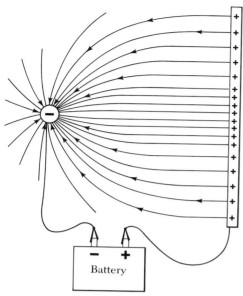

Figure 11-8
*The arrows show the direction of the electric
field at various points. That is, they show the
direction of the electric force which would be
exerted on a positively charged object placed
at those points.*

Electrical Potential and Energy

Since electric forces act on charged bodies in electric fields,
physical work has to be done in order to move them around. The
situation is quite analogous to lifting a mass in a gravitational
field, with the added complication that the electric force you
have to overcome might be in the direction of the field or exactly
opposite to it, depending on whether the charge on the object is
positive or negative. Imagine an object bearing a positive charge
q, placed in a uniform electric field that slants down toward the
left (Figure 11-9). ("Down", has gravitational connotations which
don't apply here, but it's a hard word to avoid.) Let $F_E = q\mathscr{E}$ be the
electric force exerted on the object by the field. Then let $F = -F_E
= -q\mathscr{E}$ be the force that you exert on the object in an effort to
move it in the electric field. This force is just large enough to
keep the object moving at a constant velocity along a path of
length Δx, which slants "up" against the electric field. The work
done by the force F really depends only on Δy, the distance the
object is moved in direct opposition to the electric field.[5] (Just

(5) The use of Δy here is consistent with the use in Chapters 3 and 7. The field
there was gravitational, and Δy was always vertical. Here, electrical "up" and
Δy can slant, or be in any direction relative to our local gravitational field.

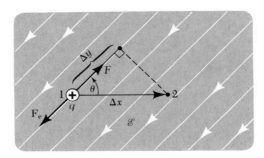

Figure 11-9
*F_e indicates the direction of the force exerted
on the positive charge q. F indicates the
external force applied to q, doing work on it.*

as the work done on a massive object in a gravitational field
(page 158) depends on how high you lift it.) Therefore, ΔW, the
work done, becomes

$$\Delta W = (F_{avg})(\cos \theta)(\Delta x) = (F_{avg})_x(\Delta x)$$

$$= (-q\mathscr{E}_{avg})(\cos \theta)\Delta x \qquad \text{Equation (11-3)}$$

$$= (-q\mathscr{E}_{avg})(\Delta y)$$

If $\mathscr{E}$ is uniform ($\mathscr{E} = \mathscr{E}_{avg}$) the expression becomes

$$\Delta W = -q\mathscr{E}\Delta y.$$

In other words, the work done, or the change in potential energy
is equal to minus $q\mathscr{E}\Delta y$.

If we substitute an object having twice as much charge on it,
twice as much work must be done in order to move it the same
distance in opposition to the electric field. The increase in potential energy would also be twice as great. To obtain a quantity
which depends only upon the field and the distances moved, we
can again divide by the charge on the body:

$$\frac{\text{work}}{\text{charge}} = \frac{\Delta W}{q} = \frac{\Delta(P.E.)}{q} = -\mathscr{E}_{avg}\Delta y \qquad \text{Equation (11-4)}$$

The quantity $\Delta(P.E.)/q$ is given a special name. It is called the
electrical potential difference, ΔV, between the points where
Δx begins and ends.

$$\Delta V = V_2 - V_1 = \frac{\Delta(P.E.)}{q}. \qquad \text{Equation (11-5)}$$

It is measured in *volts*, (V). When a joule of work is required to
move a coulomb of charge between two points in a field, the potential difference between the two points is said to be 1 volt. Or:

1 volt = 1 joule/coulomb

Since charges on the order of one electronic charge are often moved around in physical systems, a much smaller unit of energy, *the electron volt, eV*, has come into use. It is the work required to lift 1 electronic charge (1.6×10^{-19} coulombs) through 1 volt of electrical potential difference.

$$1 \text{ electron volt} = (1 \text{ volt})(1 \text{ electronic charge})$$

$$= (1 \text{ volt})(1.6 \times 10^{-19} \text{ coulombs})$$

$$= 1.6 \times 10^{-19} \text{ joules,}$$

since

$$1 \text{ joule} = (1 \text{ volt})(1 \text{ coulomb}).$$

In the nuclear reactions discussed in Chapter 9, the energies are normally expressed in million electron volts (meV) rather than in joules:

$$1 \text{ meV} = 1.6 \times 10^{-13} \text{ joules.}$$

Equation (11-5) is always true, no matter what the nature of the electric field. If the field is uniform, then from Equation (11-4) we obtain:

$$\Delta V = -\mathscr{E}_{avg}\Delta y$$

$$= -\mathscr{E}\Delta y$$

or

$$\mathscr{E} = -\frac{\Delta V}{\Delta y} \qquad \text{Equation (11-6)}$$

The minus sign in Equation (11-6) just reminds us that the field points downhill from high potential to low (Figure 11-10).

If the electrical potential, V, is analogous to height in a gravitational field, then the field intensity, $\mathscr{E}$, represents the *downward slope* of the potential. $\mathscr{E}$ tells us how rapidly V is changing with distance (not time). $\mathscr{E}$ can therefore be measured in units of volts/meter as well as in newtons/coulomb. One unit emphasizes the idea of force per unit charge. The other unit emphasizes the change in potential per unit distance. The two sets of units are equivalent. (Can you show this to be true?)

One more thing should be said to help make clear the distinction between electrical potential and potential energy at a point in space.[6] From Equation 11-5, $(\Delta V = \Delta(P.E.)/q)$, the difference in potential energy of a body between two points in an electrical field is equal to the electrical potential difference between the two points times the charge on the body: $\Delta(P.E.) = (\Delta V)q$. From this, the potential energy of a body at a single point in an elec-

(6) When you speak of the electrical potential, V, at a point in an electric field, you are focusing your attention on the point itself. When you speak of the potential energy of a charged object, qV, at the same point, you are including that particular object in your attention as well.

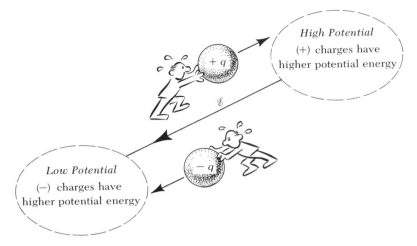

Figure 11-10
By definition, a region of high electrical potential is one where positively charged objects have high potential energy. Positive work must be done on positively charged objects to get them there. A negatively charged object will have low potential energy in a region of high potential!

tric field can be spoken of as $P.E. = (V)q$. Operationally, it represents the work required to bring an object bearing the charge q from a place very far away to the point in the electric field. Note that a positively charged body will possess high (positive) potential energy when it is in a region of high potential, but that a negatively charged body will have a low (negative) potential

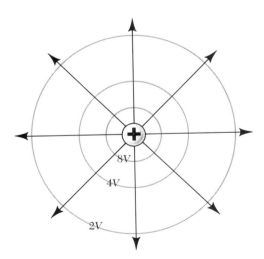

Figure 11-11
The equipotential surfaces for the field of a point (or spherical) charge are spherical. A small charged object will have the same potential energy everywhere on a given equipotential surface.

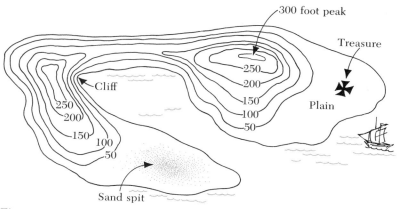

Figure 11-12
Electrical potential is something like the contours on a topographic map.

energy at the same point. This makes sense when you realize
that the electrical forces at work pull negative charge toward
regions of high potential, but that positively charged bodies have
to be *lifted* to regions of high potential against the prevailing
electric forces. If you have to do work on something to get it
somewhere, it will have higher potential energy there.

From Equation (11-3), $(\Delta W = -q\mathscr{E}_{avg} \cos \theta \, \Delta x)$ it can be seen
that it is possible to move a body in an electric field along a
path, such that θ is always 90 degrees, or $\cos \theta = 0$. Along such a
path, the work done on the body will be equal to zero; its electrical potential energy will remain constant. Such paths or surfaces are called *equipotentials*. All points on them are at the same
electrical potential or voltage. For any electric field, equipotential lines or surfaces can be constructed. They will be everywhere
perpendicular to the electric field. The equipotentials for a
point-charge electric field are shown as shaded lines in Figure
11-11. They are concentric circles. If a real spherical object were
placed so that its surface coincided with a four-volt equipotential
surface in Figure 11-11, then all parts of its surface, and the object itself, would be at a potential of four volts. Physically, the
equipotential lines of an electric field are analogous to the lines
of constant altitude shown as contours on maps of the earth's
surface as illustrated by Figure 11-12.

Electrical and Gravitational Analogies

In several instances we have used gravitational analogies to help
visualize electric field concepts. Table 11-1 presents a concept-
by-concept comparison of gravity and electricity, showing that
once you have learned to understand a concept in gravitational
field theory, you don't have to learn it a second time in electrical
field theory.

Table 11-1

Gravity Field	*Electric Field*

Intensity

Gravitational force per unit mass: $"g" = F_g/m$, measured in N/kg, or m/sec². For a *point*, or spherical mass,	Electrical force per unit charge: $\mathscr{E} = F_E/q$; measured in N/C, or volts/m. For a *point*, or spherical charge
$$g = Gm/r^2$$	$$\mathscr{E} = Kq/r^2$$
Page 113	Page 278
$$G = 6.67 \times 10^{-11} \frac{\text{N·m}^2}{\text{kg}^2}$$	$$K = 9 \times 10^9 \text{ N·m}^2/\text{C}^2$$
$"g" = 9.8$ N/kg at the earth's surface.	

Charge

Gravitational forces are always attractive. Gravitational charge appears to be of just one sign. There seem to be many "basic" lumps of mass: electrons, protons, neutrons, and others.	Electrically charged bodies can attract or repel. Although charging can be explained in terms of transfer of a single kind of charge, deficiencies as well as excesses in charge give two sorts of charged bodies. Until recently, it was thought that charge always comes in lumps amounting to 1.6×10^{-19} coulomb. (There is now a theoretical possibility of "quarks", fundamental particles having smaller lumps of charge.)

Potential Energy

ΔW, Work done $= \Delta(P.E.)$	$\Delta W = \Delta(P.E.)$
$$= mg\Delta y$$	$$= -q\mathscr{E}\Delta y.$$
Equation (7-4), pg. 158. (In a uniform gravitational field.)	Equation (11-3) pg. 280. (In a uniform electric field.)
Δy is the distance moved "up".	Δy is the distance moved in opposition to the field.

Potential

(Not mentioned thus far.)	$\Delta W/q = \Delta V$, measured in volts.
$\Delta W/m = \Delta(\text{grav. potential})$	Equations (11-4) and (11-5), pg. 280.
$$= \Delta\Phi$$	In a uniform field,
In a uniform field,	$$\Delta V = -\mathscr{E}\Delta y$$
$$\Delta\Phi = -g\Delta y;$$	$$\mathscr{E} = -\Delta V/\Delta y.$$
$$g = -\frac{\Delta\Phi}{\Delta y}$$	

Table 11-1 (*continued*)

Gravity Field	Electric Field
	Equipotential
A line, or surface, along which the gravitational potential energy is constant. That is, a line along which the altitude is the same. Always perpendicular to the maximum slope.	A line, or surface, along which the electrical potential energy is constant. That is, a line along which voltage is the same. Always perpendicular to the electric field.

Some More Problems

Two large metal sheets are placed parallel to each other and 6 cm apart. A 1.5 volt flashlight battery is then connected to the sheets as shown in Figure 11-13. A small wax ball, bearing a charge of -2×10^{-6} coulomb, is put in the space between the sheets.

(a) What is the electric force on the wax ball? First,

$$\mathscr{E} = \Delta V/\Delta y = 1.5 \text{ V}/.06 \text{ m} = -25 \text{ V/m}.$$

The field is directed from right to left between the sheets. Therefore, the electric force on the *negatively* charged ball will be toward the *right*. The magnitude of the force will be:

$$F_E = q\mathscr{E} = (-2 \times 10^{-6} \text{ C})(-25 \text{ V/m})$$
$$= 50 \times 10^{-6} \text{ J/m}$$
$$= 50 \times 10^{-6} \text{ N}.$$

(b) How much work must be done to move the wax ball from the right sheet to the left sheet? (Δx is negative because the ball will be moved toward the left. For the negatively charged ball this is like lifting it uphill.)

$$\Delta W = F\Delta x = -F_E\Delta x = (-50 \times 10^{-6} \text{ N})(-6 \times 10^{-2} \text{ m})$$
$$= (50 \times 10^{-6} \text{ N})(6 \times 10^{-2} \text{ m})$$
$$= 300 \times 10^{-8} \text{ N}\cdot\text{m}$$
$$= 3 \times 10^{-6} \text{ J}.$$

Alternatively,

$$\Delta W = (\Delta V)q = (-1.5 \text{ V})(-2 \times 10^{-6} \text{ C})$$
$$= 3 \times 10^{-6} \text{ J}$$

(c) If the wax ball has a mass of 0.5 gm, and it is released from rest at the left sheet, what will be its speed when it arrives at the right sheet?

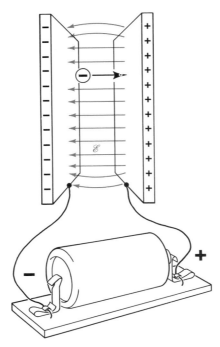

Figure 11-13
The force acting on a charged wax ball placed in a region where the electric field is uniform.

First, as is shown in (b) above, if 3×10^{-6} J of work is needed to move it from the right sheet to the left, the ball has a potential energy equal to 3×10^{-6} J when it is at the left sheet. (This is assuming a potential energy equal to zero at the sheet on the right.) As the ball accelerates;

$$-[\Delta(P.E.)] = +[\Delta(K.E.)] = 3 \times 10^{-6} \text{ J}.$$

$$\tfrac{1}{2}mv^2 = 3 \times 10^{-6} \text{ J}$$

$$v^2 = \frac{3 \times 10^{-6} \text{ J}}{(\tfrac{1}{2})(0.5 \times 10^{-3} \text{ kg})}$$

$$= \frac{3 \times 10^{-6} \text{ J}}{0.25 \times 10^{-3} \text{ kg.}}$$

$$= 12 \times 10^{-3} \text{ J/kg.}$$

This equals 12×10^{-3} m²/sec² since a joule equals one kg·m²/sec². Solving for v by taking the square root of this quantity gives:

$$v = \sqrt{12 \times 10^{-3} \text{ m}^2/\text{sec}^2} = \sqrt{1.2 \times 10^{-2} \text{ m}^2/\text{sec}^2}$$

$$= 1.09 \times 10^{-1} \text{ m/sec} = 10.9 \text{ cm/sec}.$$

(d) Now suppose we rotate the sheets so that they are horizontal, as shown in Figure 11-14, with the positively charged sheet six cm above the other and replace the battery with a higher voltage power

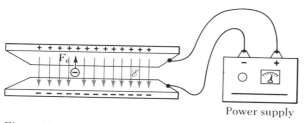

Power supply

Figure 11-14
*Electric force acts in the upward direction on the oil drop
and weight acts downward.*

supply. We now have an apparatus for doing the Millikan oil drop
experiment. Suppose that we have caught between the sheets an oil
droplet whose mass is 10^{-13} kg, and which carries 20 extra electrons.
Its charge will then be

$$q = 20(-1.6 \times 10^{-19}) \text{ C}$$
$$= -3.2 \times 10^{-18} \text{ C}.$$

What voltage must the battery furnish in order to just suspend the
oil drop between the sheets? We assume that the drop is not moving,
so that it experiences no air drag. First, we require the upward force
of the electric field F_E, to be equal to the downward weight, mg, of
the drop:

$$F_E = q\mathscr{E} = mg$$
$$= (10^{-13} \text{ kg})(9.8 \text{ N/kg}).$$

The requisite field;

$$\mathscr{E} = \frac{mg}{q} = \frac{9.8 \times 10^{-13} \text{ N}}{-3.2 \times 10^{-18} \text{ C}}$$
$$= 3.06 \times 10^5 \text{ N/C or } 3.06 \times 10^5 \text{ V}/m.$$

$\mathscr{E}$ should be written as negative since it points *down*. The requisite
potential difference, ΔV, is given by

$$\Delta V = -\mathscr{E}\Delta y,$$

where Δy equals 6 cm, or 6×10^{-2} m, the distance between the sheets.

$$\Delta V = -(-3.06 \times 10^5 \text{ V/m})(6 \times 10^{-2} \text{ m})$$
$$= +18.4 \times 10^3 \text{ V}.$$

Electric Current and Power

Work cannot be obtained from an electric charge unless the
charge moves. If work is done by electric generators in "lifting"
charge to regions of high potential (that is, in separating positive
from negative charge), then by conservation of energy, we should
be able to get that work back as useful energy by letting the

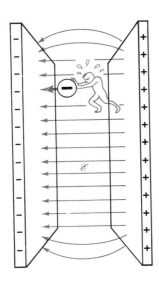

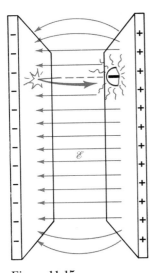

Figure 11-15

As negative charges are accelerated toward the positive plate, they lose potential energy, gain kinetic energy, and then deliver it to the positive plate in the form of heat.

charge "run back downhill" through controlled channels. Electrical engineering and electronics are almost entirely concerned with making charge move through such things as transistors, wires, or motors. In order to get some feel for the energy released when charge moves back to a place of lower potential, let us reconsider the wax ball between the parallel, electrically charged sheets on page 286. We said that work equal to $q\Delta V$ had to be done to move the ball from the right to the left sheet. Suppose we have gotten the charge (and ball) over to the left sheet, and have just let go of it. It will now accelerate back toward the right sheet, losing its electrical potential energy and gaining an equal amount of kinetic energy (Figure 11-15).

When the charge bangs into the right sheet; it is slowed down again, its kinetic energy is dissipated into heat, and the charge wanders slowly out the wire leading from the sheet back to the positive terminal of the battery. Now suppose we let N (N is just a whole number) such charges accelerate from the left to the right sheet every second. They no longer need to be attached to wax balls. The heat given to the right-hand sheet, per second, would then be equal to $q(\Delta V)N$. This can be called the *power* dissipated at the right sheet. (see page 172).

At this point it is useful to bring in a concept representing the "traffic flow" of charge through the space between the sheets. Such a flow of charge is called an electric *current*, symbolized by the letter "I" and measured in *amperes* (A). If a coulomb of charge passes through a device every second one ampere of current is said to be flowing in the device. By convention, (shades of Ben Franklin) the flow of current is taken as being the direction of flow of *positive* charge. That is, a flow of *electrons* toward the right constitutes a current flowing toward the left. This convention is consistent with taking the direction of the $\mathcal{E}$ field as being the direction of the electric force exerted on a positive charge.

In the parallel-sheet example above, the current, I equals qN; N charges pass through per second and each charge amounts to q coulombs. Thus qN coulombs pass through each second. The power, in watts, is

$$\text{Power} = \Delta V q N = \Delta V I \qquad \text{Equation (11-7)}$$

$$\text{watts} = (\text{volts})(\text{amperes})$$

Consider a numerical example: a high-voltage generator pumps a microcoulomb of charge from one charge terminal to another in one second, while the potential difference between terminals is maintained at 40,000 volts. What is the power generated? From Equation (11-7),

$$\text{Power} = \Delta V I = (4 \times 10^4 \text{ volts})\left(\frac{10^{-6} \text{ C}}{1 \text{ sec}}\right)$$

$$= 4 \times 10^{-2} \text{ VA} = 4 \times 10^{-2} \text{ watts.}$$

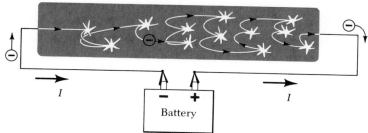

Figure 11-16

*When the battery is connected, negatively charged electrons travel
bumpily from left to right through the heating element, "falling downhill". As they pass through the battery, they are pumped "uphill" again.*

The charges that were discussed in the parallel-sheet example
"fell" freely between the sheets, accelerating uniformly (since
$\mathscr{E}$ and F_E were constant between the sheets) and gave up most of
their energy to the sheet, heating it up. If we were determined to
convert electrical energy from the battery into heat a more usual
plan would be to let the charges fall downhill through a heating
element. In this case the charge bumps its way through the
material, giving up its energy in a series of encounters with
atoms in the material, something like a ball making its way
through a pinball machine as shown in Figure 11-16. Thus the
whole element is heated.

Standards of Electrical Measurement

Until now we have not mentioned any *standard* to which electrical measurements are referred. When charge was first discussed
we actually were introducing a fourth quantity or dimension
that goes along with mass, length, and time; and there ought to
be a standard for it. Early experience showed that electric charge
is not easy to measure accurately in large quantities. A far more
accurately measurable quantity is the *force* between two wires
carrying a certain electric current. The next chapter will discuss
this force further. But for now, we can say that the standard of
electrical measurements is the ampere, and that the coulomb is
operationally defined as one ampere·second! It is the amount of
charge which passes through a wire in one second when there is
a steady current of one ampere in the wire.

Commercial Electric Power

The voltage which is generated and distributed in most parts of
the world differs from that which batteries produce in that it *alter-
nates* or oscillates between plus and minus. In many areas, one
side of the power supply (or electric outlet) is positive and the

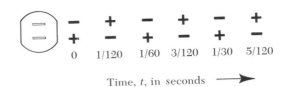

Time, *t*, in seconds ⟶

Figure 11-17
*60-cycle AC electrical outlet, and a schematic representation
of the alternating current through time.*

other negative. Then 1/120 second later the first side is negative
and the other positive. After another 1/120 second, they are back
where they started, having completed a plus-minus swing. The
voltage swings up and down like a teeter-totter 60 times per
second as shown in Figure 11–17. (In some power systems, one
wire in the plug is grounded and the other goes positive and
negative relative to it. However, the potential difference between the wires varies with time in an analogous manner.) Such
a source of electric power or energy is called 60 cycle AC, for
alternating current. The alternating current feature permits
changes of voltage by the use of devices called transformers.
Before the power leaves the central generating plant, it is usually
raised to a high voltage so that energy losses in the transmission
lines will be reduced. (For given power, Equation (11-7) indicates that increasing the voltage will reduce the current required. Heat losses are less if current can be reduced.) When
the lines approach the area in which the power is to be used,
the voltage is "stepped down" in a series of transformers until
it reaches 110-115 volts for use in the ordinary wall plug. Some
portions of Europe have 220-230 volt electric power, so that
travelers using electric razors or other appliances must be careful lest they burn out their 115 volt devices.

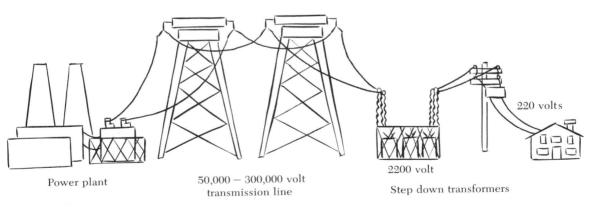

Power plant   50,000 − 300,000 volt   2200 volt
                      transmission line   Step down transformers   220 volts

Figure 11-18
Transporting electric power.

How much current do ordinary electrical appliances use? From Equation (11-7) we can quickly find how much current a 100 watt light bulb draws when it is plugged into a 115 volt socket.

$$I = \text{power}/\Delta V = 100 \text{ watts}/115 \text{ V}$$

$$= .87 \text{ A}$$

The electric wiring in most buildings is divided into separate circuits which are wired in a "parallel" scheme. Figure 11-19 shows a schematic diagram of a typical circuit. The main current branches out, passing through the various parts of the circuit simultaneously, then returning to a common wire. (It is also possible to wire devices in "series", so that they are in line, with the current passing through them sequentially.) An ordinary family dwelling will usually have four or more such circuits. The total current I in the circuit is roughly equal to the sum of the currents I_1, I_2, and I_3 and, in this case, is limited to 15 A by the fuse. It is inserted as an intentional weak link that will heat up and melt if the total current flowing in the circuit exceeds 15 A, preventing overheating of the wires going up through the walls of the house. This is particularly true when *short circuits* occur. A short circuit results when charge jumps directly from a high to a low potential without spreading its energy safely through a power consuming device of some kind. For instance, in Figure 11-17 considerable energy will be released at the place where the two wires can touch. If they were to touch, sparks and a possible fire could result unless the current is interrupted by the broken fuse.

The 15 A fuse in the circuit above will permit the dissipation of power in the amount ΔVI equal to (115 V)(15 A), or 1725 watts. The circuit as shown is dissipating a total of about 830

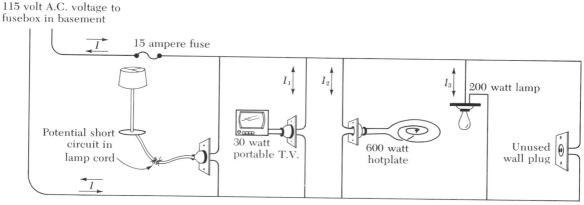

Figure 11-19

A typical American household's electric circuit.

watts. If a 1150 watt toaster were to be plugged into the empty plug, the fuse would burn out because the total wattage being used in the circuit would exceed 1725.

We are speaking freely of power here, but what people really pay for, when they send in the monthly check to the power company, is electrical *energy*. The watt, W, is a unit of power and must be multiplied by a time unit to obtain a certain amount of energy. The units in which electric energy is usually sold are *kilowatt-hours*, kWh, and they cost three cents, more or less, depending on how many of them you buy. To leave a 100 watt light bulb burning for eight hours consumes:

$$(100 \text{ watts}) (8 \text{ hr}) (3600 \text{ sec}) = 288 \times 10^4 \text{ joules},$$

or

$$(0.1 \text{ kilowatt}) (8 \text{ hours}) = 0.8 \text{ kWh}$$

and costs about two and one-half cents.

Environmentalists are now pointing out that the people who really pay for electric power are not exclusively the people who buy and use it. Those who live downwind from the power plant also pay. But this is changing. Smoke abatement is clearing the air—and increasing the cost of electric power for the consumer.

Electrical Properties of Materials

It was mentioned earlier that charges can move freely around in some materials but not in others. One way to represent the response of a material when a potential difference is applied across it is to make a graph of the current versus the applied voltage, as shown in Figure 11-20. Good conductors, such as salt water and metals, display a steep slope compared to poor conductors. At

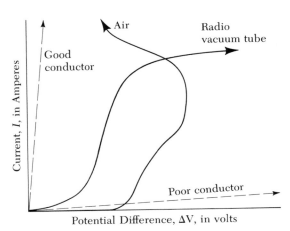

Figure 11-20
Electrical conducting properties of various materials.

very low temperatures, some materials become "superconduct-ing", in which case the resistance to the flow charge disappears entirely and the graph would be essentially vertical. That is, even if you reduce the potential difference to zero—remove the battery—the current keeps flowing! For a poor conductor (good insulator) such as rubber, most plastics, dry wood and *pure* water the graph is nearly horizontal. For air, the graph would be horizontal up to a point where the *field intensity* is about 3×10^6 V/m, and then would rise sharply. Once the spark jumps, a lower voltage will maintain the current. Thus, the curve doubles back on itself. The field required to cause a significant current in a material is called the *breakdown strength*.

Some materials, such as the good or poor conductors in Figure 11-20 show a straight-line graph between current and applied voltage if the current through them is not too great. In such cases;

$$I = (\text{constant})\,\Delta V.$$

When the relation between current and voltage is this simple, the material is said to obey *Ohm's law*. The constant in the equation is called the material's *conductance*, and is represented graphically by the slope of the I *vs* ΔV graph. The reciprocal of the conductance is called *resistance*. Resistance, R, is measured in *ohms*.

$$I = \frac{\Delta V}{R}$$

amperes = volts/ohms

or;

$$\Delta V = IR. \hspace{3cm} \text{Equation (11-8)}$$

The latter form of the equation tells us that if the two ends of a device having a resistance, R, are connected to a power supply furnishing a potential difference, ΔV, then a current, I, will pass through the device. The power dissipated in such a device, from Equation 11-7 will amount to

$$\text{power} = I\Delta V = I(IR) = I^2R. \hspace{2cm} \text{Equation (11-9)}$$

From this we can see now how it is that the reduced current in the high-voltage transmission line referred to on page 290 decreases the energy lost to electrical heating in the line.

What must be the resistance of the 100-watt lamp spoken of on page 291? We know that when a voltage of 115 volts is applied to the bulb, a current amounting to 0.87 A flows in it. Then from Equation (11-8),

$$R = \Delta V/I = 115 \text{ V}/0.87 \text{ A}$$

$$= 132.3 \text{ ohms}.$$

The Earth's Electric Field and Thunderstorms

MAN, AM I TURNED ON!

200 V

100 V

$\mathscr{E}$

0 V

Measurements show that an electric field having a strength of more than 100 V/m exists in the air at the surface of the earth. Does this mean that we walk around with our heads at a potential 200 volts positive or negative with respect to our feet? Not quite. The equipotentials drape up over us like blankets. We are part of the earth, electrically. But the equipotentials *are* crowded together over our heads, and this increases the field at that point. [See Equation (11-6). It shows that as the Δy for a given ΔV decreases, $\mathscr{E}$ increases.] As we'll see in the next paragraph, this can be bad news.

We said on page 278 that a charged sphere of radius r should produce a field at its surface equal to Kq/r^2. It would then follow that the earth, having an electric field, must carry a charge. One might expect that if r, the radius of the earth, is equal to 6.45 $\times 10^6$ m and the field intensity $\mathscr{E}$ equals 100 V/m, then the total charge on the earth:

$$q = \frac{\mathscr{E}r^2}{K} = \frac{(100 \text{ V/m})(6.45 \times 10^6 \text{ m})^2}{9 \times 10^9 \text{ Nm}^2/\text{C}^2}$$

$$= 4.62 \times 10^5 \text{ C}$$

$$= 462,000 \text{ C}!$$

(The electric field points down, or towards the earth, so the charge is negative. See Figure (11-6), and change the sign of the charge.) The total charge thus calculated is based on the assumption that the electric field at the surface of the earth is 100 volts per meter, pointed downward. This is not actually always true, because the charge on the earth may be positive below regions of stormy weather. There are indications that the total charge on the earth is considerably less than 460,000 coulombs.

But what about the whole earth system? How does the earth get its charge? There are large exchanges of charge within the earth-atmosphere system, driven largely by electric storms.[7] Thunderstorms can be visualized as being giant generators, driving negative charge (in most cases) to the earth by means of lightning strokes and forcing positively charged ions (atoms stripped of their outer electrons) up into the ionosphere. Estimates put the potential difference between the top and bottom of a thunderstorm at several millions of volts![8]

The average electric field intensity in the air below a thunderstorm can rise to many times the 100 V/m mentioned earlier. For

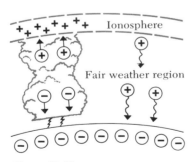

Figure 11-21
Thunderstorms pump positively charged ions up and negatively charged electrons down.

(7) McDonald, James E., "The Earth's Electricity", *Scientific American 189*, April, 1953, pg. 32.
(8) Lightning may also occur in sandstorms, snowstorms, or in the clouds over erupting volcanoes. Any mechanism which separates large amounts of charge can lead to atmospheric breakdown (lightning). The same principle—blowing particles or droplets through air—was at work when Millikan squirted oil into his apparatus as indicated on page 273.

a given potential difference, ΔV, between a cloud and the earth, lightning is likely to strike where the field intensity first exceeds the breakdown strength of air. Walking on a treeless plain while an electric storm is approaching puts you in danger because, as the electric field increases above your head, the electric field there may be the first area to reach the breakdown value. Carrying a pointed umbrella would be especially foolish, since it would further intensify the field.

Local folklore says that cows in the mountains of Ecuador have learned to lower their horns toward the ground when storms approach. Whether they really do this or not, it would be a good idea if they did. The tendency of pointed objects to intensify electric fields, page 278, and thus "attract" lightning is the basis for the lightning rod, invented by Benjamin Franklin.[9] It provides an easier and harmless path for charge to travel between thunderclouds and the earth.[10]

A lightning flash can involve many individual strokes, each carrying thousands of amperes of current. However, the strokes are brief, lasting less than a thousandth of a second. An average lightning flash, including all the strokes, lasts a few tenths of a second and transfers about 25 coulombs of negative charge to the earth. The averaged-out current in a flash is thus 100-200 A while it lasts.[11] The averaged-out current in a *thunderstorm* is about three-quarters of an ampere, since time elapses between flashes of lightning. The total negative charge coming to the whole earth is about 1500 coulombs per second, or 1500 A. This means there must be about 2000 active thunderstorms on earth at a given time.[12]

The violent influx of negative charge is balanced by a more benign downward drift of positive ions in regions of fair weather.

Figure 11-22
Sometimes lightning doesn't do what it's supposed to do.

What *Is* Charge?

Physicists have given electric charge its name. Operationally, they conceive of it as the cause of certain kinds of effects. They

(9) I. B. Cohen, in the *Journal of the Franklin Institute*, May, 1952, pg. 393, reports that prejudice kept the rods from being used on church steeples in some countries for almost 150 years after they were invented in the 1750's. The particular obstacle was the belief that ringing church bells kept lightning away. Several priests and sextons died in church steeples while applying this mistaken belief.

(10) The rods neither cause nor prevent strokes of lightning, unless they are attached to buildings more than 100 meters high. However, by intensifying the electric field around them, they make it likely that a lightning stroke in the vicinity will come to them because that's where the breakdown strength of the air (3×10^6 V/m, mentioned on page 293) will first be exceeded (usually). If the rods are *not* properly connected to a grounding wire, they constitute more of a danger than protection.

(11) Uman, Martin A., *Understanding Lightning*, Bek Technical Publications, Carnegie, Pa., 1971.

(12) McDonald, footnote 7.

know how it behaves and how to use it. Nevertheless, charge cannot be seen. For these reasons, electric charge furnishes a good example of both the power and limitations of scientific understanding. The gravitational influence sent out by mass should be equally mystifying, but we see and feel the mass directly and this tends to stop our questions. Charge, on the other hand, remains forever invisible.

Is charge just a manifestation of the work done in rubbing two things together, or in stripping them apart? The evidence says no. Unlike heat, which is discussed in Chapter 15, electric charge is always embodied in a tiny bit of mass.[13] In Chapter 9, it is pointed out that the electron is a basic constituent of matter. You can get lots of charge out of an object, but there *is* a limit, posed by the structure of atoms and the forces between charges. When you charge something, you are adding (or removing) a tiny piece to (or from) the object.

In the next chapter, magnetism, the other side of the electrical coin, is discussed; as is the puzzle of how electricity, magnetism, gravity, and other physical influences reach out through (empty?) space.

Summary

Objects become electrified by adding charge to (or withdrawing charge from) them. This can be accomplished by rubbing them with the right material, or by touching them with a charged object. The charging process will be most effective if a material which loses charge easily is brought together with a material which gains charge easily. Charge is never created or destroyed: just transferred. Not only is charge conserved; it is invariant.

Sparks occur when charge enters, or leaves, an object rapidly. Charges move freely through conductors but not through insulators.

Charge is quantized. The basic lump is minus 1.6×10^{-19} coulomb and it is attached to a particle having a mass of 9.1×10^{-31} kg (the electron).

Similarly charged objects repel. Oppositely charged objects attract. A charged object can also attract an uncharged object by polarizing it. To show with no uncertainty that an object is charged, another charged object must repel it.

Coulomb's law: the force between spherical (or point) charges

(13) It is possible to pass through the boundary between energy and charge (or matter) by means of a phenomenon called *pair production*. Under the right conditions, energy in the form of a pair of gamma rays (see Table 13-2 on page 339) are convertible into a pair of equal and oppositely charged particles: an electron and positron, for example. Net charge is still not created because the total charge of the "system" remains at zero. To make the reaction possible, the gamma rays must possess energy, E, sufficient to supply the mass of the two charged particles according to Einstein's equation $E = mc^2$.

dies off as the inverse square of the distance between their centers, just as in gravitational interactions:

$$F_E = Kq_aq_b/r_{ab}^2.$$

$$= (9 \times 10^9) \, q_aq_b/r_{ab}^2 \qquad \text{Equation (11-1)}$$

The electric field intensity at a point in space $\mathscr{E} = F_E/q$, measured in N/C, or in V/m. $\mathscr{E}$ is a vector having the direction of the force exerted on a positively charged object.

Near a point or spherical charge, $\mathscr{E} = 9 \times 10^9 q/r^2$, but in the cases of other charge distributions the expression for electric field intensity would be more complex or more simple. Between two oppositely charged parallel conducting plates, $\mathscr{E}$ is uniform, not depending on position at all.

In *any* electric field,

$$\Delta W = -q\mathscr{E}_{avg}\Delta y. = \Delta(P.E.) \qquad \text{Equation (11-3)}$$

Difference of potential, in volts,

$$\Delta V = \frac{\Delta W}{q} = \frac{\Delta(P.E.)}{q}$$

$$= -\mathscr{E}_{avg}\Delta y \qquad \text{Equations (11-4), (11-5)}$$

$$\mathscr{E}_{avg} = -\frac{\Delta V}{\Delta y} \qquad \text{Equation (11-6)}$$

$\mathscr{E}$ points from high potential toward low potential. It gives the slope of the potential, $-\Delta V/\Delta y$.

$$1 \text{ volt} = 1 \, \frac{\text{joule}}{\text{coulomb}}, \text{ so } 1 \, \frac{\text{newton}}{\text{coulomb}} = 1 \, \frac{\text{volt}}{\text{meter}}$$

The electric potential energy of a charged object is equal to its charge times the electrical potential at that point.

Equipotentials are lines (or surfaces) over which the electric potential energy of a given charged object is the same. Equipotentials always cross the lines of force in an electric field at right angles.

Charge in motion is called current, I. It is measured in amperes, A.

$$\text{Electric power in watts} = (\Delta V)(I) \qquad \text{Equation (11-7)}$$

Current, not charge, furnishes the standard for electrical measurements. A coulomb can be thought of as an ampere-second.

In America, commercially generated electric voltage oscillates at 60 hertz (cycles per/sec, abbreviated Hz) and is delivered to houses, buildings, and businesses at a potential difference of either 115 or 230 volts.

Electric energy is sold in units of kilowatt-hours, kWh,

$$1 \text{ kWh} = 3.6 \times 10^6 \text{ joules.}$$

Some conductors follow Ohm's law:

$$\Delta V = IR \qquad\qquad \text{Equation (11-8)}$$

where R is the resistance of the material in ohms.

For ohmic materials, power then is

$$I\Delta V = I^2R, \qquad\qquad \text{Equation (11-9)}$$

measured in watts.

The earth is negatively charged. Therefore, it has an electric field pointing in toward its surface. Thunderstorms pump negative charge to the earth by means of lightning. They also pump positive charge up into the ionosphere at the same time. In regions of fair weather the positive charges float down to the earth, preventing build-up of excessive negative charge on the earth.

Further Reading

Benedict, Manson, "Electric Power From Nuclear Fission", *Technology Review 74*, Oct-Nov., 1971, pg. 32. (Discusses some of the problems mentioned in Chapter 9, with the focus on the electrical energy which is produced.)

Ehrenreich, Henry, "The Electrical Properties of Materials", *Scientific American 217*, Sept., 1967, pg. 180. (Materials differ by a factor of 10^{23} in their ability to conduct electricity. Modern physicists try to tell why and to utilize a wider range of electrical conductivities.)

Feynman, Richard P., *Lectures on Physics*, Vol. II, Chap. 9, pp. 4–11, Addison-Wesley, Reading, Mass., 1963. (Informative discussion of charging and electrical storms.)

Jefimenko, Oleg, "Motors Using the Earth's Atmospheric Electric Field", *American Journal of Physics 39*, July, 1971, pg. 776. (Small motors can be made to run simply by connecting them to an antenna run up into the air.)

Millikan, Robert A., *Electrons (+) and (−)*, University of Chicago Press, 1939. (An outline of the physics of particles, built around the author's experiments to determine the charge on the electron.)

Moore, A. D., "Electrostatics", *Scientific American 226*, March, 1972, pg. 47. (Discusses modern applications of electrostatics, such as xerography.) See also his book with the same title, Doubleday Anchor Book, 1966.

Roller, D. and Roller, D. H. D., "The Development of the Concept of Electric Charge: Electricity From the Greeks to Coulomb", Harvard Case Studies in Experimental Science, Case 8, 1954.

de Santillana, Giorgio, "Alessandro Volta", *Scientific American 212*, Jan., 1965, pg. 82. (Volta invented the battery, conquering Luigi Galvani's theory of animal electricity. Electrically, there is no difference between living and non-living things.)

Uman, Martin A., "Everything You've Always Wanted to Know About Lightning But Were Afraid to Ask", *Saturday Review*, May 13, 1972, pp. 36–41.

Problems

11-1   An uncharged metal sphere hangs on an insulated thread. A second body, positively charged, is brought near but does not touch the metal sphere. Is it repelled, attracted, or uninfluenced? What if the uncharged sphere is an insulator?

11-2   Small cork particles are attracted to charged rubber, or glass rods. But if you watch closely, you will see some of the particles shoot away from the rod after awhile, especially when the air is moist. Can you think of an explanation for this observation?

11-3   A plus (3×10^{-6}) coulomb charge is situated one meter from a plus (1×10^{-6}) coulomb charge. Where would you have to put a third positively charged body so that the electric force on it would be zero? Suppose a minus (3×10^{-6}) charge is placed one meter from a plus (1×10^{-6}) coulomb charge. Where would you put a positive charge NOW so that the electric force on it would be zero?

11-4   A conducting body is charged negatively at the rate of 10^{-10} C/sec. How many electrons reach it per second? What is the charging current in amperes? If the body has a radius of 0.1 m, what will its potential be after ten seconds of charging? (The electrical potential of a sphere bearing a charge q is given by Kq/r, where r is the radius of the sphere.)

11-5   Two similar pith balls 0.2 m apart bear charges of plus $(5.0 \times 10^{-7}$ coulomb and minus 3.0×10^{-7} coulombs, respectively. They are allowed to touch and then are placed 0.2 m apart again.

   (a) Will the force now be attractive or repulsive?
   (b) What is the final charge on each ball?
   (c) What is the magnitude of the initial and final electric force between the balls?

11-6   Calculate the electric potential at the origin $(x = 0, y = 0)$ if a negative charge of (5.0×10^{-7}) coulomb is located at $y = 0$, $x = 0.5$ m and a positive charge of (3.6×10^{-7}) coulomb is located at $y = 0.3$ m, $x = 0$.

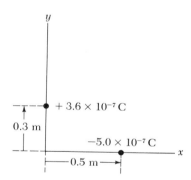

11-7   Two objects, each bearing a charge of minus (2×10^{-8}) coulomb are located 15 cm from each other.

   (a) Calculate the electric potential at the point A.
   (b) Find the electrical potential energy in joules of a third object bearing a charge of minus (5×10^{-8}) coulomb and placed at the point A.
   (c) Find the work, in joules, required to move the charged object from point A to point B, situated midway between the other two charges.

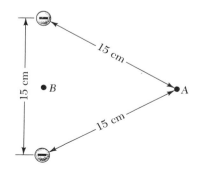

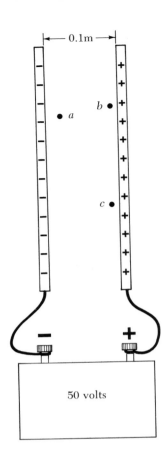

|← 0.1m →|

b •

• a

c •

50 volts

11-8 Suppose the point a is .02 m from the left sheet, and points b and c are .01 m from the right sheet.

(a) Which points are at the higher potential?
(b) What is the direction and magnitude of the field intensity between the plates?
(c) Suppose a body carrying a charge of plus 10^{-9} coulomb is put into the field between the plates. Will it have a higher energy at a, b, or c? (Ignore gravity.)
(d) Calculate the work done in moving the charged object from b to a, and from c to a.

11-9 Calculate the velocity of a 500 electron volt electron.

11-10 Calculate the accelerating potential required to produce a one-percent relativistic mass increase in an electron.

11-11 Two points are at the same potential. Does this mean that no work is done in moving a charged object between the two points? Does it mean that no force need be exerted on the object? Does the path followed from one point to the other make any difference?

11-12 Electric charge is not used up when current flows through a toaster. What electrical quantity *is* used up? Explain.

11-13 Salt water conducts electricity. When people tell lies, their palms get sweaty. Building on these two facts, design the electric circuit for a simple lie detector.

11-14 H. R. Crane, formerly a professor of physics at the University of Michigan, suggested the following set of problems. Suppose you buy two flashlight batteries for 60 cents and perform an endurance test with them. You hook them up so that they push current through a light bulb. You measure the voltage (V) and current (A) as time passes. After 2.5 hours the light is too dim to be of use as a flashlight.

Table T

Time, Hrs	Volts, V	Current, A
0	3	0.55
0.5	2.52	0.50
1.0	2.16	0.45
1.5	1.86	0.42
2.0	1.56	0.38
2.5	1.26	0.35
3.0	1.02	0.32
3.5	.78	0.30
4.0	.6	0.28
4.5	.48	0.26
10.0	.12	0.18

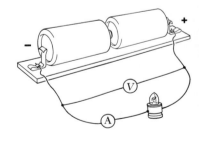

(a) First, calculate the total energy delivered to the light bulb, in watt hours, and in joules, during the useful life of the batteries (2.5 hours) What is the cost of this energy per kilowatt hour?

(b) If a cordless vacuum cleaner using five amperes at 110 volts were to be run by flashlight batteries, how much would it cost to run for one hour?

(c) How much would it cost to run the vacuum cleaner using commercial power costing 3 cents per kWh?

(d) How does the resistance of the filament in the light bulb change with time? Can you account for this?

11-15 How many seven-watt, 115-volt light bulbs can be put on a Christmas tree if the strings of lights are all plugged into a house circuit with a 15 ampere fuse in it?

11-16 A 60 volt battery is connected across a resistance R. The resistor dissipates 600 watts of power. What is the value of R? Electrons leave the resistor with less energy than they entered. If the resistor is 0.1 m long, what must be the magnitude of the average force acting on the electrons in the resistor?

11-17 A 15 ohm resistance is connected to a battery and ammeter. If the ammeter reads 0.5 A, what voltage is the battery providing?

11-18 If energy costs $.03 per kilowatt-hour, how much is a gram of mass worth, assuming that you could convert it all to energy?

11-19 If the average charge carried down by a lightning flash is 25 coulombs and the average current in a thunderstorm is 0.5 A, how often does lightning strike in an average thunderstorm?

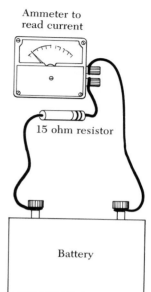

Ammeter to read current

15 ohm resistor

Battery

12
MAGNETISM AND FIELD THEORY

Magnets and Poles

Magnets are probably more familiar to you than are charged rubber rods. You know that magnets exert forces on other magnetic materials (iron and some kinds of steel, but not on copper, aluminum, lead). You might have heard the words "magnetic field" as an attempt to explain the forces. But even though these phenomena are familiar, it is by no means easy to explain why one material is magnetic and another is not. The answers to such questions are beyond the scope of this book. Nevertheless we can find out some interesting things about magnetic fields, and how they are related to electrical phenomena. Going on from there, we will examine some past and present models for explaining how one object influences another object at some distance from it, regardless of the type of physical force involved.

If you lay a piece of paper on a magnet and sprinkle some iron filings on the paper, the filings will clump together and line up in directions which will give you some idea of the shape of the field surrounding the magnet. Better yet, move a small compass

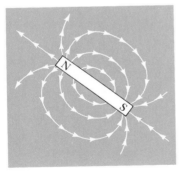

Figure 12-1

The magnetic field of a bar magnet. If a small magnet is moved around in the space surrounding the main magnet, its north pole will always line up facing in the direction indicated by the arrows.

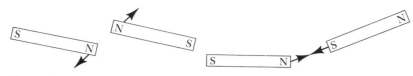

Figure 12-2

Two north poles (or south poles) repel one another. A north and a south pole attract.

around on the paper. It is a tiny magnet itself, and will point in various directions, depending on where it is with respect to the big magnet. A simple bar magnet shows a symmetrical field which doubles back on itself. The arrows in Figure 12-1 show how the compass lines up, at various points, in the space around the magnet. By definition, the *direction* of the magnetic field is the direction in which the *north pole* of another magnet points when it is placed in the field. (This assumes the magnet is free to rotate.) A compass needle (or any other magnet) has a north pole (usually marked with an N or an arrowhead) and a south pole, which experiences a force opposite to the direction of the field in which it is put. Therefore, a compass needle lines up parallel to the magnetic field at any point in space. The term "north pole" arose in the early days of navigation with the magnetic compass. It denotes the end of the compass (or magnet) which tends to point north when you carry it from place to place. Later in the chapter we will discuss the earth's magnetic field and the fact that compasses do not point directly toward the earth's geographical north pole. William Gilbert (page 105) was the first to use the word "pole" in connection with magnets in general. (For our purposes, "plus" and "minus" would work just as well as "north" and "south".)

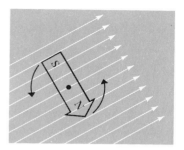

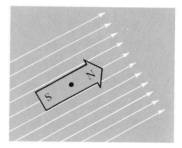

A south pole attracts a north pole of another magnet, so, following our convention, the field must point *in* toward the south pole, as shown in Figure 12-2.

The field of a bar magnet, in at one end and out at the other, is called a *dipole field*, and is similar to the electric field set up by a pair of opposite charges of equal magnitude, as shown in Fig. 12-3. This raises the possibility of some kind of equivalence between magnetic "poles" and electric "charges". But, the peculiar thing about magnetic fields is that they are *always* dipole fields or some combination or warping of them (Figure 12-4). That is, magnets always have both a north and a south pole. Theory indicates that isolated magnetic poles called "magnetic monopoles" could exist, but so far there has been no experimental evidence for them. If you break a magnet into pieces, each will have a north and a south end. The nearest thing to an analog of the radial field around an electric charge is one end of a long needle-shaped magnet (Figure 12-5).

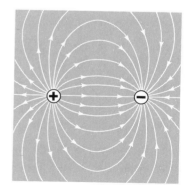

Figure 12-3

The electric field around a pair of equal but opposite charges is similar in shape to the magnetic field around a bar magnet.

Magnetism and
Field Theory

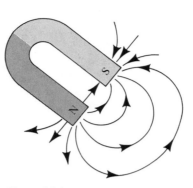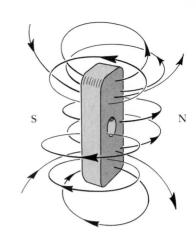

Figure 12-4
Fields produced by other magnets.

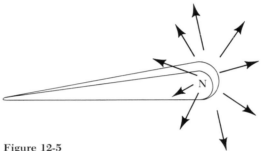

Figure 12-5
The field is almost radial near the end of a long magnet.

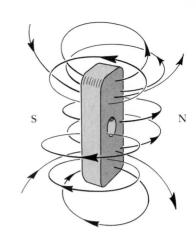

Right hand

Lines of
magnetic field

Figure 12-6
*If you grasp a current-carrying
wire with your right hand, thumb
pointing in the direction of
positive current, your fingers
wrap around the wire in the
direction of the magnetic field
produced by the current.*

The Magnetic Effects of Electric Currents

For centuries magnetism and electricity were thought to be independent, spooky phenomena. With the advent of a systematic investigation of electricity and magnetism, separate concepts and units of measurement were developed for each. We can see now why no connection between the phenomena was detected. There is no interaction between *static* electric and magnetic fields, and until Alessandro Volta (1745–1827) invented the battery in 1799, scientists had no means for maintaining a steady *motion* of electric charge. In 1802, an Italian named Romognosi found that an electric current would deflect a magnet. He wrote a letter to a local newspaper concerning his observation but evidently did not pursue the matter further, and few European scientists knew of his discovery. It was not until 1818 that Hans Christian Oersted (1777–1851) noticed that when a compass needle is held close to a wire carrying a current it rotates. He began a detailed investigation of the phenomenon. Because he did this, he became a famous man. Several streets and a park in

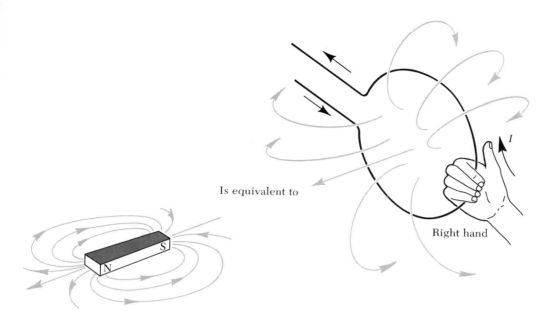

Is equivalent to

Right hand

I

S

N

Figure 12-7

Copenhagen are named after him and his picture is on the Danish 100-kroner bill. He measured the strength of the magnetic field set up by an electric current in a wire, and found that it died off with distance from the wire. This was no surprise. What *was* surprising was that the magnetic field seemed not to point out from the wire carrying the current. Instead, it was wrapped around the wire in a circular fashion! (Figure 12-6.)

If a wire is bent into a loop and an electric current is sent through it, the field produced at some distance from the loop is the same as the dipole field of a bar magnet, as illustrated in Figure 12-7. If such a current loop is placed in a larger magnetic field, it will line up, behaving precisely as a bar magnet or compass would.

Electromagnets

If the loop of wire above is replaced by a very long wire wrapped into a coil having very many turns, and an electric current is made to flow through it, a powerful electromagnet can be the result; especially if the space inside the coil is filled with iron. Such devices are put to wide use in science and industry whenever a temporary or strong magnet is needed. Large ones are used to lift steel scrap around junk yards and small ones are used to make telephones and doorbells ring.

Figure 12-8

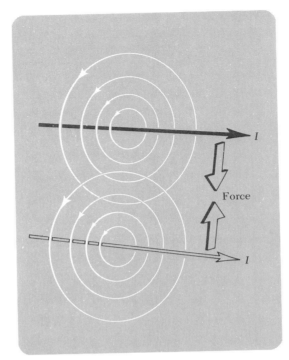

Figure 12-9
*Two wires carrying a current in the same direction
attract each other magnetically.*

Defining the Ampere

If electric currents produce magnetic fields that affect magnets (compasses), it follows that two currents, each setting up magnetic fields, will affect each other. Experiment shows that two currents flowing in the same direction, in two parallel wires, exert an attractive force on each other (Figure 12-9). This force can be measured very accurately, and is the basis for the standard of electric measurement, the ampere, as was mentioned on page 289. *One ampere is the unvarying current which will produce a force of 2×10^{-6} newtons per meter of length between two parallel wires placed one meter apart in vacuum.* The definition of the ampere operationally determines, in turn, the *coulomb*, or ampere·second.

Forces on Charged Particles Moving
in Magnetic Fields

When Oersted made his investigations in the 1820's, he had no clear picture of what electric current was. Physicists now usually visualize it as being discrete lumps of charge marching along. If such a current sets up a magnetic field, then it must be the individual charges in motion which create the field. If two current-carrying wires experience forces toward, or away from, each other then a single charge moving through a magnetic field, should experience a force, regardless of the source of the field. The interesting thing is that the force exerted on a moving charged particle by a magnetic field is at *right angles* to the field and to the velocity of the charge. The magnitude of the force is found experimentally to be:

$$F_M = qvB \sin \theta \qquad \text{Equation (12-1)}$$

where F_M is the force, measured in newtons, v is the velocity of the charge, in meter/sec, B is the magnetic field or "flux density", in Webers/m^2, and θ is the angle between v and B. For a particle moving in a direction perpendicular to the magnetic field, Equation (12-1) becomes $F_M = qvB$. Equation (12-1) *defines* magnetic flux density, for the purposes of this book. There is also no need to puzzle over the strange units in which magnetic fields are measured. For our discussion all that needs to be remembered is that a magnetic field has an intensity of one weber per m^2 if a force of one newton is exerted on one coulomb of charge injected into the magnetic field at a velocity of one meter per second perpendicular to the field.

Since the force is always at right angles to the velocity, the particle will not speed up or slow down. A steady magnetic field thus does not increase or decrease the kinetic energy of the particle. It does no work on it. (We will see later that a changing magnetic field induces a voltage or electric field which can do work on a charged particle.)

There are several ways to remember the direction of the force exerted on a moving charge—or a wire carrying a current—in a magnetic field. One is called the *right hand rule*. If your hand is held as shown in Figure 12-11, the rule is as follows: if the magnetic field is in the direction of the thumb and the current or movement of *positive charge* is in the direction of the three bent fingers, the force will be in the direction of the straight index finger. IMPORTANT: for a negative charge moving in the same direction, the force will be in the opposite direction, (Or, for negative charges, you can use your left hand.)

The second scheme for determining the direction of the force could be called the "slingshot" effect. It utilizes the physical fact that the magnetic force tends to move the charge or current-carrying wire in the direction of decreasing magnetic field. In order to explain this statement suppose we look at the positive charge or current-carrying wire in Figure 12-6 head-on, as shown in Figure 12-12. The magnetic field is still *up*. The charge (now coming out of the paper) will produce a local magnetic field which wraps around it in a counter-clockwise sense. (Up on the right side and down on the left.) This means that on the right side

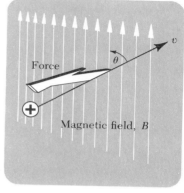

Figure 12-10
The magnetic force exerted on a moving charged particle is in a direction which is perpendicular to the plane containing the velocity vector v *and the magnetic field vector* B.

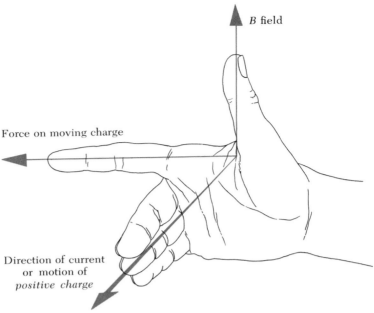

Figure 12-11
*The right hand rule: thumb up, index finger to the left, the other fingers
pointing out of the paper.*

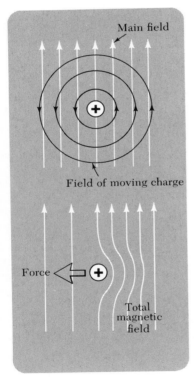

Figure 12-12
*The "slingshot effect". If the
current in the wire is out of the
paper and the magnetic field
points toward the top of the page,
the magnetic force on the wire
is to the left.*

of the charge or wire, the local field and the main magnetic field
add together, producing a stronger total magnetic field. The lines
of force representing this total field would be crowded closer
together. The total field on the left of the charge or wire would
be weakened because the local and main fields oppose. The lines
of force on the right try to "straighten out", and the wire or charge
experiences a force toward the weaker field on the left. Perhaps
this model treats lines of magnetic field too much like rubber
bands, but it describes the direction of the force in a slightly
more physical way than does the right hand rule.

You can illustrate these magnetic forces for yourself if you can
get an unfrosted light bulb and two magnets. When the light bulb
is turned on, the current in it oscillates back and forth 60 times
per second. If you bring the magnets up close to the bulb, the
filament carrying the current will feel a sideways force, first in
one direction and then in the other. The glowing filament will
vibrate back and forth 60 times per second (Figure 12-13).

The same force is at work in electric motors. Wire coils in the
motors are designed to carry current while they are in magnetic
fields, in such directions as to guarantee that the coils experience
forces which always tend to rotate them in the same direction.

If a charged particle is shot into a magnetic field with velocity,
v, perpendicular to the field, it will travel in a circular path.

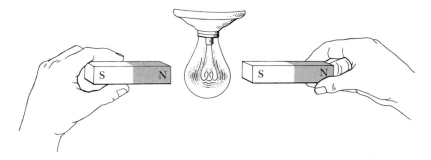

309

*Forces on
Charged
Particles Moving
in Magnetic
Fields*

Figure 12-13
A lamp filament carrying a positive current down *toward the bottom of the
page will experience a magnetic force* into *the paper if the magnetic field
points from right to left. When the current in the filament reverses, the force
will be directed outward from the paper.*

From Chapter 3, we can find the radius of the path in which the
particle must travel (Figure 12-14).

Centripetal force $= qvB$

$$\frac{mv^2}{r} = qvB$$

Solving this for r gives:

$$\frac{mv}{qB} = r \qquad\qquad \text{Equation (12-2)}$$

If a group of charged ions, all having the same charge, q, are injected with the same velocity, v, into a magnetic field, B, then
they will travel in circles of different radius, depending upon

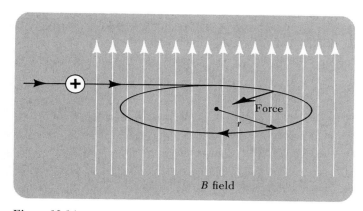

Figure 12-14
*A charged particle entering a uniform magnetic field will
travel in a circular path if its initial velocity is perpendicular to the magnetic field.*

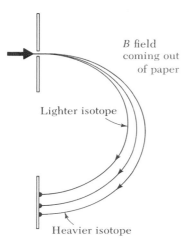

Figure 12-15

The larger the mass for a given charge, the larger the radius of the path followed by a charged particle in a magnetic field.

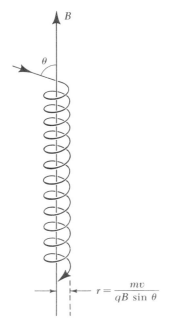

Figure 12-16

If a charged particle enters a magnetic field in a direction such that θ is not 90°, it will travel in a spiral path.

their masses, *m*. This principle was the basis for one of the methods of separating uranium isotopes mentioned on page 238. If uranium atoms are ionized (charged, by removing one or more electrons from each one) and injected into a magnetic field, the various isotopes will follow circular paths having different radii. Each can be collected separately from the others after one semi-circle of travel as shown schematically in Figure 12-15.

A particle which enters a magnetic field at some odd angle θ will travel in a spiral of radius:

$$r = \frac{mv}{qB \sin \theta},$$

Equation (12-3)

with a line of magnetic force along its center as shown in Figure 12-16.

The Earth as a Magnet

What has made magnetism of practical interest for so long is that the earth itself is a huge magnet, with a dipole field analogous to that which would be produced by a "south" magnetic pole buried beneath the Queen Elizabeth Islands, west of northern Greenland, and a "north" pole underneath Antarctica. Although the lines of magnetic force spring upward from the southern hemisphere and plunge into the northern hemisphere, they always have a component parallel to the earth's surface. This component points north (more or less), and magnetic compass needles line up parallel to it. For centuries, compasses have thereby given travelers a sense of direction when other orientation cues are absent.

As figure 12-17 shows, the earth's magnetic field is tilted and off center with respect to the earth's axis of rotation. Therefore, compasses do not point toward the geographic north pole. The earth's magnetic field also varies with time. Indications are that the "south" pole has wandered over wide portions of the Pacific ocean. Paleobiologists, paleontologists, and archeologists are beginning to speculate on what changes in (or the temporary disappearance of) the earth's magnetic field might have had on life in the past.[1]

The source of the earth's magnetic field is uncertain. One possibility is that electrically charged, molten rock in the earth's interior rotates slowly with respect to the earth's crust.

During the late 1950's a series of satellite experiments were conducted by Dr. James Van Allen, of the State University of

(1) Cox, Allan; Dalrymple, G. Brent; and Doell, Richard R., "Reversals of the Earth's Magnetic Field", *Scientific American 216*, Feb, 1967, pg. 44.

Iowa, and others. The purpose of the experiments was to find out what kind of electrically charged particles, if any, might be in the regions of space out beyond the edge of the earth's atmosphere; not just hundreds, but thousands of miles above the earth's surface. The sensors in the satellites showed that there *were* charged particles out there, but whenever the satellites passed through either of two altitude bands, the counters went dead. This was very strange. Finally Carl McIlwain, a member of Van Allen's research group, noted that if the rockets had gone through a region with a very *high* charge density, the counters would not be able to count fast enough, and would jam. The group designed new counters which could handle a higher density of charge, and the result was the discovery of two belts of charged particles, now called the Van Allen radiation belts. One of these belts is centered about 2500 miles out from the earth's surface, has a bean-shaped cross section, and contains mostly protons. The other is centered about 11,000 miles out, contains mostly electrons, and has a banana-shaped cross-section[2] (Figure 12-18). As the earth moves in its orbit, it intercepts charged particles from the sun and other sources. When some of these particles encounter the earth's magnetic field, they begin moving in spiral paths along the field lines. As they approach the top or bottom of the belt, the magnetic field intensity increases (the lines are closer together) and the particles travel in tighter spirals. (In Equation (12-3) r gets smaller.) The turns of the spiral are also closer together, and the particle can actually be bounced back by the magnetic field, so that its path doubles back on itself. The result is a trapping of charged particles in the earth's magnetic field. Some of the particles do leak on down to the earth's atmosphere, but new particles coming in from the outside maintain the high density of particles in the belts (Figure 12-19).

Figure 12-17
The magnetic field around the earth. Theoretical view prior to the 1950's.

Figure 12-18
The Van Allen belts around the earth, discovered in the 1950's. (From "Radiation Belts Around the World" by J. A. Van Allen. Copyright © 1959 by Scientific American, Inc. All rights reserved.)

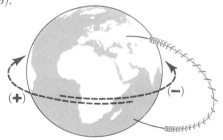

Figure 12-19
Charge particles trapped by the earth's magnetic field. In addition to spiralling up and down, the particles drift slowly around the earth. (From "Radiation Belts Around the World" by J. A. Van Allen. Copyright © 1959 by Scientific American, Inc. All rights reserved.)

(2) Van Allen, James A., "Radiation Belts Around the Earth", *Scientific American 200*, March, 1959, pg. 39. Offprint 248.

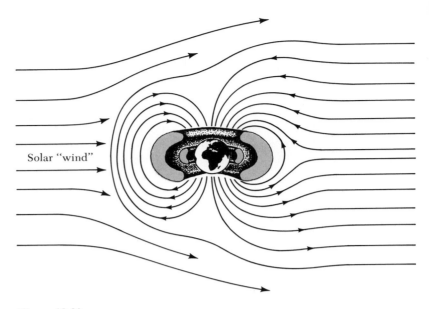

Solar "wind"

Figure 12-20
The magnetic field around the earth is now seen as the sum of the earth's field plus fields due to charged particles streaming from the sun past the earth.

Superimposed on the dipole field and the Van Allen radiation belts is a variable "wind" of particles from the sun. This wind compresses the earth's magnetic field on the side facing the sun and stretches it out on the dark side. The resulting overall shape of the field is that of a long teardrop.[3]

Although the detailed mechanisms are still in doubt, there is a connection between the solar wind and the aurora borealis, or northern lights. It seems as though some of the very energetic particles in the solar wind which just get caught at the edge of the compressed magnetic field of the earth (not in the Van Allen belts) come crashing into air molecules at points which form a rough oval around the earth's magnetic poles — north and south. The resulting excitation and ionization of the air molecules produces visible light. At times when the solar wind is weak, northern (and southern) lights are not visible.

Changing Magnetic Fields: Induction

Now let's complete the picture of interactions between electric and magnetic phenomena. In 1831, Michael Faraday (1791–1867) in England, and Joseph Henry (1797–1878) in America, found (independently) that when a magnet is moved past a con-

(3) Cahill, Laurence J., Jr., "The Magnetosphere", *Scientific American* 212, March, 1965, pg. 58.

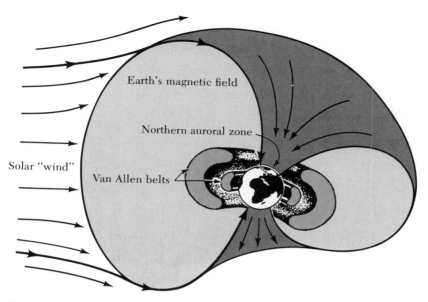

Figure 12-21

Magnetic fields and the Northern lights. (Adapted from "The Aurora", by
Syun-Ichi Akasofu, copyright © 1965 by *Scientific American, Inc.* All rights
reserved.)

ducting wire, an electric current is induced or generated in the
wire. Physicists now know that *any* relative change in a magnetic
field in a particular frame of reference will produce an electric
field in that frame of reference. If a conducting wire is present,
the voltage produced will push a current around it. The "relative
change" can be obtained in many ways: by moving the magnetic
field; moving the wire relative to the magnetic field; by changing
the magnetic field; or, by bending the wire into a loop and changing the shape (area) of the loop in the magnetic field. Any of these
changes will cause a current to flow in the wire. *The induced
current is in a direction which produces a magnetic field in opposition to the change of field which induced it.*

Let's see how this works in the case of the magnet and coil in
Figure 12-22. As the magnet is moved to the left, the leftward
magnetic flux increases in the coil. An induced current will then
flow around the turns of the coil in a direction such that it produces a magnetic field to the right. Looking at Figure 12-7, you
can see that the current must flow in a counterclockwise direction, that is, from right to left in Figure 12-22. If the magnet is
now pulled away (to the right), the magnetic flux in the leftward
direction in the coil decreases. The induced current will now
flow in the opposite direction (clockwise) in the coils.

Work must be done when a magnetic field is changed in the
presence of current carrying wires, or when a current carrying
wire is pulled through a magnetic field. Mechanical energy is

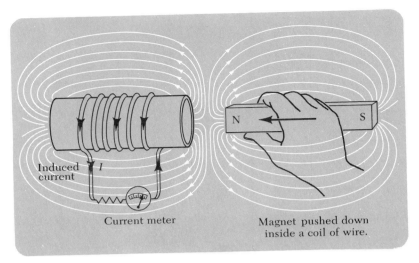

Induced current

Current meter

Magnet pushed down
inside a coil of wire.

Figure 12-22
*Movement of the magnet relative to the coil of wire produces an electric
current.*

converted to electrical energy. This is the basis of the electric
generator. The same principle was at work in the electric locomotives running downhill described on page 181. It is also used,
on a far more delicate scale, in magnetic tape players. A varying
magnetic field previously recorded along the length of the tape
induces a variable electric current as the tape moves through
the playback head. The current is amplified (electronically
magnified) and the amplified current is sent into loudspeakers
which change the electric variations back into mechanical vibrations, producing sound waves.

Field Theory and Mechanism

We have discussed in some detail the characteristics of and interactions between electric and magnetic phenomena. Although we
have included a historical dimension, our perspective has, in
reality, been modern. For this reason, one problem which was
paramount when physical theory had to be more literally mechanical than it is now has not been faced. The problem was
simply this: how does one object exert forces on another object
which it does not touch?

How *does* the earth attract the moon? What reaches out from
an electrically charged phonograph record to attract a particle
of dust floating overhead? Down through the centuries, there
have been two ways of answering these questions:

(1) One object simply acts at a distance upon the other object, pulling it in (or pushing it out) along a line joining the centers of the two
objects: *action-at-distance*.

(2) The force is transmitted across the space between the objects by some unseen but mechanically potent and continuous medium. In other words, the two objects are really still connected by an invisible third body: *action by contact* or *direct action.*

One of the first to think about the problem of force transmission was Roger Bacon. He suggested that one object, such as the earth, might attract another object by means of "species," a communicated tendency that made the object want to move toward the earth. There were species to explain electric, magnetic, and other, more occult attractions as well. Bacon's theory required that the space between the two interacting objects be filled with a medium to ensure continuity. (Like Aristotle, he abhored a vacuum.) The species were then communicated as pulses through the medium. Bacon did not attempt to describe how the pulse in the medium might take place, but his speculations do hint at modern wave and field concepts. In the early 1600's, some 350 years later, *Francis* Bacon theorized that whereas gravity and magnetism can be transmitted without the aid of a medium between objects; odors, light, heat, and sound all required a material, or immaterial, medium (the difference is not clear) for their transmission. Another century after this, Newton did not see how his gravitational forces could simply leap across the space between attracting bodies:

> That gravity should be innate, inherent and essential to Matter, so that one Body may act upon another at a distance thro' a *Vacuum,* without the Mediation of any thing else, by and through which their Action and Force may be conveyed from one to another, is to me so great an Absurdity, that I believe no Man who has in philosophical Matters a competent Faculty of thinking, can ever fall into it. Gravity must be caused by an agent acting according to certain Laws; but whether this Agent be material or immaterial, I have left to the Consideration of my Readers.[4]

He felt that the interactions must be transmitted by chains, or streams, of corpuscles. However, he was wise enough to realize that the real meat of his gravitational theory, the mathematics, was independent of how the attraction might be transmitted.

In the discussion on page 107, it was pointed out that Descartes saw all space as being filled with a swirling fluid whose whirlpools helped push the planets around the sun. He decided, unlike Newton, that the medium was primary.

Later, in 1782, G. L. LeSage hypothesized that space is filled with tiny particles moving in all directions. They do not collide with each other, but when they hit a large piece of matter, they are absorbed, and in so doing, exert a force on the object which absorbs them. An isolated object would be hit by the particles from all sides, so no net force would result. However, if two objects were in the same vicinity, they would partially shield each

(4) From a letter written to Richard Bentley, an English clergyman.

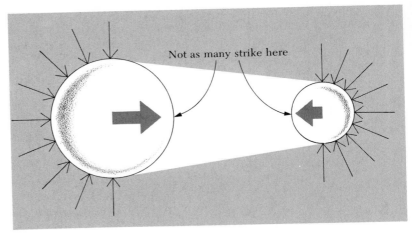

Not as many strike here

Figure 12-23
LeSage's model to explain gravitational attractions.

other; the number of particles striking them on their near sides would be decreased.

The net result would be an attraction between the two objects, and the attraction turned out to have the inverse-squared distance dependence which Newton's theory called for. Thus, it presented a semi-mechanical model to explain gravitational forces. However, the model did not work well when more than two objects interacted. Furthermore, the steady rain of particles would impede the motion of planets through space, if they really operated the way LeSage said they did. LeSage's model was never seriously considered.

By the middle of the eighteenth century both the action-at-a-distance theory and the continuous medium, direct action theory had their supporters, and criteria for testing them had been proposed. At that time natural philosophers thought that if the direct action (medium) interpretation were correct, then the interactions would:

(1) require a transit time;

(2) be affected by changes in the medium separating the objects; and,

(3) require energy to be located in the space between the objects.

For the supporters of both theories, continuity (smoothness) became an important philosophical consideration. Many scientists supported the direct-action point of view for this reason; the space between the object was then filled with something. However, in 1758, Roger J. Boscovich (1711–1787), a Serbo-Croatian mathematician, theologian, and one of the earliest supporters of Newton's theories, found *support* for *action-at-a-distance* using the same principle. He said that if matter were composed of *point* atoms sending out forces through empty space, then there would be no break—no discontinuity—where the material ended and the space or medium began, since a true point occupies no space. Although the idea of empty space was repugnant to the German philosopher Immanuel Kant (1724–1804), both he and the English philosopher John Locke (1632–1704) favored action-at-a-distance ideas because they related most directly to what was actually observed. No invisible material medium had to be postulated. On the European continent action-at-a-distance principles prevailed for most of the nineteenth century.

However, the work of Michael Faraday (1791–1867) in England, based on meticulous and prolonged experimentation, began to raise serious problems for the action-at-a-distance school. By 1851, he had concluded that light, heat, electricity, and magnetism all must involve a medium of transmission, *according to the three criteria above*. Gravity did not seem to. As evidence for the continuous medium theory collected, mathematicians with experience in describing the flow of heat and fluids were trying to figure out what kind of substance must exist in the seemingly empty space between objects that were interacting electrically, magnetically, or in other ways. The "fluid" came to be called "aether", as was pointed out in Chapter 8, and it had to be a very unusual "fluid" mechanically. It had to be quite rigid—like jelly—so that light would travel through it at high velocity. At the same time, it had to be thin enough to permit the passage of moving masses with no resistance. Worst of all, it had to be capable of exerting magnetic forces at right angles to a line connecting the two interacting objects. A great deal of time was spent by many brilliant scientists in the nineteenth century trying to spell out the characteristics of the aether. All this time, Faraday, who had little mathematical training—perhaps an asset at the time—preferred to picture the interactions most simply in terms of "lines of force".

His visualization made the forces an attribute of the space around the charged object or magnet, without worrying so much about what set up the influence. He is thus the originator of the modern concept of *field*. As far as Faraday was concerned, the lines of force gave all the answers one could want. Earlier, he had tried to picture the field more mechanically, seeing the lines of forces composed of closely spaced but separate and invisible particles. However, these ideas were not an integral part of his model.

Maxwell (see page 193) was at his creative peak in the 1860's. In contrast to Faraday he was a sophisticated, mathematical physicist. (He was also an experimenter and a builder of scientific toys.) He understood Faraday's thinking and was able to give it mathematical form. Furthermore, he felt the responsibility to spell out the mathematical characteristics—the quantitative behavior—of the aether that transmitted electricity, magnetism and light. For awhile, he too tried a mechanical approach, speaking in terms of "tubes of flow" and "molecular vortices". He thought this would establish the direct-action, rather than action-at-a-distance theory as correct. However, by 1864, he had realized something important: the aether was always seen as "that which caused or carried the interaction". If the aether were a material medium, and not just a mathematical artifice, then it ought to be detectable in some additional way. But it never was. Maxwell realized that the aether was an unnecessary concept. He came to see "field" as a fundamental concept in itself, and that nothing was gained by trying to liken a field to some other more familiar and mechanically describable fluid. His equations, or pictorially, Faraday's lines, told him everything he could find out. The system needed no mechanical scaffolding.

The aether concept, tied so closely with familiar, mechanical experience and with the idea of absolute motion and rest, lived on another two generations. In fact, as was mentioned on page 194, Maxwell's equations, showing that light travels with a certain speed in empty space, were used by other scientists as evidence for the possibility of detecting absolute motion. Absolute motion and the aether concept finally received their death blow from Einstein's special theory of relativity.

Maxwell's work provides an important landmark in the evolution of scientific thinking. Before him, physicists felt that all phenomena must ultimately be explainable by models; actual devices that could be seen, with the oscillations, tensions, gears and cables of familiar machinery. Since his work, physics has learned to find reality in equations—in correct predictions—without requiring mechanical models to form connections with familiar phenomena. Scientific reality had begun its switch from materialism-mechanism to more symbolic (mathematical) abstraction. Visualization and literal models were deemphasized. Newton's words on page 121 now seem almost 200 years ahead of their time. Once physicists gave up describing the medium which carries "the message", then the nineteenth century criteria, and the arguments separating action-at-a-distance from direct-action collapsed. Einstein's rule that no disturbance could be transmitted instantaneously ruled out action-at-a-distance. However, the lack of necessity of a material medium to carry the disturbance did away with the direct-action theory. They were *both* replaced by the *field theory*, which limits itself to the mathematical description of the space in which the object is located.

Twentieth Century Field Theories

The decline of mechanism as a basic principle in explaining interactions through space has been traced. Let us now consider two modern approaches to understanding fields. One is Einstein's general theory of relativity and the other is the quantum theory of fields. The first is a geometrical theory; the second, an exchange theory, involving the emission and absorption of particles.

Certain details in the motion of the planets around the sun indicate that either the force between the sun and the planets is not quite proportional to $1/r^2$, or that Newtonian mechanics is not quite correct, or both. For example, the elliptical orbits of the planets rotate about the sun (Figure 12-24) in a manner that Newtonian mechanics could not completely account for. Einstein's *general* theory of relativity, which has already been mentioned in connection with non-inertial frames of reference and gravity in Chapter 8, gave an explanation. Although it is phrased in highly abstract mathematical language, it can be partially understood as reducing Newton's pushes and pulls to a geometrical problem of space "warpage". According to Einstein, when mass exists at some place in space, it gives curvature to the space around it. Matter and curved space can be interpreted as being synonymous. A mass, moving from A to B (as shown in Figure 12-25), which passes by a second one, travels along what is, in its frame of reference, the shortest line between the two points A and B. This shortest distance is not a straight line near the second mass. The shortest distance between A and B is actually a curved line. (This is what is meant by a curved space.) Another way to describe what is happening would be to say that the masses accelerate toward each other. As George Gamow has suggested in one of his popularizations of the general theory of relativity,[5] gravitational forces act on an object something like the wrinkles in a rug would act on a marble lying on the rug. By tugging on the edge of a rug and making its wrinkles move around, you could manipulate the ball around in much the same manner as if you were attracting it gravitationally, or with a magnet. Whether Einstein's general theory of relativity constitutes a better *explanation* than Newton's law of gravitation is a moot point. It is certainly further removed from common sense observations and experience. However, in the sense that Einstein's theory has wider application and predicts more correctly, it is a better theory.

Two other well-known predictions resulted from Einstein's general theory of relativity. One is that there should be interactions between light waves and gravitational fields. This was checked experimentally during a solar eclipse in 1919, when it was found that the path of star light on its way toward the earth was bent when it passed the edge of the sun. The third prediction is that time should pass more slowly in regions where the gravitational field is stronger. The most recent verification of this[6] was a highly interesting experiment because it gave evidence

(5) Gamow, George, "Gravity", *Scientific American 204*, March, 1961, pg. 94.
(6) Hafele, J. C. and Keating, R. E., *Science 177*, pg. 166, 1972.

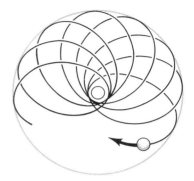

Figure 12-24

Planets travel around in elliptical orbits but the orbits themselves rotate slowly around the sun.

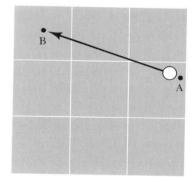

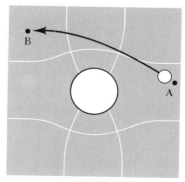

Figure 12-25

Gravitational attractions can be seen as resulting from the curvature of space which mass causes.

which supports both the special and general theories of relativity. J. C. Hafele and R. E. Keating synchronized a pair of atomic clocks (page 14). Then they took one of the clocks on two plane trips around the world. One trip was in an easterly direction, while the other, to the west, was also made at a higher altitude. During the trip eastward, the plane's motion was aided by the turning of the earth, so it traveled faster. According to the special theory of relativity (page 205), the eastward-traveling clock should have lost time relative to the clock at home. The westward-moving clock, traveling against the earth's rotation, moved more slowly than its twin at home, and should have gained time relative to it. Along with all this the clock in the airplane should have gained a smaller amount of time on both trips, according to the *general* theory of relativity. The earth's gravitational field is weaker further from the earth, so time should be slowed less than on the surface of the earth. (The higher altitude on the Westward trip should have made the time gain somewhat larger.) To summarize, the eastward-traveling clock should have lost a small amount of time (as predicted by special relativity). But, some of this loss should be offset by the gain in time (predicted by general relativity). The predicted result was an over-all *loss* of 40 *nano*-seconds. Hafele and Keating measured a loss of 59 nanoseconds, providing verification well within experimental error. The westward-traveling clock should theoretically gain time from both effects, resulting in a predicted overall gain of 275 nanoseconds. Experimentally, they measured a 273 nanosecond gain!

Page 71 states that the inertia of an object can be interpreted in terms of its interactions with the rest of the mass in the universe. In a similar way the gravitational interactions between two objects seems to depend upon the rest of the mass of the universe. In particular, it has been argued[7] that the universal gravitational constant, G, depends upon the average mass per unit volume (density) of the universe: Since the universe seems to be expanding, as evidenced by motion of the distant galaxies of stars *away* from us, G might be decreasing. If G had been greater in the past then the earth would have been closer to the sun and its past history would have been affected accordingly. Evidence for and against these ideas is still sketchy. The lesson to be learned from all this is that the full understanding of even a simple phenomenon, such as the falling of a ball, might require knowledge about the constitution of the entire universe.

More recently, the quantum theory of fields has been proposed to explain physical forces of all kinds. It has been especially successful in explaining the intense but short-range forces that seem to hold the nucleus together. It can perhaps best be described by an analogy. Suppose two children were ordered to play catch in their front yard. They might wander closer to or farther away from each other, but as long as they toss the ball back and forth

(7) Sciama, Dennis, "Inertia", *Scientific American 196*, Feb., 1957, pg. 99.

("exchange" the ball) they can not be separated. The exchange situation, in effect, exerts an attractive force between the children. If the children now are thought of as being the size of nuclear particles, and the ball as a quantum of energy, you have a model for nuclear forces. (Of course, the distance over which they can now toss the ball, and thus the range of the force holding them together, is quite small: on the order of 10^{-15} m.) The quantum theory of fields suggests that two interacting nuclear particles exchange "virtual" quanta with one another, "tossing" them back and forth. The theory indicates that the virtual particles exchanged might have mass. The greater the mass, the more intense and short range is the force field. The concept of binding by the exchange of virtual quanta might sound unreal, but it does make numerical predictions which match experimental observations. For example, when two nucleons are torn apart by a third energetic particle a new particle is often formed. The mass of this new particle matches the mass that an exchange particle (meson) would have had, calculated on the basis of the intensity of the field which held the two nucleons together initially. In other words, a sort of nuclear "glue" tends to lump itself into quanta when two particles are rent asunder.

By using Einstein's mass-energy equation $E = mc^2$, the "lump of glue" is expected to have a mass equal to the initial interaction energy divided by c^2. Very roughly, the energy of interaction, or the work required to separate the two nucleons, should be equal to the average field intensity times the range of the interaction. However, this simple-minded model overlooks one of the main mysteries in particle physics: that is, why the lumps only have certain specific masses.

The same rough model works for gravitational, electric, and magnetic interactions as well. In these cases the quanta exchanged have no rest mass, and the interaction takes on the familiar inverse of the distance squared dependence.

A Summary and Comparison of Physical Forces

Physicists have been able to account for all "physical" phenomena in terms of just four force fields. We have already discussed gravitational fields, associated with mass; and encountered electric and magnetic fields, which are associated with charge. We also just encountered a third force field; that which is found between, or is postulated as holding nucleons together in the nucleus. The intensity of this force, called the nuclear force, was attested to by the large nuclear reaction energies described' in Chapter 9. In the last few years, a fourth force, with the odd name of *weak*[8] has been proposed to account for the fact that

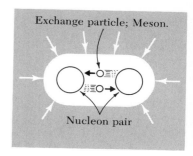

Figure 12-26
Nuclear attraction resulting from the exchange of virtual particles.

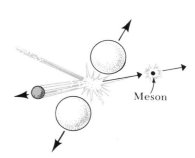

Figure 12-27
As the binding "glue" disappears, a new particle — a meson — appears.

(8) Marshak, R. E., "The Fourth Force of Nature", *The Physics Teacher* 9, Nov., 1971, pg. 435; Dec., 1971, pg. 516.

certain clumps of nuclear particles hold together longer than expected. Table 12-1 (below) compares their relative intensities, and lists the exchange particle that seems to provide the attraction or repulsion for each of the four forces. The quantum of energy in the weak interaction, the W particle, as well as the quantum of gravitation, the graviton, have not been detected experimentally.

To give an illustration of the relative strengths of gravitational and electrical interactions, consider a pair of electrons placed one cm apart. Each has a mass of 9.1×10^{-31} kg and a charge of 1.6×10^{-19} coulomb. The electrical force, F_E, acting between them is given by Kq_aq_b/r_{ab}^2, and the gravitational force, F_G, between them is given by Gm_am_b/r_{ab}^2. The ratio of electrical to gravitational force, for a given distance of separation, would thus be:

$$\frac{F_E}{F_G} = \frac{Kq_aq_b}{Gm_am_b},$$

where $K = 9 \times 10^9$ N·m²/C² and $G = 6.67 \times 10^{-11}$ N·m²/kg²

so

$$\frac{F_E}{F_G} = \frac{9 \times 10^9 \ (1.6 \times 10^{-19})^2}{6.67 \times 10^{-11} \ (9.1 \times 10^{-31})^2}$$

$$= \frac{(9 \times 10^9)(2.56 \times 10^{-38})}{(6.67 \times 10^{-11})(82.8 \times 10^{-62})}$$

Table 12-1   *Comparison of Force Fields and Their Quanta.*

Force Field	Role	Relative Intensity	Range	Quantum
Nuclear	Binding force holding the nucleus together	1	10^{-15} m	Meson
Electro-magnetic	Binding force in atoms, molecules; Large scale Electromagnetic phenomena	10^{-12}	Infinite	Photon (Quantum of Light)
Weak	Certain subatomic particle decay processes	10^{-14}	10^{-15} m	W Particle
Gravity	Weight; astronomical, and geological processes	10^{-48}	Infinite	Graviton

$$= \frac{23 \times 10^{-29}}{552 \times 10^{-73}}$$

$$= .0418 \times 10^{44}$$

$$= 4.18 \times 10^{42}.$$

The electrical interaction between two electrons is 10^{42} times as strong as the gravitational! Protons (hydrogen atom nuclei) have a mass about 1836 times as great as that of electrons. The gravitational attraction F_G, between them is thus $(1836)^2$ or 3.37×10^8 times greater than for electrons. Nevertheless, the ratio F_E/F_G is still greater than 10^{34}.

It might seem odd, then, that gravity plays such a large part in our lives. One reason for this is that we spend our lives on the surface of a very large mass, the earth. A second reason for the predominance of gravitational forces in the large-scale universe is that they are all *attractive*. Electric forces, in contrast to this, can be either attractive or repulsive, and on the average cancel out or neutralize. Furthermore, electric forces in the nucleus tend to be overcome by nuclear forces, as we shall see below.

Planets and stars are spherical, or nearly so, because of gravitational forces pulling them together. In fact, the fusion processes which are the source of stellar energy can start only in the crushing pressure and heat resulting from a giant cloud of mass gathering itself together gravitationally. The size and shape of a shining star such as the sun reflects a balance between an inward gravitational pressure and an outward pressure created by the heat released in the star's fusion processes. (See page 242.) As their nuclear fuel is consumed, stars the size of the sun expand into a huge "red giant", and then collapse into a "white dwarf" about the size of the earth. The dwarf gradually cools and the star dies. According to the theory, stars whose mass is greater than approximately 1.2 solar masses collapse more violently. They collapse through the white dwarf stage without pausing, and then explode into a "supernova" (page 64). During this explosion, much of the material in the star is sprayed outward, perhaps to take part in the formation of some new star in the distant future.[9] The core which remains, collapses into a super dense (neutron) state, wherein the material is literally compressed by gravitational attraction into one big "nucleus". A piece of such neutron material the size of a ping pong ball would have the same mass as an earth type rock more than 100 miles in diameter.[10] If the *core* still has a mass several times the sun's mass (or greater) theory indicates a further collapse into an exceedingly strange state. The huge mass falling in on itself creates

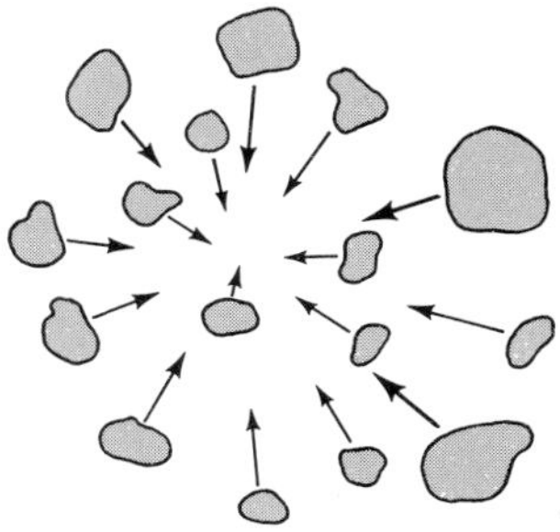

Figure 12-28a
Matter falls together gravitationally.

Figure 12-28b
If the mass is large enough, tremendous pressures and temperatures build up and nuclear fusion processes may begin at the core.

(9) Thorne, Kip S., "Gravitational Collapse", *Scientific American 203*, Nov., 1967, pg. 88. Offprint 316.

(10) Penrose, Roger, "Black Holes", *Scientific American 226*, May, 1972, pg. 38.

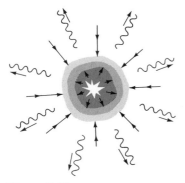

Figure 12-28c
*The inward gravitational pull is
balanced by outward pressures
from the core.*

more and more gigantic gravitational field intensities ($g = Gm/r^2$, and r is getting smaller all the time) so that, according to the general theory of relativity, space becomes so "warped" that radiation emitted by the star cannot escape. A "black hole", the location of a large mass, but the source of no radiation, is formed. Thus far, the observational supports for the existence of such "black holes" is ambiguous.

This discussion of the relative intensity of forces brings us back to the thought posed at the end of Chapter 2: that the forces acting at the surface of the earth have in many ways determined the minimum and maximum size that an intelligent being can have. Our lives are dominated by gravitational forces. The life of the ant is dominated by surface cohesive forces that are electromagnetic in origin. Some types of ants are able to walk upside down on a ceiling not because they have stickier feet than we do but because they weigh so much less than we do. (Others *do* have additional hooks or "suction cups" to supplement the surface cohesion.) As was mentioned in Chapter 2 the same predominance of surface forces would also make books with pages an impossibility for an ant.

The *range* of forces as well as their intensity is important. Two protons repel each other with a large electric force. However, if they can be forced together closely enough, they begin to experience nuclear attractions which are intense enough to continue holding them together. The process would be analogous to two particles separated by a spring, but held together by a latch:

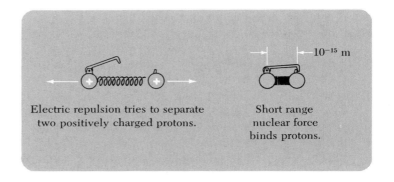

Electric repulsion tries to separate
two positively charged protons.

Short range
nuclear force
binds protons.

Figure 12-28d
*After the nuclear "fuel" is
consumed, stars either collapse
into a white dwarf; or, if the
mass is sufficient, undergo a giant
explosion, becoming a supernova.
The remaining core collapses
further to form a neutron star; or,
if its mass is sufficient, may
collapse even further to form a
"black hole". In this latter case,
gravitational fields around the
star may be so great that no
radiation can escape.*

Force Fields and Potential Energy

Having compared the four types of physical force fields, it would be well to recall, from Chapters 7 and 11, that with each force field you can associate a potential energy. This is true whether the energy is concerned with arrangements of particles in the nucleus (nuclear energy); arrangement of electrons around the nucleus, or atoms within the molecule (chemical energy); with distance between molecules (elastic energy, or acoustical energy); or distance between a man and the earth (gravitational potential energy). The scale, the intensity of the force, and the possibilities for energy release vary widely. Nevertheless, in each instance we're interested in the work required to separate two objects in an attractive force field, or in the energy released when they fall back together.

Extrasensory Perception and Astrology

Do the four force fields listed previously exhaust the possibilities for interactions in the physical world? Two things could be said in reply. First, physicists should not make claims as to the completeness of their world picture. There might be other sorts of forces at work which will eventually be consistently measurable and mathematically describable. Second, what does "physical" mean? What does the term encompass? The structure of science does not necessarily require a completeness of world view, or of necessity a denial of the existence of "supernatural" phenomena. Simply put, phenomena in the "supernatural" realm do not yet lend themselves to the scientific criterion of disprovability. They are outside the (present) realm of physical science. But what of extrasensory perception (ESP)? The intent of people who are working on ESP would seem to be to show that the phenomena *are* reproducible and that they belong inside the "physical" boundary.

"Extrasensory perception" includes four effects:

(1) telepathy, where two minds communicate;

(2) clairvoyance, where a mind can see and describe events far away, or at least not in view;

(3) precognition, where future events are seen; and

(4) psychokinesis, wherein the mind affects the motion of an object such as a rolling die.

J. B. Rhine of Duke University has studied these effects in a laboratory setting for many years, and claims to show that if the effects *are* physical they do not die off with distance, as do most physical force fields. (We know from our discussion of electric fields between sheets that a physical influence need not die off with distance, especially if the source itself is spread out.) Rhine finds that subjects seem able to get messages from Europe about as intensely as they do from other minds in the same room. A difficulty with ESP is that the effects are not always observed, and a statistically based statement is weak from the disprovability point of view. The investigator can never be certain an effect has stopped. It is within the realm of statistical possibility that nothing happen for a number of trials, even when a valid effect is being studied. In a typical experiment, the subject is asked to identify 25 cards dealt onto a hidden table. The subject may identify 15 correctly when, on the average, random guessing would lead to just five correct answers. The investigator, who tends to believe in ESP, points out that 15 correct answers is highly unlikely statistically. A statistician points out, on the other hand, that *anything* can happen statistically. The 15 correct answers could indeed be statistical, rather than physical in origin. Another problem is that psychically sensitive subjects are not always sensitive, so the problem of when to start recording their responses arises. Workers in the field of ESP have tended to statistically emphasize successes and ignore the times when nothing interesting happens. All of these factors have served to discredit the laboratory study of extra-sensory perception.

Of all ESP phenomena, precognition is most directly at odds with present physical theory. We have seen that special relativity denies us the privilege of knowing that something happens *now*, let alone in the future. One serious supporter of ESP[11] denies true precognition, but suggests that certain fields unknown to us guide human behavior so that a set of conditions, leading inevitably to a later result, might be sensed. It would be something like a dog sensing a change of air pressure ahead of a storm which is still hidden from view.

As a science investigating natural phenomena physics does not, or should not, deny the possibility that "psychic powers" *might* be manifested in certain individuals. The evolution of physics away from mechanistic thinking toward field theory allows for more possibilities for including ESP among physical effects. But most physicists are waiting for honestly consistent data and a testable model before accepting ESP as "science".

Astrology presents a different sort of problem for the sciences. It claims that the relative positions of the stars and planets at the time of an individual's birth play a large part in the life that the person subsequently leads. The art of astrology is more than 2500 years old. Historically, astronomy came first, developing among farming cultures because of the importance of time keeping in the planting of crops. The Assyrians who moved in on Babylon around 800 B.C. were primarily soldiers, so their interest in the stars had a different basis. It became the source of secret "classified" information, hopefully useful for predicting or influencing present and future military events. With this, astrology began. Insofar as it became secret, it was no longer science, according to Ziman's criterion.[12] Still, scientists right up to Kepler's times saw the universe astrologically, as much as they

(11) Wasserman, G. D., in *Forces and Fields*, Mary B. Hesse ed. Philosophical Library, New York, 1961, pp. 300–303.
(12) See footnote 22, page 126.

did astronomically. France employed a state astrologer until late in the 1600's.

Physical science brings with it several basic attitudes toward the universe. They have been verified repeatedly and they make astrology very difficult to accept. One attitude is that the universe is *uniform*. There is no center, no preferred place. All positions are relative. A bunch of stars here has the same effect as a bunch of stars there, as long as you are the same distance from them. All stars are made out of the same stuff. There are as many stars in one direction as in another, when you look out from any point in the universe. All of these statements make it hard to believe that the relative positions of a few stars in constellations could exert much of a changing effect on an individual's life. It should not make any difference whether we are born under Pisces or Aquarius.

The planets are fewer in number than the stars, but they are closer to us and undergo greater changes of position relative to us. However, the sun, because of its mass and the moon, because of its nearness, would have to be the main source of "cosmic" influences. It is quite possible that their changing gravitational fields and radiation affect us in ways we don't know about, or understand, yet. Or, there may be influences other than gravity and the solar wind which they send toward us. In any case, there is evidence aside from astrology, or old wive's tales, that human behavior undergoes cyclic changes which match movements of the sun, moon and planets.[13] The evidence is sketchy and the system (the human being) is very complex.

Looking at astrology more psychologically, (which might be the best way to view it anyway) it could certainly have an effect. If half the investment brokers in New York City were believers in astrology and said that next Thursday has some bad signs attached to it, their *belief* in astrology could make the market change. Belief in ESP or astrology is further evidence for the intense desire humans feel to find causes for the events they experience. This basic feel for cause and effect is, of course, at the bottom of the irrepressible questioning which has led (among other things) to science.

Summary

Magnets (compasses, for example) line up parallel to any magnetic field into which they are put. The direction of the magnetic field is defined as the direction in which the north pole of a magnet points when it is placed in the field. North and South poles always seem to occur in pairs.

(13) Gauquelin, Michel, *The Cosmic Clocks*, Henry Regnery, 1962. (After giving evidence that astrology is largely superstition or fakery, he goes on to list a large amount of evidence that cyclically varying extra-terrestrial influences do affect the behavior of inanimate as well as animate objects.)

Electric currents produce magnetic fields.

Two current carrying wires exert magnetic forces on each other. From this comes the definition of the ampere (and of the coulomb):

One ampere is the steady current which will produce a force of 2×10^{-6} N per meter of length between two parallel wires placed one meter apart in vacuum.

A charged particle moving with velocity v in a magnetic field experiences a force

$$F_m = qvB \sin\theta \hspace{3cm} \text{Equation (12-1)}$$

where B is the magnetic flux density, measured in webers/m² and θ is the angle between v and B. The force is perpendicular, not parallel to, the B field.

The earth has the magnetic field which would result from a large South pole under the Queen Elizabeth Islands and a large North pole under Anarctica. These poles have wandered widely during the earth's history.

The earth's magnetic field deflects and traps charged particles which reach it from outer space, creating belts of charged particles centered at about 2500 and 11,000 miles above the earth's surface.

When the magnetic field in a region is changed, an electric voltage is induced. This voltage, if given some charge to act upon, will create a current. The current will be in a direction such that the magnetic field associated with it will oppose the original change in magnetic field.

When wires are mechanically moved through a magnetic field, currents are generated in them. This is the principle behind the electric generator.

Changes in electric and magnetic field thus affect each other.

Until the mid-Nineteenth century, disputes raged between advocates of the action-at-a-distance and direct-action explanations for interactions over large distances. These two viewpoints were finally replaced in the 1860's by a field theory, which focused attention on the region around the object experiencing the force. The acceptance of this theory ended the principle that all of physics must be understood in literal terms of mechanical objects, such as gears and pulleys.

Modern field theories utilize concepts such as a geometrical "warping" of space, or the exchange of a virtual particle between the interacting objects.

The forces thus far proposed to explain the physical world, in order of decreasing intensity, are nuclear, electromagnetic, "weak" and gravity. Gravity forces are weakest, yet are of most importance in the large-scale universe. This is because they extend outward over long distances and they are always attractive.

Whenever two objects are pulled apart while they are attracting each other, their potential energy increases.

Extrasensory perception and astrology are two examples of the belief that force fields, in addition to those listed above affect us, or that the physical forces affect us in ways that we don't understand yet. They are further examples of humanity's desire to find causes for its behavior.

Further Reading

Akasofu, Syun-Ichi, "The Aurora", *Scientific American 213*, Dec., 1965, pg. 54. (Explains the northern lights.)

Blin-Stoyle, R. J., "The End of Mechanistic Philosophy and the Rise of Field Physics", pp. 5–29 in *Turning Points in Physics*, Harper Torchbook TB 535, pp. 5–29, 1961. (Readable, non-mathematical.)

Dyson, Freeman, "Field Theory", *Scientific American 188*, April, 1953, pg. 57, Offprint 208. (Discusses the classical and quantum theory of fields.)

Einstein, Albert and Infeld, Leopold, *The Evolution of Physics*, Simon and Schuster, N.Y., 1942. (First 160 pages trace the rise of the principle of mechanism, then its replacement by field theory.)

Jammer, Max, *Concepts of Force; A study in the Foundations of Dynamics*, Harvard University Press, Cambridge, 1957. (First 50 pages give historical background.)

Kondo, Herbert, "Michael Faraday", *Scientific American 189*, Oct. 1953, pg. 90. (Faraday's discovery of electromagnetic induction led to the electric generator, but his main contribution to physical theory was the concept of field.)

Newman, James R., "James Clerk Maxwell", *Scientific American 192*, June, 1955, pg. 58. (Account of the life and works of the greatest theoretical physicist of the nineteenth century.)

Park, David, "Fields", Chapter 3 in *Contemporary Physics*, Harcourt Brace Jovanovich Harbinger Book, New York, 1964. (Shows the misleading character of mechanical models and discusses the decline of the materialistic interpretation of the physical world.)

Parker, Eugene N., "The Solar Wind", *Scientific American 210*, April 1964, pg. 66. (The sun sends out streams of particles. They can be thought of as part of the sun's atmosphere, and go out as far as the planet Jupiter.) also, "Universal Magnetic Fields", *American Scientist 59*, Sept./Oct., 1971, pg. 578. (Strong magnetic fields are found throughout the universe and are responsible for many exotic effects.)

Sciama, D. W., *The Physical Foundations of General Relativity*, Doubleday Anchorbook, Garden City, N.Y., 1969. (Discusses inertia, the principle of equivalence, and the general theory of relativity.)

Tricker, R. A. R., *The Contributions of Faraday and Maxwell to Electrical Science*, Pergamon Press, London, 1966.

Wheeler, John A., "Our Universe: the Known and the Unknown", *The Physics Teacher 7*, Jan. 1969, pg. 24. (A very readable discussion of gravitational collapse, the expansion of the universe, and other cosmological problems.)

Problems

12-1 If you had been in Oersted's place how would you have "measured the magnetic field set up by a current"? (page 305).

12-2 In what direction does a compass point when it's at the geographic north pole? What *is* the geographic north pole? In what direction does a compass (try to) point when it's at the magnetic north pole?

12-3 How does the earth's magnetic field affect the electron beam in a TV set, in the northern hemisphere?

12-4 Suppose the vertical component of the earth's magnetic field is 10^{-5} W/m² and directed down. A horizontal powerline carries 50 A of current south. Find the direction and magnitude of the magnetic force per meter of powerline. (Note: you need to show first that qv is equal to the current I times length, ℓ.)

12-5 An electron enters the region outlined by the dashed lines (as shown in the margin) and is deflected by a magnetic field having a flux density $B = 0.1$ W/m². The velocity of the electron is 5×10^8 cm/sec. If the path taken is as shown,

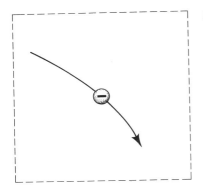

(a) show the direction of the force exerted on the electron.
(b) what must be the direction of the *B* field within the dotted region?
(c) calculate the magnitude of the force exerted on the electron.

12-6 Uranium ions are shot into a region containing a magnetic field of intensity $B = 1.2 \times 10^{-2}$ W/m². The ions have a speed of 1.46×10^4 m/sec and were injected in a direction perpendicular to the magnetic field. They have a charge of 4.8×10^{-19} C and travel in a circle having a radius of 100 cm. Are the ions $^{235}_{92}$U or $^{238}_{92}$U?

12-7 A magnet is thrust into a coil of wire, the north pole first, as shown in the margin. Explain the direction of the current which is induced in the coil of wire.

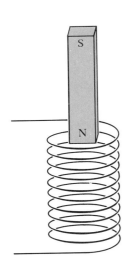

12-8 A magnet, with its South pole down, is dropped through the center of a ring of wire. Describe the forces which act on the magnet.

12-9 What happens when a particle is suddenly created, á la quantum theory of fields? Is the field in question established instantly everywhere or does it take time to propagate?

12-10 Explain in your own words what a mechanistic point of view is.

12-11 Suppose the positions of the stars do influence our lives. Is it a denial of Einstein's special relativity to say that the position of the constellations at the moment of birth determines the course of our lives?

13

WAVES

Much of physics is concerned with *movements* of energy. If all of the energy in the universe were to stay in the same place, the universe would be far less interesting. For one thing, there would be no life on the earth. Let's consider some of the possible ways of moving energy around.

Disturbances That Travel

Suppose you want to send a signal to someone who's asleep in a boat in the middle of a pond. To send a signal, you must get a bit of energy to the person in a controlled fashion. First, you could throw a stone and hit the person on the head.

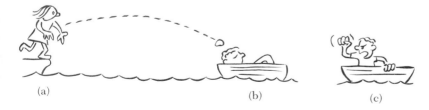

(a)　　　　　　　　　　(b)　　　　(c)

You would then have sent the energy by *transfer of mass* from one place to another. Second, you could slap the water with an oar. Then a water wave would travel across the pond, eventually rocking the boat.

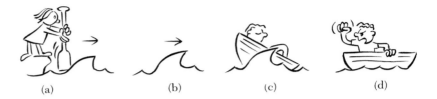

(a)   (b)   (c)   (d)

Third, you could yell at the person and a sound wave would move through the air, reaching the person in the boat in a fraction of a second.

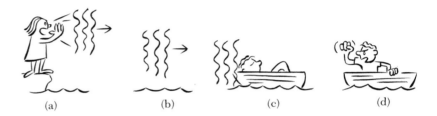

(a)   (b)   (c)   (d)

The interesting thing about the latter two signals is that they involve the transfer of *no* matter. Nothing moves from shore to boat except a *disturbance*. This peculiarity of waves should be kept in mind throughout the chapter. They are unlike tides, wind or spreading odors. Waves are simply disturbances moving through space, whether the disturbance is a jostling of water molecules, air molecules, or electromagnetic field. Waves are something that happens to something else. After the wave has passed, everything is back where it was at first. Everything, that is, except a bit of energy.

A wave may be a one-shot pulse or a continuing oscillatory (back and forth) disturbance. Most of our discussion concerns oscillatory waves. There are several parameters which are useful in describing periodic waves. One is the *wavelength*, λ, (lambda). Figure 13-1 shows λ as the distance between two peaks in a wave moving along a string. It could also be the distance between two "valleys", or, for that matter, between any two nearest similar events at a given time. At a point past which the wave is moving the disturbance will seem to repeat itself with some *frequency*, f, That is, while we're watching a wave pass through one part of the string, the number of peaks (or valleys) passing per second will be f. Frequency is measured in units of cycles per second, now called *hertz*, after Heinrich R. Hertz

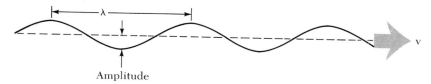

Amplitude

Figure 13-1

Wavelength is the distance between two similar parts of a repetitive wave.

(1857–1894), the German physicist who pioneered in the study of electromagnetic wave transmission.

Now imagine the wave in Figure 13-1 to be moving to the right with a velocity, v. It will cover a distance, λ, in the time it takes to complete one cycle: $\underset{|\leftarrow\lambda\rightarrow|}{\smile}$. When a certain peak reaches the point in the string that we are watching, the previous peak will have moved a distance, λ, to the right. The time required to complete one cycle is called the period, τ, of the oscillation (page 54). If it has a repetition frequency of f hertz; that is, if it takes $1/f$ seconds to complete one cycle, then the velocity of the wave,

$$v = \frac{\text{Distance Traveled}}{\text{Time Elapsed}} = \frac{\lambda}{1/f}$$

or

$$v = \lambda f. \qquad\qquad \text{Equation (13-1)}$$

If we watch the string for one second, f peaks will pass by and the very first peak will have moved a distance, λf, to the right.

To illustrate this further, suppose you are on a dock, watching waves come in from a passing boat. You look at the waves and see that there is about one meter between crests. You then listen and notice that three crests slap the pilings of the dock every two seconds. From these observations you know that the wavelength, λ is one meter and the frequency, f, is 3/2 or 1.5 hertz. The wave must then be moving with velocity:

$$v = \lambda f = (1\text{m})(1.5 \text{ hertz}) = 1.5 \text{ m/sec.}$$

One further wave concept is the *amplitude*, or maximum disturbance set up by the wave (shown in Figure 13-1). The energy carried by a wave is proportional to the square of the amplitude of the wave. But, except for certain water waves, the *energy* transported by the wave has nothing to do with the *velocity* of the wave. A loud sound wave travels no faster than a soft one: it just has a larger amplitude. Sound waves do not slow down as they fade out. The same relation holds for bright versus dim light waves.

For simplicity we have pictured a wave in which the disturbance is up or down, that is, perpendicular to the direction in

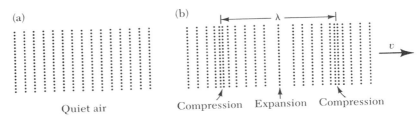

Figure 13-2
(a) Schematic representation of the molecules in quiet air. (b) Molecules in the air with a longitudinal sound wave passing through them from left to right.

Figure 13-3
Cross section of an ocean wave traveling from left to right. Circles are orbits of water particles in the wave. At the surface their diameter equals the wave height.

which the wave travels. Such waves are said to be *transverse*. Electromagnetic waves have this character. On the other hand, sound waves in air involve a back and forth disturbance of air molecules along a line parallel to the direction in which the wave moves (Figure 13-2). Such waves are called *longitudinal*. Waves on the surface of water require a rolling disturbance which combines longitudinal and transverse disturbances (Figure 13-3). In solids, *torsional waves* are also a possibility.

The important thing to remember is that the wave velocity is not the same as the velocity with which things shake as the wave passes through the medium.

Sound Waves

When you whistle, snap your fingers, talk, or sing, a part of your anatomy is shaking back and forth rapidly. Near the vibrating surface, whether it be lips, finger skin or vocal cords, the air follows the surface and also starts oscillating. The oscillations spread from one molecule to the next, and a sound wave moves outward from the vibrating surface. By everyday standards, sound waves move pretty fast: 331 m/sec or 1090 ft/sec—about a mile in 5 seconds—in air at 0°C (32°F). The time required for sound to

travel is easily detected from echoes. If you stand a hundred feet or so from a large reflecting wall and clap your hands, the sound comes back, delayed by the time required for it to travel from you to the wall, and back again. Dividing the total time elapsed in half, and multiplying by 1090 gives the distance to the wall, in feet. In more dense media the speed of sound is much greater, as shown in Table 13-1.

Our hearing apparatus intercepts vibrations having a frequency of greater than about 20 hertz as a steady tone. We can also hear sounds up to 20,000 hertz, depending upon our age. As we grow older, beginning for most people in the early 20's, our ear mechanism stiffens and there is a steady decline in the highest frequency we can hear. Hi-fi buffs in their 50's may be paying a lot of money to reproduce sounds they cannot hear!

Sounds waves above the audible frequency range are called *ultrasonic*, and have been applied to such diverse tasks as sending "silent" signals to dogs (in air) and cleaning jewelry (in liquids).

The *pitch* of a sound wave depends upon its frequency. The higher the frequency, the higher the pitch. When the frequency doubles, we say the pitch goes up one *octave*. High C on the musical staff is two octaves above middle C. Thus, its frequency, 1050 hertz, is two times two, or four, times the frequency of middle C, 264 hertz.[1] The frequency interval between 20 hertz and 20,000 hertz is slightly less than 10 octaves.

Table 13-1

Medium	Longitud. Wave Velocity in m/Sec.	Medium	Longitud. Wave Velocity in m/Sec.
Water	1500	Cork	500
Benzene	1295	Brick	3650
CO_2	1259	Lucite	2680
Paraffin	1300	Pyrex glass	5640
Wood (Oak)	3850	Mild steel	5960

(1) Diatonic scale, with A taken to be standard pitch at 400 hertz.

Knowing the speed of sound and its frequency of vibration you can figure out how large sound waves are; that is, the wavelength, or how far it is between successive compressions or expansions in the moving wave. For a 20 hertz sound wave in air,

$$\lambda = \frac{v}{f} = \frac{331 \text{ m/sec}}{20} = 16.5 \text{ m}.$$

For a 20,000 hertz sound wave,

$$\lambda = \frac{v}{f} = \frac{331 \text{ m/sec}}{20,000} = .0165 \text{ m} = 1.65 \text{ cm (about } \tfrac{2}{3} \text{ inch)}$$

Suppose a tuning fork, train whistle, violin, soprano, and clarinet were each to give out a sound having the same pitch, in turn. The basic frequency would be the same in all cases but you would have no trouble distinguishing one from the other. They all have different quality, or *timbre*. How can this be? To answer this question, let us begin by diagramming a sound wave in a different way. We draw a stop-motion diagram as usual, but instead of trying to sketch the longitudinal motion of sheets of molecules, we plot the movement up and down: up corresponding to a compression of the molecules, and down to an expansion or movement away from each other. The tuning fork produces a smooth sine curve (see Appendix F). It's pure, but it's not very interesting to listen to. The violin, the human voice, and the clarinet, on the other hand produce more complex waveforms. They can be thought of as the result of superimposing various amounts of *"overtones"* upon the *fundamental* wave which determines the pitch. Overtones have a frequency which is a whole number times the frequency of the fundamental tone. Any sound source can be synthesized by adding together the right recipe of fun-

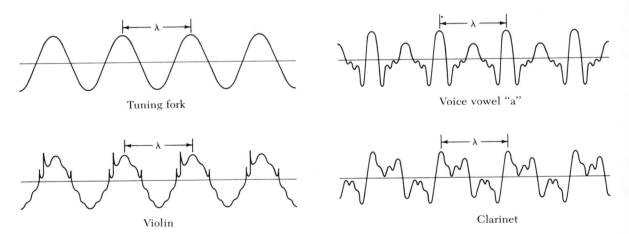

Tuning fork

Voice vowel "a"

Violin

Clarinet

Figure 13-4
Waveforms of various sound sources, all having the same pitch.

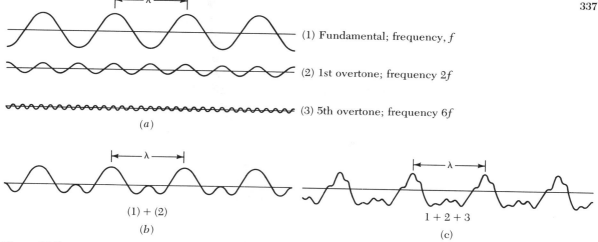

(1) Fundamental; frequency, f

(2) 1st overtone; frequency $2f$

(3) 5th overtone; frequency $6f$

(a)

(1) + (2)

(b)

$1 + 2 + 3$

(c)

Figure 13-5

The production of more complex wave forms by adding together various simple sine waves. The pitch of the complex wave's tone is the same as that of the fundamental wave.

damental and overtone waves.[2] Figure 13-5 shows an example of this.

The first overtone has twice the frequence and half the wavelength of the fundamental wave. It's therefore an octave higher in pitch. If the fundamental and first overtone are mixed in the proportions shown in Figure 13-5a (that is, with the overtone having about half the amplitude of the fundamental), the composite wave shown in Figure 13-5b is the result. If some fifth overtone is also added, with an amplitude about one quarter the amplitude of the fundamental, then Figure 13-5c is the result. According to a theorem of the French mathematician Jean B. J. Fourier (1768–1830), *any* repetitive phenomenon, such as a wave, can be built up or synthesized by adding together the right amounts of sine or "tuning fork" waves having various frequencies. *Noise* is a mixture of waves, with a variety of wavelengths, which add together randomly.[3] We will look into some other consequences of adding waves together in Chapter 14.

There are other ways in which a singer or instrumentalist can add interest to the musical tones that they emit. One technique is called *vibrato*, wherein the performer varies a fundamental pitch up and down rapidly. Violinists do this by moving their fingers back and forth a short distance along the string while they

(2) The synthesis of musical tones by electronic means has made rapid progress in the last 20 years. Workers are now realizing that the "attack" or way in which a tone begins also has an important effect on the quality of the tone.

(3) Beranek, Leo L., "Noise", *Scientific American 215*, Dec., 1966, p. 66. Offprint 306.

bow, thus changing the length of string which is vibrating. On page 359 it is shown that changing the length of a vibrating object varies its frequency of vibration.

Electromagnetic Waves

The last chapter showed that motion of electric charge produces magnetic fields, and that changes in magnetic fields make electric charges accelerate. Putting it another way, changing electric fields produce magnetic fields, and changing magnetic fields produce electric fields. As the frequency of change increases the fields tend to peel off the wires and circuits which set them up and move out through space. Maxwell showed in the 1860's that just as a stone, dropped on the surface of a pond, sends out a disturbance; a change in either an electric or magnetic field does not just shimmer in one place, but rushes outward at 186,000 miles/sec or 3×10^8 m/sec. Figure 13-6 is an attempt at illustrating what seems to be going on. When the electric or magnetic field is building up at a point, a magnetic or electric field is also generated at the same point in space, but at right angles to the other field. The electric field is shown as being in a vertical plane in Fig. 13-6. It increases upward, decreases, increases downward, decreases, etc, moving along rapidly to the right like some immaterial water wave. The magnetic field does the same, and is shown here in the horizontal plane. The vibrating electric charges which generated the wave in the first place are somewhere off to the left.

With this model of the electromagnetic field Maxwell joined electricity, magnetism, and light, three phenomena which previously had been thought to be independent of each other.

Electromagnetic waves of every imaginable frequency seem to be produced in the natural or man-made world. Even the current which oscillates back and forth in a 60 hertz AC lamp cord

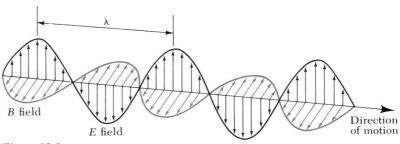

Figure 13-6
An electromagnetic wave moving toward the right.

Table 13-2   The electromagnetic spectrum

	Wavelength, in Meters
Cosmic rays	10^{-14} and smaller
Gamma rays	$3 \times 10^{-14} - 3 \times 10^{-12}$
X-rays	$3 \times 10^{-12} - 10^{-9}$
Ultra-violet	$10^{-9} - 4 \times 10^{-7}$ $(1 - 430$ nm$)$
Visible light	$4.3 - 6.9 \times 10^{-7}$ $(430 - 690$ nm$)$[4]
Infra-red	$6.9 \times 10^{-7} - 10^{-3}$ $(690 - 10^6$ nm$)$
Microwaves	$10^{-3} - 10^{-1}$
UHF TV	$.64 - .34$
FM, short wave	$1 - 1.6 \times 10^2$
AM radio	$1.8 \times 10^2 - 6 \times 10^2$
Extremely long wave (used for underwater communication)	$10^8 - 10^{10}$

radiates some power outward. At this low frequency, the amount of power in the wave is very small, but it can still be detected. Radio or television sets can pick up a 60 hertz hum from nearby fluorescent lamps. The wavelength of a 60 hertz electromagnetic wave is gigantic:

$$\lambda = \frac{v}{f} = \frac{c}{f} = \frac{3 \times 10^8 \text{ m/sec}}{60 \text{ hertz}} = 5 \times 10^6 \text{ m}$$

$$= 5 \times 10^3 \text{ km}$$

$$= 3105 \text{ miles.}$$

Warm objects also give off electromagnetic waves. Most of the wave energy, in this case, consists of infra-red radiation, heat waves which are feelable but not visible. When objects become "red-hot", it means that some of the electromagnetic radiation has a sufficiently small wavelength for us to see (Figure 13-7). An incandescent lamp is simply a wire protected from the oxygen in the air, and heated hot enough so that much of its radiation is visible to us. *Light* is *visible* electromagnetic radiation. Table 13-2 outlines some of the radiation within the electromagnetic wave *spectrum*. All of it travels with a velocity of 3×10^8 m/sec in vacuum.

(4) Our vision fades gradually as deep red and deep violet wavelengths are approached, so that at 430 and 690 nm, our visual sensitivity has dropped to about one percent of what it is in the middle of the visible region.

Light

We live in a sea of electromagnetic radiation, some of it coming from man-made sources and some of it from natural sources. Large or small, these *waves* are not directly visible. We can't see them go by like water waves or waves in a rope. They represent oscillations in electric and magnetic fields, and when you get right down to it, our only evidence for their existence comes from the forces they exert on charged particles.

We are blind to all electromagnetic waves except for a narrow "visible light" region of the spectrum lying between 430 and 690 nanometers. (Other wavelengths affect us physically but are not visible: microwaves and infra-red warm, ultra-violet tans, and x-rays can damage tissue. Medical science is beginning to find other effects of electromagnetic radiation in the mending of bones and healing of wounds. Some kinds of radiation from electrical machines might be harmful without our knowing it.[5]) Our eyes respond to less than one octave in the spectrum. However, we have broadened our detection of electromagnetic radiation by using instruments such as radios, and infra-red sensors.

If a source of light gives off all the wavelengths in the visible region at the same time, in the same amounts, the sensation that we see is of *white light*. If the light is then separated into its component wavelengths, (we will see two ways to do this later) the various wavelengths will produce the various color sensations, as listed in Table 13-3.

The color of most objects is determined by the frequencies of the light that they reflect. If white light illuminates a blue blanket, most of the other colors are absorbed but much light in the region of 450 nanometers (6.7×10^{14} hertz) is reflected. However, a very hot object itself, can become a source of light as has been mentioned. In the case of very hot glowing solids, all colors are

MAN, HAVE I GOT THE 450 NANOMETERS!

Table 13-3   The visible spectrum

Color	λ in nm	f in hertz
violet	430	7×10^{14}
blue	450	6.7×10^4
green	550	5.5×10^{14}
yellow	600	5×10^{14}
orange	630	4.8×10^{14}
red	650	4.5×10

(5) Becker, R. O., "Electromagnetic Forces and Life Processes", *Technology Review* 75, Dec., 1972, pg. 32.

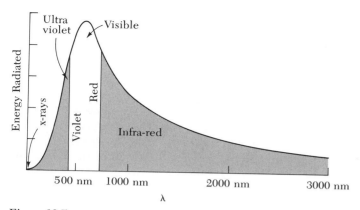

Figure 13-7

The continuous spectrum emitted by a hot object such as the sun.

emitted simultaneously. Figure 13-7 shows how the energy radiated by an object at 6000 degrees Celsius (the temperature of the sun) is spread among the various wavelengths. Most of the energy is in the infra-red region, but visible light, and even x-rays are emitted.

On the other hand, some sources (such as gases) give off only certain colors, not a continuous spread of colors. If the energy that they emit is concentrated near just one wavelength, they are called *monochromatic* sources.

As was outlined in previous chapters, it took physicists a long time to accept the idea that electromagnetic radiation can travel through space without the aid of a material medium, or an aether. It can also travel through material media, but unlike sound waves, it usually travels more slowly in more dense surroundings. We think of glass as transparent because electromagnetic waves which are visible to us can get through it. However, other materials such as metals may be "transparent" to waves that are not visible. (For example, the visibly opaque metal silicon is exceedingly transparent to infra-red (heat) radiation). We see materials like silicon as opaque because they don't let visible light waves through.

Gravity Waves

From what has been discussed thus far, we might expect that a change in *any* field quantity should propagate outward as a wave. Does this mean that changes in a gravitational field occurring during the collapse (or explosion) of a star or in the rotation of two stars around each other, will produce gravitational waves? The theoretical answer is yes. Einstein's general theory of relativity published in 1916, showed that gravitational effects must

travel *at* the speed of light, but no faster. A much later paper[6] showed that gravitational waves could exist and that they should be transverse in nature, just like electromagnetic waves. No one had seriously looked for such waves until the late 1960's when Joseph Weber, a physicist at the University of Maryland, began assembling a large gravity wave detector. His first apparatus used a large cylinder of aluminum which would flex slightly when a gravity wave passed through it. Exceedingly sensitive detectors were used to record the flexing. With this equipment, Weber has not only detected gravitational waves experimentally, but seems to observe an embarrassing wealth of them. In the first few years of his study he observed at least one and sometimes two waves per week. His results were raising many questions about the possible sources of so many gravitational waves.[7] Numbers of other physicists have become interested in the subject and are trying to reproduce Weber's results.[8] Not everyone agreed that Weber's effects were really caused by gravity waves.

Reflection of Waves

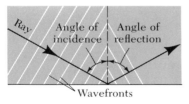

Figure 13-8
Reflection of a pulsed wave in a rope.

Whenever a wave strikes the boundary between two media in which the wave velocity is different, part of the wave is *reflected*. Part of the wave's energy always bounces back.

When a wave strikes a two-dimensional surface the direction of the reflected wave is always such that the *angle of reflection* is equal to the *angle of incidence*. (It is convenient to define a *ray* as giving the direction of travel of the wave and a *wavefront* as the surface over which the disturbance is the same at a given instant.) The tricky thing is that you can't always be sure what the angle of incidence *is*. For a smooth, flat reflecting surface, the incident rays come into the surface at the same angle of incidence, and they reflect in a common direction. This kind of reflection is called *specular*, after the Latin word for mirror. However, if the reflecting surface is rough, as in the case of a river or lake, the angle of incidence of a ray will depend upon where the ray hits the water. The light undergoes *diffuse* reflection, with light coming off in various directions (Figure 13-10). If you look at the moon reflected from a lake or river with small ripples on the surface the reflection appears to be stretched out vertically (Figure 13-11). Most surfaces reflect less diffusely at glancing angles. If you hold a piece of paper up so that light from a lamp bounces off the paper, as shown in Figure 13-12, the paper will look shiny. The light which reaches you is just skipping off the tops of the rough spots, so the whole surface looks smoother.

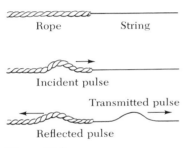

Figure 13-9
Reflection of a wave from a smooth surface.

(6) Einstein, A. and Rosen, Nathan, "Gravitational Waves", *Franklin Institute Journal 223*, pp. 43–54, Jan., 1937.

(7) Weber, Joseph, "The Detection of Gravitational Waves", *Scientific American 224*, May, 1971, pg. 22.

(8) Logan, Jonathan L., "Gravitational Waves—a Progress Report", *Physics Today 26*, March, 1973, pg. 44.

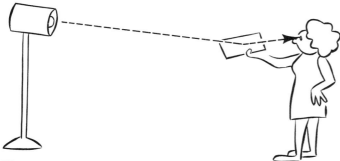

Figure 13-12
Surfaces look shinier when the light glances off them.

Aside from roughness or smoothness, one material can be a better reflector of waves than another. Metals are good reflectors of light because less light gets into them. *Hard* materials are better sound reflectors than soft materials.

Architects use these principles to focus or break up sound waves in large halls. A hard, smooth surface mounted above the stage at the correct angle will project more wave energy out into the audience. To prevent too much reverberation of sound, brick walls can be roughened by intentionally sinking in an occasional brick or by letting an occasional brick stick out. The reflected sound will be scattered, and spread out more evenly.

Figure 13-10
Reflection from a rough or matte surface.

Bending of Waves

Up to this point the emphasis has been on the wave which always bounces back when a wave strikes the boundary between two media. In many instances a portion of the wave is able to enter the second medium and travel through it. However, some changes in this transmitted wave occur which produce interesting visual effects. For example, when you look down into a pool, objects appear to be nearer the surface than they really are. If you stick a pencil into a glass of water, it appears to be bent. When people walk along a hot, dry desert and peer toward the horizon, they think they see a lake ahead. These phenomena — and many more — are not hallucinations, but a direct result of the properties of waves passing from one medium into another. First, an explanation for these phenomena will be proposed. Then, further consequences of this explanation (known as Snell's Law) will be discussed. These examples will involve light waves because the results are easy to see, but the same principles apply to sound, water, radio, and other sorts of waves as well.

People normally think of a wave as traveling in a straight line from its source to whatever place it goes. However, if a wave passes from one region into another where its velocity is different, it may bend. (A portion is also reflected, but we will neglect

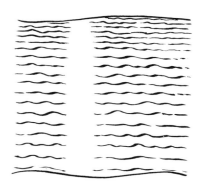

Figure 13-11
Surface roughness spreads out the moon's reflection.

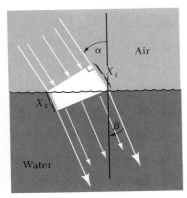

Figure 13-13
*Waves are bent when they enter
a medium in which their speed
is different.*

that fact for the next few pages.) Consider a light wave passing from air into water, where the wave travels only three-fourths as fast. Let the incoming ray make an *angle of incidence, α,* with a line perpendicular to the air-water interface and let the refracted ray make an *angle of refraction, β,* with the perpendicular. Let the wave velocity be v_1 in the first medium (air, in this case) and v_2 in the second medium (water). Note that the left end of the wave front reaches the interface at a time when the right side of the wavefront still has a distance X_1 to go. An instant later the left side of the wavefront enters the second medium. Therefore, in the time, Δt, during which the right side of the wave front has traveled the distance X_1 in air, the left side will have penetrated a distance X_2 into the water. Since v_2 is less than v_1, the distance X_2, which equals $v_2\Delta t$, is smaller than the distance X_1, which is equal to $v_1\Delta t$. (Remember, Δt is the time required for the *wave-front* to make the complete transition from one medium to the other.) Quite clearly, the direction of the wave is changed because one side of it spends a longer time in the slow medium than the other side does. If the wave moves from a slow to a faster medium, the bending is in the opposite direction. Such bending of waves is called *refraction.* Note that a wave entering a slower medium is bent toward the perpendicular and that a wave entering a faster medium is bent away from the perpendicular, as shown in Figure 13-14.

How much does the wave bend? If we enlarge Figure 13-13 and concentrate on the two right triangles formed by the two wave fronts, and on the distances X_1 and X_2, we can compare the angles $α$ and $β$.

First, we need to convince ourselves that the old and the new wavefronts make angles $α$ and $β$ with the interface itself. The interface of the two media forms the common hypotenuse, h, of the two triangles. Therefore:

$$\frac{X_1}{h} = \sin\alpha \text{ and } \frac{X_2}{h} = \sin\beta.$$

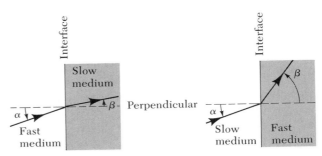

Figure 13-14
*These phenomena hold true no matter what sort of wave
is being discussed.*

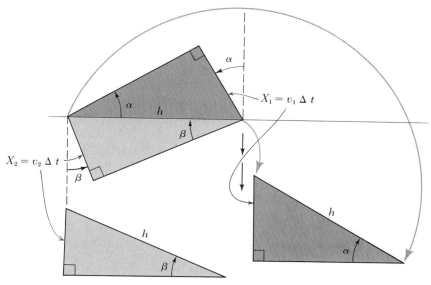

Figure 13-15
The distances X_1 and X_2 can be seen as the heights of two triangles having the same base, h. The heights are in the same ratio as the sines of the opposite angles.

For this to be true;

$$h = \frac{X_1}{\sin\alpha} = \frac{X_2}{\sin\beta},$$

or

$$\frac{X_1}{X_2} = \frac{\sin\alpha}{\sin\beta}.$$

Since X_1 equals $v_1\Delta t$ and X_2 equals $v_2\Delta t$, then:

$$\frac{\sin\alpha}{\sin\beta} = \frac{X_1}{X_2} = \frac{v_1\Delta t}{v_2\Delta t} = \frac{v_1}{v_2}.$$

Therefore:

$$\frac{\sin\alpha}{\sin\beta} = \frac{v_1}{v_2}. \qquad\qquad \text{Equation (13-2)}$$

This is called *Snell's law of refraction*. It applies to *any* wave passing obliquely from one medium to another.

It is also possible to interpret the path of a wave ray, through a region where its velocity changes, in terms of an "economy principle" called the *principle of least action*, or *Fermat's principle*, after the French lawyer and mathematician Pierre de Fermat (1601–1665). It says, in effect, that between two points the wave will always follow the *quickest* path (the path that takes the least time). Thus the ray will go further in the medium where its velocity is greater, thereby reducing the distance it has to go in the slower medium.

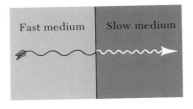

When a wave changes velocity, is this the result of a change in frequency, f, or in wavelength, λ? (Remember that $v = f\lambda$). If you stop to think of it, if two waves are to match up at the interface, they must vibrate together with the same frequency. Therefore, it is wavelength which changes when the wave crosses the boundary into the new medium.

Index of Refraction

The relative velocities of a light wave in two different media is sometimes expressed as the relative *indices of refraction*, η_1 and η_2, of the two media:

$$\frac{v_1}{v_2} = \frac{\eta_2}{\eta_1},$$

Snell's law is thus often expressed as:

$$\frac{\sin \alpha}{\sin \beta} = \frac{\eta_2}{\eta_1} \qquad \text{Equation (13-3)}$$

The index of refraction, η, of a single medium is defined to be the ratio of the velocity of light in vacuum to the velocity in the medium:

$$\eta = \frac{c}{v} \qquad \text{Equation (13-4)}$$

The more slowly the wave travels in the medium, the larger is its index of refraction.

From Table 13-4, you can see that the index of refraction of water relative to air is 1.333. (Light travels 1.333 times faster in air than in water.) This causes some interesting illusions to result when you look at something in the water from the air. First, as

Table 13-4 *Indices of Refraction*

Medium	η	Medium	η
Air	1.00027	Mercury	1.6 − 1.9
Benzene	1.498	Paraffin	1.43
Diamond	2.42	Quartz	1.46
Ethyl ether	1.352	Rock salt	1.54
Flint glass	1.6	Vacuum	1.00000
Gelatin	1.53	Water	1.333

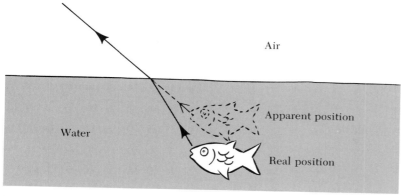

Figure 13-16
The fish appears to be higher than it really is.

you look down into water, refraction makes objects appear to be shallower than they really are. (Our brain notes the direction from which the ray comes and assumes the object lies in that direction as shown in Figure 13-16. A stick thrust into the water (Figure 13-17) appears bent in a direction opposite to that of the refraction of the light rays! If you have a chance to visit an aquarium and look through a window which goes above and below the level of the water, seals or other animals sticking out of the water will appear broken in two. The parts above water appear to be 1.333 farther away than the parts below water because light travels about 1.333 times faster in air than in water.

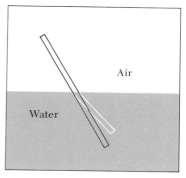

Figure 13-17
You can't trust your perceptions. The stick appears to be bent.

Atmospheric Refraction

Using Table 13-4, it can be determined that the speed of light in air is almost the same as in a vacuum, but not quite. It is also true

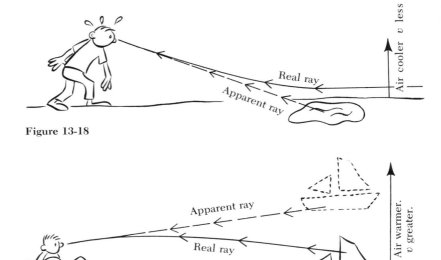

Figure 13-18

Figure 13-19
We cannot mentally account for the fact that the light ray bends as it passes through the air. The ship appears to loom above the water.

that the speed of light in air depends slightly upon the temperature, decreasing with decreasing air temperature. This physical fact also results in some interesting illusions. When a man walks along a hot desert, rays of light coming up to him from the hot floor of the desert will curve *up* (Figure 13-18). As the rays move upward they enter air which is somewhat cooler (a medium with a slower speed of light). Thus, they will bend toward the perpendicular. Then as he looks down at what he thinks is desert floor ahead of him, he will really be seeing rays from objects *above* the desert floor; in particular, from bits of blue sky. It will seem to him that a shiny lake lies in the desert ahead; a *mirage*. The same phenomenon can be observed when you're traveling in a car on a black asphalt highway on a hot day.

An opposite illusion, called *looming*, can occur on the ocean on days when the water is much cooler than the air above it. In this case the wave velocity increases as it departs from the surface of the water. Thus, a ray coming from an object, such as a ship, will bend *toward* the water; away from the perpendicular. The apparent path of the ray makes it appear that the ship is floating in the air above the surface of the water (Figure 13-19).

Total Reflection

Figure 13-14 showed that when a wave crosses the boundary into a faster medium, it is bent away from the normal to the interface. Since the angle β is greater than the angle α there must be a value of angle α such that angle β is 90°, or greater. The meaning of this statement is that there must be a *critical* angle, α_c, above which *no wave gets through* the surface. The whole wave is reflected back into the first medium. This phenomenon is called *total reflection*. Since the sine of 90° equals one, from Equations (13-2) and (13-3);

$$\frac{\sin \alpha_c}{1} = \frac{v_1}{v_2} = \frac{\eta_2}{\eta_1},$$

$$\sin \alpha_c = \frac{\eta_2}{\eta_1}, \qquad \qquad \text{Equation (13-5)}$$

The statement that, "when the critical angle is exceeded no light gets out, but all is internally reflected," can be checked by doing a simple experiment with a triangular prism. If a light ray is projected into a prism and the prism is rotated as shown in Figure 13-21, you can see the reflected ray brighten just at the angle for which the refracted ray disappears. The light energy which is shared by the refracted and reflected rays, for angles less than the critical angle, is carried entirely by the reflected ray when the critical angle is exceeded and total reflection occurs.

If medium 1 is water and medium 2 is air,

$$\frac{\eta_2}{\eta_1} = \frac{1.00027}{1.333} = 0.75$$

$\sin \alpha_c = 0.75$, so angle $\alpha_c = 48.5°$. This means that a fish looking up through a still water at the sky can only see out through a cone having a half-angle of 48.5°. As it looks up at the surface of the water at angles greater than this, the surface will act as a 100 percent reflecting mirror. Nevertheless, the fish does see everything in the air above the water. Its view will be distorted, however, with objects near the horizon appearing to be shrunk in size. Photographs taken with a wide-angle or "fish eye" lens on the camera show the same distortion (Figure 13-22).

From what we have said, the larger the index of refraction of a material, the smaller will be the critical angle for light rays to escape; that is, the closer the ray must be to the perpendicular to escape. For diamond, with an index of refraction equal to 2.42, the critical angle is only 24.4°. Such a small critical angle makes it relatively difficult for a ray to escape from the diamond once it's inside. This physical property is what makes diamonds sparkle; light which gets in, bounces around until it happens to approach a facet with a sufficiently small angle of incidence. The light then escapes, sometimes coming at the viewer from an unexpected direction (Figure 13-24).

The same principle has been applied to the manufacture of "light pipes", or *fiber optics* devices which will carry light around corners.[9] A light pipe is made by bundling together

(9) Kapany, Narinder S., "Fiber Optics", *Scientific American 203*, November, 1960, pg. 72.

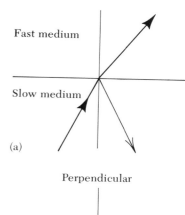

(a)

Fast medium

Slow medium

Perpendicular

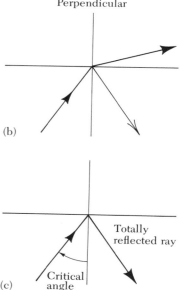

(b)

(c)

Totally reflected ray

Critical angle

Figure 13-20
When the angle of incidence reaches a critical value, total reflection occurs.

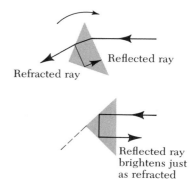

Reflected ray

Refracted ray

Reflected ray brightens just as refracted ray cuts out.

Figure 13-21
Total internal reflection. No light escapes to the left from the prism.

Photograph by Michael Simon.

Figure 13-22
A fish eye view of the author poking a stick at a fish.

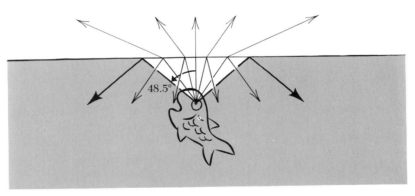

48.5°

Figure 13-23
The fish's view is limited to a cone with a 48.5° half-angle.

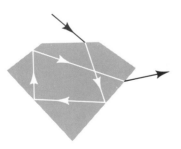

Figure 13-24

hundreds of thin glass fibers, each coated with a substance having an index of refraction between that of glass and of air. To get into one of the fibers a ray must travel nearly parallel to the axis of the fiber. This means that whenever it strikes the side of the fiber from the inside, the angle of incidence is far greater than the critical angle. The ray "rattles" along the fiber, never escaping until it reaches the end of the fiber, as shown in Figure 13-25. By using a large bundle of fibers, whole pictures or images can be carried from one end of the bundle to the other, even when the bundle is tied into a knot! Fiber optics have made it possible to illuminate and inspect objects which are very hard to get to, such as the interiors of hearts, or stomachs, prior to surgery or other medical treatment.

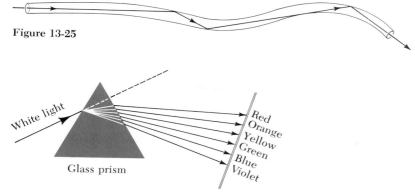

Figure 13-25

Glass prism

Red
Orange
Yellow
Green
Blue
Violet

White light

Figure 13-26

*Various wavelengths are slowed by different amounts
when they enter glass from air. Thus, they are refracted by different amounts. If the sides of the glass
are parallel, then these differences in refraction are
cancelled out when the light re-emerges into the air.
However, if the sides of the glass are not parallel, as in
a prism, the various wavelengths come out in different
directions and are thus separated.*

Dispersion

Why is it that when white light passes through the beveled edge
of a mirror or a prism, it spreads out in various colors? It is because various wavelengths of visible electromagnetic radiation
travel with different velocities in a transparent medium such as
glass. The shorter violet or blue wavelengths are slowed down
more when they enter glass from air than are the longer orange
and red wavelengths. This means the violet and blue are refracted more than the red. What enters as white light, made up
of many colors, is spread out into the separate colors. The phenomenon whereby wave velocity (any wave, not just electromagnetic waves) depends upon wavelength is called *dispersion*. It
provides one means for spreading a wave containing many wavelengths out into its component parts, or *spectrum*.

Atmospheric Absorption; Pollution

Several factors can prevent a wave from penetrating very far into
a particular medium. First, it can be *absorbed*. In this case the
atoms or molecules in the medium take energy away from the incoming wave, converting it to heat or extra electron energy. Some
of the energy is re-radiated, but it goes off in all directions, and
will not always have the same wavelength as the original incoming wave. The material in a medium will usually have a few
specific wavelengths (frequencies) which it responds to, and

these will be the ones that it absorbs most effectively. It follows from this that a transparent medium is one whose atoms and molecules just don't respond much to the incoming wave.

A wave can also be *scattered* by molecules or very small particles suspended in the medium. Here the disruption of the wave is closer to a direct bouncing from the molecules, but there is still a wavelength-dependent amount of scattering. For light waves in air, the molecules are "seen better" by the shorter wavelengths so more blue light is scattered than red — and the sky is thus blue. Sunsets are red because the light coming to us slants down through more air. Most of the blue light is scattered in other directions, so that the light reaching us is largely red. On the other hand, if the particles that are suspended in a medium are distinctly larger than the incoming wavelength, then ordinary *reflection* (as from a flat smooth mirror) can occur. The effect on the incoming wave is much the same as when scattering occurs, except that the reflection does not depend on wavelength.

White paint is made by suspending colorless but highly reflecting particles in a transparent medium. The particles reflect and refract all the light repeatedly from each other, sending back out white light (if white light was what shone on the paint in the first place).

A number of factors, some of them a result of our technology, can effect the absorption of electromagnetic radiation in the earth's atmosphere. First, *ozone* (a molecule composed of three oxygen atoms, and formed by natural processes in the upper atmosphere) absorbs most of the ultra-violet wavelengths coming from the sun. Enough ultraviolet radiation to give us sunburn gets through but essentially, we are protected. Ozone tends to react with and be destroyed by the particles which our chemical processes emit, if the particles are small enough to float up to where the ozone is formed (about 22 kilometers in altitude). Anything which might reduce the ozone content in the atmosphere markedly could make going out on a sunny day a dangerous adventure.

Carbon dioxide (CO_2), which is produced by all burning processes (such as the one shown on pg. 233) also plays an important absorptive role in the atmosphere. It lets through visible light waves from the sun, but absorbs and traps much of the longer-wavelength infra-red radiation which is re-emitted by the earth. Thus, it acts as a blanket, or like the window glass in a greenhouse; hence the name the "Greenhouse effect" for this phenomenon. Some prophets have foretold a disastrous warming of the earth if the CO_2 in the atmosphere increases even a few percent. The polar ice caps would melt, raising the level of the oceans several hundred feet and flooding all of our coastal cities.

More recent theories[10] say that the CO_2 content would have to increase ten-fold before the greenhouse effect would become

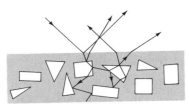

Figure 13-27
The particles which make white paint white are themselves colorless!

(10) Newell, R. E., "The Global Circulation of Atmospheric Pollutants", *Scientific American* 224, Jan., 1971, pg. 32. Offprint 894.

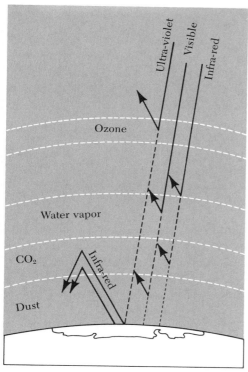

Figure 13-28
*The effect of various layers in the earth's
atmosphere on radiation passing through them.*

serious and that the oceans of the earth would absorb the CO_2 before it could reach this level.

The amount of *water vapor* in the atmosphere might also be important, because it can reflect radiation from the sun before it gets to the earth, leading to an overall *cooling*. Here, as in the case of other pollutants, the effect depends on particle size.

Dust, from both natural and man-made processes, plays a complex role in the atmospheric absorption of radiation.[11] Like water vapor, it could back-scatter incoming radiation and cause a cooling of the earth, triggering another ice age. But it can also absorb infra-red radiation, as does CO_2, and lead to a *warming* effect. On balance, the prediction is that dust will cool the earth.

The fact is that we are not sure *what* the result is going to be, if we keep polluting the atmosphere. Mixing, and inter-relationships between CO_2, water vapor content, dust, and temperature are still essentially mysteries. However, it *is* safe to say that man has the power to unwittingly upset the delicate temperature balance which has made possible the presence of life on earth as we now know it.

(11) Rasool, Ichtiaque and Schneider, S., *Science 173*, pg. 138, 1971.

Summary

A wave is a propagation of a disturbance through space. Most waves are oscillatory in nature, involving some sort of shaking back and forth at some frequency f (in hertz). If they travel with velocity, v, and if similar phases of the disturbance are separated by a distance λ (called the wavelength), then

$$v = f\lambda. \qquad \text{Equation (13-1)}$$

The energy of a wave is more closely related to its amplitude than to its velocity. The disturbances themselves can be longitudinal, transverse, rotational, or even torsional.

Sound waves are longitudinal disturbances and require a material medium for their transmission. In air, they travel a mile in about 5 seconds, but they travel much faster in water and solid materials. Humans can hear sound as a steady tone from about 20 to 20,000 hertz.

The pitch of a musical tone is determined by its frequency of vibration. An interval of one octave corresponds to a doubling of frequency. The quality of a tone is determined by the number and amount of higher frequency overtones which are mixed in with the fundamental vibration (which determines the basic pitch).

Electromagnetic waves travel at 3×10^8 m/sec in vacuum. The disturbance involves the electric or magnetic field in a region. The oscillations are transverse, and a material medium is not required for their propagation.

We can see very little electromagnetic radiation: only the portion with a wavelength of between 430 and 690 nm. It is this range of wavelengths which gives rise to all the colors we see and which is correctly called "light". X-rays, ultra-violet, infra-red, radio, and television waves all involve the same phenomenon, but (depending on the radiation involved) at smaller—or larger—wavelengths than those of light.

Sources of electromagnetic radiation typically emit many wavelengths. A source which emits just one wavelength, or color, is called monochromatic.

Recently, gravitational waves, corresponding to changes in gravitational fields in space, may have been detected. They also travel at the speed of electromagnetic waves.

When a wave encounters a boundary between two different media part of it is always reflected. When a wave strikes a smooth surface its angle of reflection equals the angle of incidence. As waves enter a new medium in which they travel faster, or slower, they are refracted respectively away from, or toward, the perpendicular to the boundary.

The degree of bending is given by Snell's law:

$$\frac{\sin \alpha}{\sin \beta} = \frac{v_1}{v_2} = \frac{\eta_2}{\eta_1}. \qquad \text{Equations (13-2), (13-3)}$$

where v_1 and v_2 are the velocities of the wave in the two media and η_1 and η_2 are the indices of refraction of the two media.

The index of refraction is a measure of how much more slowly light travels in the medium, than in a vacuum;

$$\eta = \frac{c}{v}.$$

Equation (13-4)

When a wave approaches a boundary across which it speeds up, it can be totally reflected back into the slow medium if the angle of incidence is larger than the critical angle for which

$$\sin \alpha_c = \frac{v_1}{v_2} = \frac{\eta_2}{\eta_1}.$$

Equation (13-5)

Dispersion is a phenomenon wherein the wave velocity in a medium depends on its wavelength. When a light wave enters glass, the shorter wavelengths (blue-violet) are slowed and refracted more than the longer wavelengths (orange-red). In this way white light, containing many colors, can be spread out into a rainbow-like spectrum by use of a prism.

The earth's atmosphere is important in protecting us from harmful radiations from the sun and from outer space. It does this by absorbing, scattering or reflecting the incoming radiation. Scattering gives the sky its blue color. Carbon dioxide, ozone, water vapor, and dust in various combinations can play a significant part in atmospheric absorption. Small changes in the amount of these substances in the air as a result of man-made pollution could drastically alter the protective nature of the atmosphere as well as the average temperature of the earth.

Further Reading

Benade, Arthur H., *Horns, Strings and Harmony*, Doubleday Anchor Book, Science Study Series, Garden City, 1960. (The physics of musical instruments given with minimum of mathematics. Good illustrations.)

Beranek, Leo L., "Noise", *Scientific American 215*, Dec. 1966. Offprint 306. (What noise is, and the problems it causes.)

Born, Max, "The Air and its Relations", Chapter 1 in *The Restless Universe*, Dover Publications, N.Y., 1951. (A quick run through many of the concepts we have discussed in this text. Pages 38–44 is an especially clear discussion of why the sky is blue. The book features "motion pictures" in the margins, generated by flipping the pages.)

Griffin, Donald R., *Echoes of Bats and Men*, Doubleday Anchor Book Science Study Series, Garden City, 1959. (Good discussion of radar.)

Heirtzler, James R., "The Longest Electromagnetic Waves", *Scientific American 206*, March 1962. (Charged particles streaming out from the sun oscillate slowly to produce waves having extremely long wavelength.)

Hutchins, C. M. and Fielding, F. L., "Acoustical Measurements of Violins", *Physics Today 19*, July 1968, pg. 35. (With improved acoustical design, violin makers are beginning to recapture some of the empirical secrets of the old masters.)

"Science and the Citizen", *Scientific American, 223*, Dec. 1970, pg. 41. (Objects can be seen in murky water by pulsing the source of light and receiving back only those pulses which travel out to the region where the object is located.)

Shankland, Robert S., "The Development of Architectural Acoustics", *American Scientist 60*, March-April, 1972, pg. 201. (An informative discussion of acoustical design from ancient to modern times.)

Problems

13-1 A certain guitar string has a fundamental vibration frequency of 100 hertz. What are the frequencies of the first three overtones?

13-2 The time standard, as was noted on page 14, is the 9.19×10^9 hertz vibration of energy emitted and absorbed by the Cesium 133 atom. In what portion of the electromagnetic spectrum (Table 13-2) does the radiation lie?

13-3 A police radar set sends out a pulse. The pulse is reflected by an auto and gets back to the set 2 microseconds after it was sent. How far is the car from the radar set?

13-4 A tuning fork vibrating at 450 hertz is attached to one end of a string. The other end of the string is fastened solidly to the wall. The fork is pulled away from the wall so that tension is exerted on the string and standing waves are set up in the string.

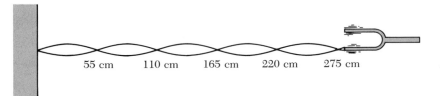

55 cm   110 cm   165 cm   220 cm   275 cm

What is the velocity of the disturbance set up by the tuning fork in the string?

13-5 What makes a tuning fork die down in vibration after it has been struck to set it going?

13-6 Find the frequency of (a) red light of 650 nanometer wavelength; (b) microwaves of 2 cm wavelength; (c) a musical tone having a wavelength of 1 m.

13-7 How big are the waves which an FM receiver picks up when it is tuned to 88 megacycles?

13-8 Someone sees a flash of lightning, then hears the thunder 15 seconds later. How far away did the lightning strike?

13-9 What is the speed of light in water and in quartz?

13-10 A ray of light enters a container of water at an angle of incidence of 40 degrees. At what angle does the ray travel in the water?

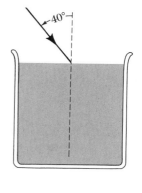

13-11 A hunter spots a dangerous man-eating tiger on a patch of desert on a hot day. To hit a particular spot on the animal should the hunter aim high or low?

13-12 A boat sits on a smooth pond which is 20 m deep. As the sun sets where does the shadow of the boat hit the sand on the bottom of the pond?

13-13 Suppose five parallel rays of light in air strike a glass sphere. Sketch what happens to the rays.

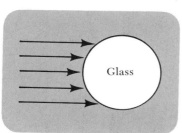

13-14 A person standing at x sees another person drowning at y. Suppose the first person can run on land four times as fast as he or she can swim in water. What is the quickest path that the person can take from x to y?

13-15 When you look up at the moon from the earth's surface, how is the moon's color affected by the atmosphere?

14

THE INTERACTION OF WAVES

In Chapter 13 it was shown just how waves of various kinds travel through regions where their speed changes, or where they are absorbed. Now we want to look into some of the interesting phenomena which result—patterns which develop—when *more* than one wave passes through a medium at the same time.

Standing Waves; Resonance

For simplicity, suppose we begin by passing just two identical waves through the same medium in opposite directions at the same time. One of the easiest ways to do this is simply to reflect one wave back into the medium. Suppose we continually send waves toward the right in a rope, and tie down the right end of the rope (Figure 14-1). If we now tie the rope down at a second specific point, but somehow manage to keep the rope vibrating, an interesting phenomenon can result: *standing waves* (Figure 14-2). The rope will oscillate up and down transversely, but at certain points, called *nodes*, the rope is not disturbed. At these points the rope has equal and opposite up and down urges from both the wave moving toward the right and the one moving toward the left. There's a node every *half* wavelength. To obtain

Figure 14-1

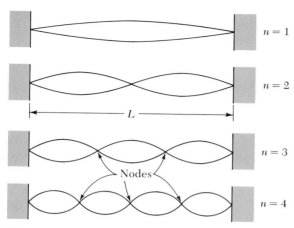

Figure 14-2
*Standing waves. In each case, the distance between
nodes is $\lambda/2$.*

a standing wave pattern the length, L, of the "medium", whether
we're speaking of light waves or sound waves or waves in a rope,
must be equal to a whole number, n, times one-half the wavelength, λ.

$$L = \frac{n\lambda}{2}$$ Equation (14-1)

Dividing both sides of Equation 14-1 by L gives $1 = n\lambda/2L$. Dividing both sides of this by λ gives $1/\lambda = n/2L$. From Equation
(13-1), $v = f\lambda$, and $f = v/\lambda$, so $f = v/\lambda = v(1/\lambda) = vn/2L$. The frequency f must increase as n increases. The first overtone ($n = 2$)
would have twice the frequency of the fundamental frequency
of vibration.

From the standpoint of wave analysis, standing wave patterns
provide a good means for observing wave phenomena which
would otherwise be too fast-moving to keep up with. Given a
wave with a certain wavelength, a reflecting surface is moved
back and forth in the direction of the wave motion until a pattern
of nodes and loops forms. The distance between nodes will, of
course, be $\lambda/2$. Alternatively, given a medium or a sounding
chamber of a certain dimension L, you could vary λ until it fits
the relation $\lambda = 2L/n$.

When the right relationship between wavelength and L exists,
a small amount of energy will keep a large vibration going in the
system. This is an example of a phenomenon called *resonance*.
A homely, but hopefully not too gross, example of this can be ex-
perienced in a booth in a public restroom with parallel reflecting
walls 2.5 to three feet apart. Start humming at as low a pitch as
possible, then slowly raise the pitch of the hum. When the sound

wave you are producing gets to the right wavelength and frequency, a standing wave pattern or resonance will develop, and you'll hear the hum get much louder. A bass will have no trouble humming the frequency corresponding to the fundamental resonance ($n = 1$ in Figure 14-2), but a soprano will have to try for one of the overtones.

Every physical system which is capable of vibrating has certain frequencies at which it *resonates*. A small amount of oscillatory energy having the right frequency will set the system vibrating with a surprisingly large amplitude. You can see or hear this at work in:

playground swings;

diving boards at swimming pools;

larynx and nasal passages of Wagnerian sopranos;

ultrasonic waves cleaning jewelry;

wine glasses about to be shattered by sound waves;

radio antennae and circuits tuned to a certain station;

loose parts in cars, which rattle when the car travels at certain speeds;

the sounding board and body of a violin;

and analogously in:

the absorption of infra-red radiation by carbon dioxide in the atmosphere (Chapter 13); and,

the absorption and emission of 9.19×10^9 hertz radiation by cesium 133 atoms in an atomic clock. (Chapter 2).

The calculation of the spectrum of resonant frequencies of systems has become a major branch of physics, particularly in the

"That's a note I don't hit very often."

(Courtesy of Joseph Zeis.)

atomic and nuclear realm. One major difference here is that the resonant frequencies are related to the possible energies of the system, rather than to its size.

The relation between wavelength and size of sounding chamber (Equation 14-1) also plays a major role in determining the size of musical instruments.[1] A piccolo designed for producing tones in the upper audible range is just inches long. An organ pipe which gives forth the "pedal tones" that set a whole auditorium vibrating can be 10 m long!

It should be re-iterated, that standing wave patterns can be produced with *all* kinds of waves. The only difference in technique would lie in the length of the medium, L, and in the means (voltmeter, microphone, visual observation of water wave ripples, etc.,) used to detect where the nodes and the regions of vibration lie.

Interference

From this point we could go on to study a rich multitude of phenomena involving the adding together of several waves in the same medium at the same time. But we must limit ourselves. Suppose first that we have two sources sending out similar waves through the same medium. We will let the waves be simple sine waves and will assume that they leave their respective sources in step (*in phase*). That is, they leave vibrating "in time" with each other. When two waves are in phase, their sum is a wave having the same wavelength and frequency, but with a larger amplitude, as shown in Figure 14-3a. If, on the other hand, the waves are exactly out of step, their net effect is zero, as shown in Figure 14-3b. In this case a disturbance from one wave is always exactly opposed by an equally sized disturbance from the other

Figure 14-3a

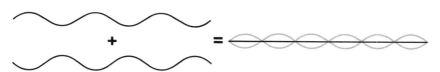

Figure 14-3b

(1) Hutchings, C. M. and Fielding, F. L., "Founding a Family of Fiddles", *Physics Today* 20, Feb., 1967, pg. 35.

Figure 14-4
*The summation of the two waves produces a wave whose amplitude
increases and decreases with time. This variation produces the
phenomenon of beats.*

wave. These effects are called *constructive*, and *destructive interference*, respectively. (The nodes spoken of, on page 358, were places where the interference was always destructive.)

If two waves with different wavelengths are now added together the composite wave will vary in amplitude, as shown in Figure 14-4. This rise and fall in amplitude can itself be heard (in the case of sound waves) as *beats* or surges. The closer together the frequence and wavelength of the two waves the less frequent will be the beats:

$$\text{beat frequency} = f_1 - f_2 \qquad \text{Equation (14-2)}$$

By listening to beats, piano tuners (and players in orchestras) can bring two instruments to the same pitch. When the beats disappear, the fundamental frequencies in the two tones are the same.

The phenomenon of beats can be simulated by looking through two combs that have different spacing between the teeth. Where teeth are directly behind each other you can see through both combs but when they're staggered, the teeth on one comb block out the spaces between the teeth on the other comb, as shown in Figure 14-5.

Interference in Thin Films

One interesting example of destructive interference that can be seen in nature—and technology—occurs in thin film coatings. The iridescent coloring of hummingbird throats, for example, is

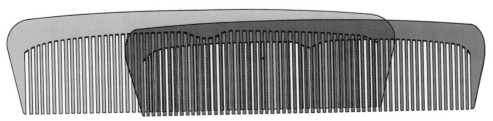

Figure 14-5

not a result of pigmented, absorptive coloring, but of interference; certain wavelengths in the light striking the feathers undergo destructive interference. We see the brilliant remainder as reflection. The same is true of the colors produced when a thin film of oil spreads over the surface of water or tarnish forms on silverware.

The principle of destructive interference can be used to reduce the reflection from camera lenses.[2] Suppose a thin transparent coating is put on a glass lens. When a ray of light strikes the outside surface of the coating part of it will be reflected and part will refract into the coating. Reflection will also occur at the surface of the glass. Some of this reflected ray will reenter the air, traveling alongside the ray which reflected from the front of the coating. The ray which made the round-trip through the coating will lag behind the other ray. If the coating is made just one-quarter wavelength thick, then the ray which penetrates it will lag behind the other ray by one-half wavelength: destructive interference will result. We must remember that "wavelength", as used here applies to the light *in the coating*. If the index of refraction of the coating is 1.4, then the light is slowed by that factor while it's in the film, and the wavelength in the coating, λ_c, is also reduced by the factor 1.4. Thus, $\lambda_c = \lambda/1.4$, where λ is the wavelength of the light in air. Therefore, the coating should actually have a thickness equal to $\lambda_c/4$ which, by substitution, equals $\lambda/(1.4 \cdot 4)$. A given coating will be most effective for one wavelength, and the wavelength is usually chosen to be one in the middle of the visual spectrum; such as that of green light—about 550 nm. A quarter-wave coating should be:

$$\frac{550 \times 10^{-9} \text{ m}}{(1.4)(4)} = 98.2 \text{ nm}$$

(about 10^{-5} cm thick), in order to reduce reflections in the green wavelength range. A lens coated in this way has a purplish appearance since red and blue light are still reflected but the green is reduced.

By now you should be convinced that light waves carry energy. If two of them interfere destructively, what happens to the energy that they each possessed? It means, in the case just discussed, that the energy stays *in* the reflective coating, or the material underneath it, and so it gets warmer. One recent proposal for trapping solar energy[3] uses a series of non-reflecting thin films. These let in energy from the sun's entire spectrum, but reflect back only a small fraction.

(2) Baumeister, Philip and Pincus, Gerald, "Optical Interference Coatings", *Scientific American 223*, Dec., 1970, pg. 59.

(3) Glaser, Peter E., "Solar Energy—An Option For Future Energy Production", *The Physics Teacher 10*, Nov. 1972, pg. 443.

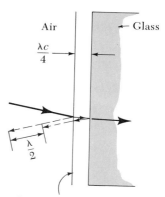

Thin transparent coating, index of refraction $\eta = 1.4$.

Figure 14-6

Cross-sectional diagram of a non-reflective coating. The wave reflected from the back surface of the coating interferes destructively with the wave reflected from the front of the coating.

Wave Interference in Two Dimensions

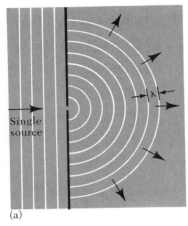

(a)

Figure 14–7a
Wave coming from a single source.

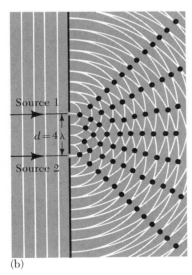

(b)

Figure 14-7b
*Coherent waves issuing from a
pair of sources. (The separation
distance d between the sources
equals 4λ.) Black dots show
locations of constructive
interference.*

Suppose now that two sources at a distance, d, apart send out
waves having the same velocity and wavelength. Suppose further (and this is important) that the sources oscillate in phase
together, or *coherently*. The wave pattern at a given instant can
be illustrated schematically, if we draw concentric circles to
represent the crests of waves issuing from the sources. There will
be points where two crests or two troughs coincide. Constructive interference will result. The amplitude of the resultant
ripple, or whatever disturbance the wave carries, will be larger
there. Such points are indicated by a black dot, in Figure 14-7b.
Between each line of dots there will be other places where the
interference is destructive. A crest from one source coincides
with a trough from the other source and the result is nothing,
assuming that the amplitudes of the separate waves are equal.
As the waves move outward the places of constructive and destructive interference sweep outward with them.

What can we say about the *location* of these points of constructive and destructive interference? Since crests and troughs
leave the coherent sources simultaneously, and travel with the
same velocity, then if you were to stand at a point which is an
equal distance from each source you would find constructive
interference. Two crests, and two troughs would always reach
you in phase. However, there are other places where the interference will also be constructive. Suppose you are 17 wavelengths
distant from one source and 16 wavelengths distant from the other.
Two crests will still reach you at the same time. They will no
longer be crests which left the sources at the same time, but since
one crest is like another, constructive interference will still result. In general, points which are 1, 2, 3, or n wavelengths farther
from one source than the other will experience constructive interference. If we designate r_1 and r_2 as the distance from any
point to the two sources, then the requirement for *constructive*
interference can be expressed as

$$r_1 - r_2 = n\lambda \qquad\qquad \text{Equation (14-3)}$$

In Figure 14-8, r_1 is 5λ and r_2 is 4λ.

Remember: it is not the distance to the sources but the *difference* in the distances which determines the nature of the interference effects at a given point.

For *destructive* interference the point must be located a whole
number of wavelengths *plus one-half wavelength* farther from
one source than the other:

$$r_1 - r_2 = (n + \tfrac{1}{2})\lambda \qquad\qquad \text{Equation (14-4)}$$

The distance difference $r_1 - r_2$ can be expressed more conveniently by first thinking of the observer as facing the two
sources from a direction given by the angle θ. As shown in Fig-

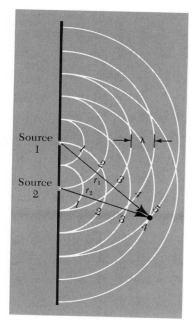

Figure 14-8

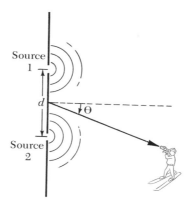

Figure 14-9

ure 14-9, θ is the angle between the line drawn from the point of observation to the intersection of a second line, drawn perpendicular to the plane of the sources and passing through the point midway between them. (This perpendicular line is also the line that connects points equally distant from both sources, the "line of equal distances".) For points a large distance from the two sources the lines r_1 and r_2 become nearly parallel. If we draw a line ℓ from *source* 2 and perpendicular to r_1 then we can specify a right triangle whose hypotenuse is d, and the quantity $r_1 - r_2$, the height of the triangle, can be expressed as $r_1 - r_2 = d \sin \theta$, since

$$\frac{r_1 - r_2}{d} = \frac{\text{opposite side}}{\text{hypotenuse}} = \sin \theta.$$

Following from this, the requirement for *constructive* interference becomes:

$$n\lambda = d \sin \theta. \qquad \text{Equation (14-5)}$$

That is, to observe constructive interference, one must look from an angle θ such that $\sin\theta = n\lambda/d$. The requirement for *destructive* interference becomes:

$$(n + \tfrac{1}{2})\lambda = d \sin \theta. \qquad \text{Equation (14-6)}$$

As an illustration, let us consider two coherent sources that are placed four wavelengths apart. ($d = 4\lambda$, as is illustrated in

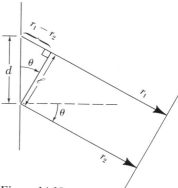

Figure 14-10

If the point of observation is far enough from the two sources, r_1 and r_2 become approximately parallel (but not equal in length).

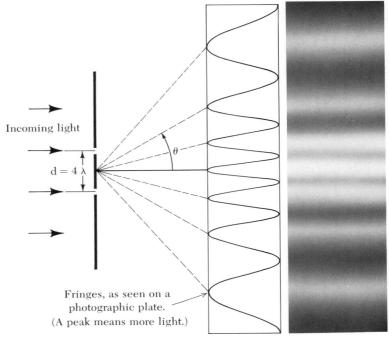

Figure 14-11
*Interference plot produced by two slits spaced a distance equal to 4λ
apart; λ/d = 0.25.*

Table 14-1

n	$\sin\theta = \dfrac{n\lambda}{d}$	θ
0	0	0
±1	± .25	±14.5°
±2	± .50	±30°
±3	± .75	±48.5°
±4	±1.000	±90°

Figure 14-7(b).) The sources could be emitting water, sound, electromagnetic, or any other sort of periodic wave. Table 14-1 and Figure 14-11 show the directions in which constructive interference should be observed, according to Equation (14-5). The diagram in Figure 14-11 is meant to show the relative wave intensities observed in the various directions. The row of "fringes" at the center portion of Fig. 14-11 shows how the interference pattern would appear on a photograph made on a flat piece of film with light waves. The portion on the right is a facsimile of such a photograph. Notice that the light intensity dies off fairly gradually when you move away from the specific angles at which constructive interference occurs.

Note also that the interference pattern will be compressed (small values of θ) when λ/d is small and that the pattern will fan out when λ/d grows larger; that is, when the distance between sources, d, decreases in comparison with the size of λ.

Stereophonic Reproduction

The brain is quite sensitive to the time lag between two waves reaching the ear from two sources which are separated from each other. This is the basis of stereophonic sound recording and re-

Sound reaches this mike sooner, and is louder.

Sound recorded in two separate "channels" of recorder.

Figure 14-12a
Recording sound stereophonically.

Later . . .

Stereo amplifier builds up the signal strength in the two channels separately and sends them to separate speakers.

Sound comes out sooner and louder from here.

Figure 14-12b
Reproducing sound stereophonically.

production. It takes more than just two speakers to make a "Hi-fi" a stereo set. The sound must be picked up by two separate microphones and each of these two signals must be recorded separately on a disc or tape. The two "channels" that are recorded can be on the same tape or disc, but the sounds received and recorded at one microphone are always separated from those which went to the other microphone. When the disc or tape is played, the electrical amplification and conversion back to sound waves at a speaker must again be done separately, though the two channels must be tied together in time: the vibrations of the two speakers must be controlled. They will not be in phase, but they will be a *controlled amount out of phase.* The sound which reached one microphone louder and sooner will come out of one of the speakers louder and sooner. The listener's brain then integrates the two waves it receives from the two (or more) speakers into a "depth" illusion.

In "quadraphonic" reproduction, four microphones, recorders, amplifiers, and speakers are used.

A Note on the Existence of Light Waves

The two-source interference phenomenon shown in Figure 14-11 was first observed with light in 1806 by the English scientist Thomas Young (1773–1829). It constitutes strong evidence that light *is* a wave. In fact, interference phenomena are *the* test for the existence of waves.

At the time Young did his experiments, Newtonian physics, which described light as a stream of particles, was in command. Whether Newton himself would have supported the particle model in the face of Young's newer counter-evidence is a moot point. The historical fact is that his *followers* were sold on the particle model and Young's crucial evidence was essentially ignored for 50 years.

One vocal opponent to Newton's particle theory of light was Johann Wolfgang von Goethe (1744–1832) the German poet. During a trip to Italy he had become fascinated with color; how it is produced and how we sense it. According to one source,[4] he was a good Romantic (see page 90), so he thought that humans should be more subjectively involved in the process of seeing, not just being an objective receiver of waves. For example, he thought that color was made in the eye, by mixing various amounts of lightness and darkness. Newton, of course, had provided experimental evidence that white light contains all the colors. Goethe's theory of color did not gain many followers but it shows an interesting attempt to bridge the gap between the mechanistic and organic world views.

The question of whether light is made up of particles, that is, energy concentrated in space; or whether it is a wave disturbance, spread out into oscillations, can't really be settled. One experiment will reveal the wave characteristics of light. Another will show its particle characteristics. Light behaves like *light* and is not completely analogous either to water waves or to a stream of bullets.

Gratings and Spectroscopy

Suppose now that *many* coherent wave sources are placed in a row, each a distance $d = 4\lambda$ from the next. (Such an array of many sources (slits) is called a *grating*.) Constructive interference will still occur in the same directions as for two sources. (See Figures 14-7b and 14-11.) However, the decrease in wave intensity as you depart from the directions of constructive interference will be

(4) Schroeer, Dietrich, "Romanticism, Physics and Goethe", pp. 101–114 in *Physics and Its Fifth Dimension: Society*, Addison-Wesley, Reading, Mass., 1972.

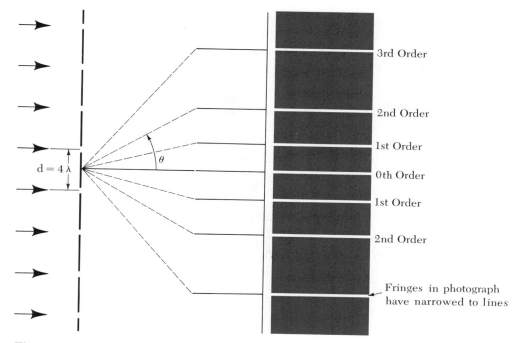

3rd Order

2nd Order

1st Order

0th Order

1st Order

2nd Order

Fringes in photograph
have narrowed to lines

$d = 4\lambda$

θ

Figure 14-13

*Interference plot produced by waves issuing from a row of many sources,
all spaced a distance, d, equal to 6λ apart.*

much sharper: The "fringes" in the center portion of Figure
14-13 narrow down to well-defined lines. These lines are usually
identified as "Zeroth order", "First order", or "Second order"
lines, depending upon whether n is equal to 0, ± 1, ± 2, and so
on. The first order interference line will be located so that, from
that angle of observation, each succeeding source in the grating
is seen as being one wavelength farther away than its neighbor.
The lines fanning out from the grating in Figure 14-13 show the
directions in which constructive interference is observed.

Up to now we have not specified what the wave sources all in
a row might be. In the case of a grating for light waves you could
not hope to put up a row of light bulbs. For one thing, you could
not get them close enough together. For another, the light bulbs
would radiate independently, not coherently. The latter problem
can be solved by placing a bright light source behind a narrow
slit. This slit lets through a part of the source's radiation which
is fairly coherent—the various parts of the outcoming wave are
more or less in phase with each other. Then this light is allowed
to pass through a set of narrow slits in a barrier. These slits become the "sources" we have been discussing. One problem still
remains. For light waves, λ ranges in size from 430 to 690 nm. To

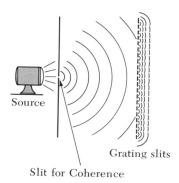

Source

Grating slits

Slit for Coherence

Figure 14-14

*The light is first passed through
a slit to make it more coherent
before it reaches the slits in
the grating.*

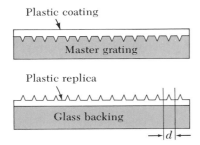

Plastic coating

Master grating

Plastic replica

Glass backing

$\leftarrow$$d$$\leftarrow$

Figure 14-15
Making a grating replica.

really observe interference effects (that is, to keep $\sin\theta = n\lambda/d$ from being immeasurably small) d should be about the same magnitude as λ. To actually cut slits this close together in a barrier is impossible. The solution to the problem came surprisingly early; during the 1500's. It was to make a series of very closely spaced parallel scratches in a smooth reflecting surface. The narrow unscratched regions then act as separate sources of light by reflection, and the device is called a *reflection grating*. Inexpensive *transmission replica gratings* are now made by pouring a plastic solution over a master reflection grating, and letting it harden. (Figure 14-15(a).) The plastic replica is then stripped off and put on a glass support Figure 14-15(b).) Light gets through the thin portions of the plastic but not the thick, rough replicas of the scratches. In this manner, gratings can be made which have 5,000 to 12,000 lines or sources per cm.

Suppose a 566 nm wavelength green light is viewed through a grating which has 6000 lines per cm. In this instance d will be equal to 1/6000 cm, which equals $(1/6) \times 10^{-5}$ m, or 1667 nm. Constructive interference will occur in directions given by Table 14-2. If observers look through the grating in a darkened room they will see a green line straight ahead ($n = 0$). They will also see three green lines that seem to come from 19.5°, 41.8°, and 82° to the left and right of the head-on direction. There will be no fourth order lines because, for them, $\sin\theta$ would have to be greater than one — a mathematical, and physical impossibility. Putting it another way, the slits in the grating are a distance, $d = 1667$ n:n/566 nm $= 3\lambda$ apart, so it is impossible to find a position such that you are four wavelengths farther from one slit than from the next.

We have assumed thus far that just one wavelength is coming through the grating. In the case of light gratings in the real world, it is more usual for *many* wavelengths to interfere. Each wavelength will produce an interference pattern like the one in Figure 14-13. All of the first order interference lines ($n = 1$) taken together are called the "first order spectrum". The total number

Table 14-2 *Directions of constructive interference.*

n	$\sin\theta = \dfrac{n\lambda}{d} = n\,\dfrac{566\text{ nm}}{1667\text{ nm}}$ $= n(.333)$	θ
0	0	0
± 1	$\pm\ .333$	$\pm 19.5°$
± 2	$\pm\ .666$	$\pm 41.8°$
± 3	$\pm\ .999$	$\pm 82°$

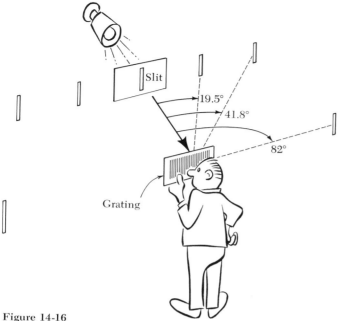

Figure 14-16
A person viewing the source of a green light through the
grating will see images of the slit in the directions indicated.

of spectra visible will depend upon the ratio of λ to d, that is, the number of possible values of n in Equation (14-5).

The real utility of the grating comes from its ability to *resolve* or separate two wavelengths that lie close to each other. The larger the number of sources (slits) in the grating, the greater its ability to separate wavelengths, since the "fringes" become narrower and narrower lines. The fact that the resolving power of gratings can be made so great makes them more useful than prisms in spectroscopy, the science of spectrum analysis (page 351).

Spectroscopy played an exceedingly important part in the birth of atomic physics near the beginning of the twentieth century and, in its many forms, is still the primary means for obtaining information about the insides of molecules, atoms, and nuclei, whether they are in laboratories or in distant stars. By measuring the wavelengths in the spectrum of an assemblage of atoms, you can find out what amounts of energy are absorbed and emitted by the atoms. This can be done, because on the scale of atoms, the changes in energy that the system undergoes are directly linked to the frequency of the wave radiated or absorbed. Sometimes theoretical models can be devised which enable resonant energies and frequencies to be predicted. In this way, theoretical and experimental spectroscopy combine to give a major scheme for finding out about the internal structure of submicroscopic systems.

For purposes of illustration we have dwelt upon gratings for light waves. The spacing, d, was a few wavelengths of light. Heat waves in the "far" infra-red, where λ can be as large as 10^{-3} m (Table 13-2, page 359) call for a source spacing of a millimeter or so. Radio waves, with wavelengths on the order of meters, require a "grating" made up of antennae spaced several meters apart and stretching several kilometers in length. The principles of interference are the same as for light grating.

In the last few decades astronomers have begun to study radiation, from stars and intersteller matter, which lies in the radio wavelength range, not the optical (visible) range. It seems that the sun and other stars emit energy stretching from one end of the electromagnetic spectrum to the other. Our atmosphere cuts off most of the radiation, but does let the visible wavelengths, along with waves in the short radio (.01 to 25 meter) range through, along with certain other wavelengths. A number of experiments have been performed, placing telescopes of various types in balloons and satellites so that they can rise above the atmospheric blanket and see what's coming in from the stars.[5] But in 1931, well before the age of flying telescopes, Karl G. Jansky, an American radio engineer, detected radiation from outer space that had a wavelength equal to about 15 m. Since that time, radio astronomy has become a sophisticated science in itself, using sensitive antennae and amplifiers to separate out the stellar radiation from the rest of the man-made radio waves that are bouncing around the earth. Radio astronomers are concerned with determining the direction from which the radiation comes, and with figuring out what is sending it. Instead of the lenses, mirrors, and photographic plates that are used in optical astronomy, the radio astronomer uses large antennae, electronic circuits, computers, and electronic recording equipment. Since radio waves are much larger than light waves one can't get sharp, focused "pictures" of radio stars, but the position of a radio emitter can be narrowed down sufficiently to allow comparison with photographs made of the same region of space. Then the astronomer can see if the radio source is also a visible star. Sometimes it is and sometimes it isn't. The planets in the solar system are also sources of radio waves.[6]

Figure 14-17 shows one scheme, among many, which radio astronomers sometimes use to pinpoint the position of radio sources in space. It consists of a "grating" made up of many receiving antennae, all spaced a distance, d, apart. They are linked

(5) Code, Arthur D., "Stellar Astronomy From a Space Vehicle", *Astronomical Journal* 65, June, 1960, p. 278.

(6) Smith, Alex G. and Carr, Thomas G., *Radio Exploration of the Planetary System*, Van Nostrand Momentum Book, Princeton, N.J., 1964

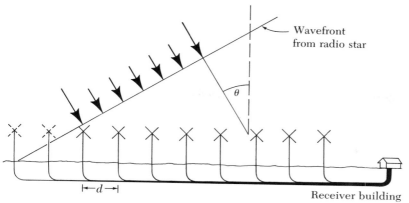

Figure 14-17

An interferometer receiving radiation from a direction at an angle θ to the vertical.

electronically to a common receiver tuned to the frequency (wavelength) of interest. Suppose that a beam of radiation arrives from a distant source, coming down in a direction which makes an angle θ with the vertical. The wavefront will reach one antenna, then the next, etc, traveling an additional distance, Δx, in each case (Figure 14-18). If $\Delta x = n\lambda$, the signals from all the antennae will add constructively and a large total signal will be recorded. As the earth turns, the source will appear to move across the sky, changing the angle θ. As $\Delta x = d \sin\theta$ changes, interference will be alternately constructive and destructive and the pattern shown in Figure 14-19 will be the result. The rises and falls in the total radio signal are similar to the fringes of light pictured in Figure 14-11. If the number of antennae is increased the broad fringes will sharpen into lines, as in Figure 14-13, and the source can be located more accurately. The central fringe in the radio pattern, corresponding to the source being directly overhead, is largest. This is true in our example because the receiving pattern for each individual antenna can be designed to give a maximum sensitivity for a source directly overhead.

As radio waves enter the earth's atmosphere they are slowed slightly, and thus at any angle other than the perpendicular they are refracted. (Recall page 348 on atmospheric refraction.) To determine the true direction of the source, the angle θ must be corrected by being increased, slightly, in order to compensate for the bending of the wave (Figure 14-20).

To really locate a radio source in space a second line of antennae must be set up, in a direction perpendicular to the original line. Then a second angle, θ', can be measured. The two angles together are needed to pin down the direction from which the radiation is coming (Figure 14-21).

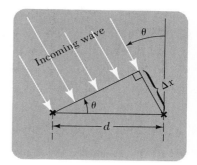

Figure 14-18

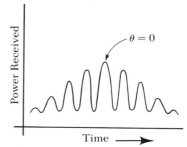

Figure 14-19

The plot of power received vs time permits very accurate measurement of when a source is directly overhead.

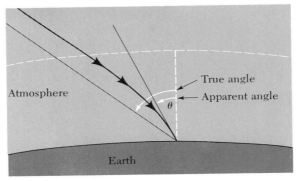

Figure 14-20
*Allowance must be made for atmospheric refraction
of radio waves from the stars.*

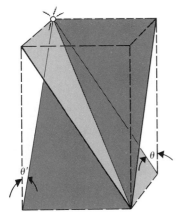

Figure 14-21
*Pinpointing a source in two
dimensions.*

If the whole array of antennae is rewired so that each becomes a *broadcast,* instead of a receiving, antenna then radio power can be beamed out in a pre-determined direction. The wave energy is concentrated in a cone so that a given signal strength will be detectable at greater distances than if it were beamed out equally in every direction. This principle has been used for long-distance radio communications, as in Radio Free Europe and the Voice of America. The longer the rows of antennae, the less the beam will spread out with distance. That is, the more radiation it will concentrate in a particular direction.

Diffraction

When a wave passes the edge of a barrier it spreads out. This behavior is called *diffraction.* Figure 14-22 shows this to be true even for light waves. Instead of a sharp change from light to dark in the shadow of the barrier, there are fringes, with shadow fringes penetrating into the "light" region. In the same manner, when a wave is sent through a narrow opening between two barriers, it spreads out. As Figure 14-23 shows, the smaller the opening in relation to the wavelength of the wave, the greater will be the spreading. You cannot make a narrow beam by sending a wave (light, or any other kind) through an opening, if that opening is smaller than several wavelengths in width.

The openings which light waves encounter are usually many wavelengths in width, so we are not used to seeing light waves undergo diffraction. Sound and water waves, on the other hand, are much larger and it is quite common for them to run into openings or barriers which are less than one wavelength across. We often hear sound that has "come around a corner". If a person stands behind a tree we can't see them but we expect to be able to hear them talk.

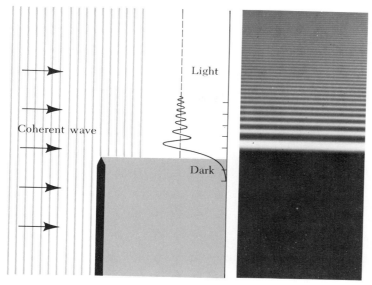

Figure 14-22

Diffraction pattern produced when coherent light passes the edge of a sharp barrier. (Photograph from M. Cagnet, M. Françon and J. C. Thrierr, Atlas of Optical Phenomena, Springer-Verlag, Berlin, 1962. Courtesy of M. Cagnet.)

(a) $\lambda = .35D$

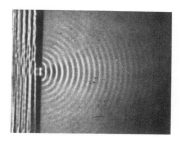

(b) $\lambda = 0.6D$

(c) $\lambda = 1.2D$

The mathematics of calculating the ways waves spread out — that is, the calculating of diffraction patterns — is complex. However, we can get some idea of what seems to be going on by looking down on a single slit through which a wave is passing (Figure 14-24). Suppose the width of the slit (broadcast antenna, loudspeaker, or harbor opening) is D. First, we divide the slit into two parts and look at it from the usual angle, θ. The portion of the wave coming from the upper half of the slit will lag behind the portion coming from the bottom half by the distance Δx, and Δx will equal $D/2(\sin \theta)$. Now, if Δx equals $\lambda/2$, the two half-opening wavelets will interfere destructively. $D/2(\sin \theta) = \lambda/2$. The angle at which the destructive interference occurs is thus given by

$$\sin \theta = \frac{\lambda}{D}.^{(7)}$$

Equation (14-7)

A more complete analysis shows that additional, very dim fringes of light do spread out to angles larger than this, but our simple treatment verifies one important thing: a wave or beam spreads out further, the more narrow is the opening D through which it passes. For example, if 550 nm green light is made to pass through

(7) For a beam coming out of a round opening with diameter, d, the relation becomes $\sin \theta = 1.22(\lambda/d)$. In either case, this represents a central diffraction fringe or illuminated region whose dark edges are located at the angle θ.

Figure 14-23

Ripple tank photographs of water waves passing through an opening in a barrier. The width of the opening, D, remains the same, but the wavelength is increased from (a) to (b) to (c) by decreasing the frequency of the waves.

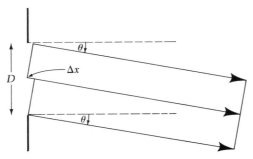

Figure 14-24
Wave issuing from a slit of width D.

a single slit one millimeter (10^{-3} m, or 10^6 nm) wide, the spreading resulting from diffraction is given by

$$\sin \theta = \frac{\lambda}{D}$$

$$= \frac{550 \times 10^{-9} \text{ m}}{10^{-3} \text{ m}} = 5.5 \times 10^{-4}.$$

The corresponding angle, θ, is almost zero, so the spreading is negligible. The slit would have produced a beam one mm wide, with fairly sharp-appearing edges. However, if the same wave passes through an opening 10^{-6} m wide,

$$\sin \theta = \frac{550 \times 10^{-9} \text{ m}}{1000 \times 10^{-9} \text{ m}} = 0.55.$$

The corresponding angle is 33.4°, so the spreading is by no means negligible. As Figure 14-25 shows, the "beam" produced is far wider than the opening through which the wave passes.

The same analysis can be applied to the water ripple waves shown in Figures 14-23(a), (b), and (c).

Table 14-3

	Wavelength (x)	$\frac{\lambda}{D}$ *or sin* θ		θ
(a)	.35D	.35	.35	20°
(b)	.6D	.6	.6	37°
(c)	1.2D	1.2(!)	$1^{(\circ)}$	90°

°Physically, when sin $\theta = 1$ the wave has spread out as far as it can go. All other larger values of λ/D become limited to a 90° spread.

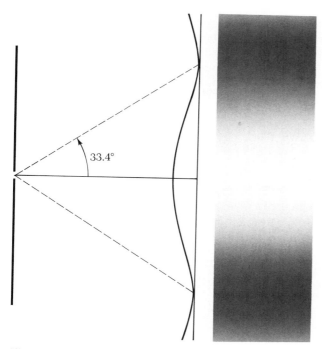

Figure 14-25
Central diffraction pattern produced when D = 1000 nm.
and λ = 550 nm.

The relative size of an object and the wavelength of a wave striking it, have been important in everything that has been discussed in this section of the book. The size comparison can also help us understand some of the differences between reflection and scattering which came up in the discussion of atmospheric absorption in Chapter 13. When a wave strikes an object much larger than its wavelength, it is *reflected* according to the rules stated on page 242. We can forget about the wave nature of light. It travels in beams or rays along straight lines. The method of description that is applicable here, and which assumes wavelength to be negligibly small is called *geometrical optics*. All that was discussed in Chapter 13 was based on this assumption. If, on the other hand, a wave encounters an object whose dimensions are about the same as the wavelength, λ, "edge effects" become more important. As can be seen from Figure 14-22, diffraction occurs when light hits the edge of an object. If the object is small relative to the wavelength, that is, if the edges of the object are close together, then diffraction and spreading will predominate. Small objects can diffract light in certain directions and they can also scatter the wave (page 352). The branch of optics which deals with the wave characteristics of light is called *wave optics*, and has been the main topic of this chapter.

Large objects reflect.

Small objects scatter and diffract in many directions.

Figure 14-26
A small object is "all edge" thus diffracts or scatters, rather than reflects, waves which strike it.

One last thing about diffraction: it affects gratings in a way which has been neglected in the discussion up to now. When many sources are placed close together, the sources themselves (slits, antennae, or whatever) must have a width, D, somewhat less than d. (In this case, d is measured from the center of one source to the center of the next.) That means that gratings involve diffraction, as well as interference, phenomena. Gratings, such as those described on pages 369–370 are usually called *diffraction* gratings even though their most interesting uses arise from interference effects. Then, what *does* diffraction do to the pattern of waves coming through a grating? This can be illustrated by imagining a grating in which the distance between the centers of the slits, d, is equal to 4λ, but whose slits each have a width, D, equal to 3λ. Figure 14-27 shows the diffraction pattern produced by each slit if it were alone and Figure 14-28 shows the interference pattern produced by the whole grating.

All of the interference fringes must fit under the diffraction "envelope" (or tent). They are still positioned according to sin θ

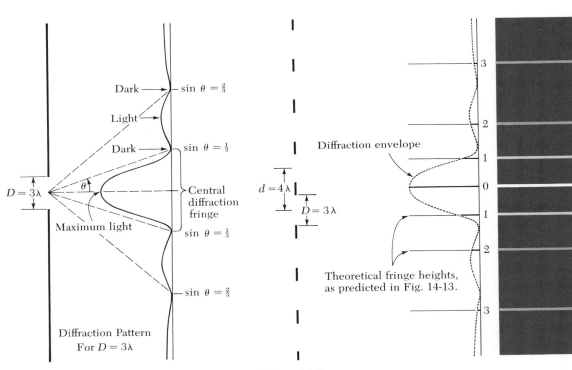

Figure 14-27
Diffraction pattern for $D = 3\lambda$.

Figure 14-28
Interference pattern produced by a wave passing through a grating made up of slits of width $D = 3\lambda$, *with a spacing of* $d = 4\lambda$ *between their centers.*

$= n\lambda/d$. but they will be squashed (dimmed) under a curve which goes to zero at $\sin\theta = \lambda/D$. While Figure 14-28 looks like a repetition of Figure 14-13, it has been changed to indicate the lines which are reduced in intensity by some factor dependent upon their position under the diffraction envelope. The same effect occurs in the radio interferometer fringe pattern diagrammed in Figure 14-19.

Lasers

Everything that has been said concerning interference and diffraction up to this point presupposes a coherent wave. Either we let parts of a single wave front re-combine after it passed through a grating, or we were confining a single wavefront by making it pass through a narrow opening in a barrier. To observe any of the diffraction and interference effects with light it must first pass through a narrow slit. In this way a portion of the emitted wave was selected which was more or less "in phase with itself". This had to be done because an ordinary light source, such as the hot filament of an incandescent lamp, is made up of many smaller independently oscillating sources. At any particular time one part of the surface of a hot filament doesn't necessarily do what other parts are doing. Of course the narrow slit that is used to make the light more coherent stops most of the wave, so we have to work with the dim remnants of the original light. A bright coherent light source would be a godsend for the study of optical effects.

A high-frequency electronic oscillator is a source of coherent radiation, but its wavelength is far too large to be visible. In the 1950's the lower size limit of electronically produced wavelengths was pushed down to a millimeter or less, by making use of smaller and smaller resonant cavities. (As shown on page 359 there is almost always a connection between the size of a system and the wavelength which it produces best, or best responds to.) However, there is a practical limit to how small these resonant cavities can be made. Late in the 1950's several scientists began looking into the possibilities of exploiting naturally occurring *atomic* or *molecular* resonances for amplifying wavelengths in the infra-red, visible, and ultra-violet range. They sought some means for synchronizing molecular resonances, making them work together. Their efforts succeeded and have led to an amazing number of applications in the years since. Like "chain reaction", the word *laser* has become a household word even though most households don't have one. Laser is an acronym for "*l*ight *a*mplification by *s*timulated *e*mission of *r*adiation." The principle upon which the laser is based dates back to 1917 and comes from the mind of—you guessed it—Albert Einstein. It is

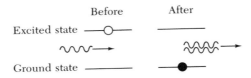

Figure 14-29
The basic laser phenomenon.

a principle which has become part of quantum mechanics, the branch of physics which physicists have developed in the attempt to describe and understand the behavior of atoms. It has no parallel in the large-scale mechanics we have concentrated upon thus far in the text. The principle can be described as follows: when an atom is excited from its ground state to an excited state (see footnote, page 229) it will then emit some energy *spontaneously* and drop back to the ground state either directly, or in steps. However, it can also be stimulated, and made to drop from the excited state to a lower excited state (or to the ground state). This can be accomplished by hitting it with a wave whose wavelength is equal to that of the energy the atom would normally have emitted spontaneously. Two interesting things can be said about this *stimulated emission process*: first, it happens more quickly than a typical spontaneous emission and second, the stimulating and the stimulated emission are radiated *in phase* — as one coherent wave! (Figure 14-29.)

In most materials the incoming stimulating energy would normally excite atoms in the ground state and be absorbed, rather than helping atoms to de-excite and being amplified. Thus, to produce "lasing" the atoms in the material must first be "pumped" so that a majority are in the excited state. To achieve this the laser material, whether gas, liquid, or solid, needs to have a "meta-stable" excited state in which the atoms spend a short time before de-exciting spontaneously (Figure 14-30). Then the stimulated emission phenomenon has time to build up.

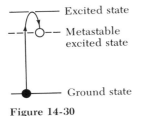

Figure 14-30

The material in a laser is typically *pumped* with light having a shorter wavelength than the laser wavelength. In addition the critical meta-stable state of a laser material is not always reached directly from the ground state. The atoms or molecules can also get there by dropping from a higher excited level. Once the atoms or molecules have reached the meta-stable state some of them will begin to de-excite spontaneously. When the light they emit strikes another excited atom it will also be stimulated to emit light. An avalanche wave process will start in the same direction as the original emitted energy was traveling. In most cases the built-up wave will pass out the sides of the medium before it reaches very high intensity. However, if the ends of the chamber containing the medium are parallel and made of reflecting mirrors, a wave will be sent back and forth through the medium many times, building up to trememdous intensity if it is traveling close to the axis of the medium. The wave travels at a very high speed (c/η, from page 346) so the buildup process does not take long. Even though the chamber is millions of wavelengths long, it is resonating, producing a standing wave between the reflecting ends. If one of the end mirrors is semi-reflecting, then part of the intense wave along the axis of the medium can escape as a laser beam. In some lasers the beam comes out continuously, whereas in others it is pulsed or shuttered to give intermittent pulses.

A laser beam is a very special sort of wave.[8]

(1) It is extremely monochromatic; that is, the energy is concentrated into a much narrower range of wavelengths, than is the case for ordinary monochromatic light.

(2) It is intense. Other beams are the result of letting a small fraction of the light from an ordinary source escape through an opening.

(3) It is highly directional; the beam spreads out very little as it travels out from the laser.

(4) It is a beam of coherent radiation; and, as such, it will produce interference effects without first having to pass through a narrow slit before reaching a grating, edge, or what have you.

Characteristics **(3)** and **(4)** are interrelated. If a wave *is* coherent; that is, if it can be treated as having one smooth wavefront, then the spreading of the beam is limited by the diffraction condition in Equation (14-7). For example, if the wavelength of the beam is 632.8 nm, as in the case of the red helium-neon laser; and, if the initial diameter of a circular beam is two millimeters (2×10^{-3} m), then the spreading of the beam will be at an angle given by the equation:

$$\sin\theta = 1.22\lambda/D$$

$$= \frac{(1.22)(632.8 \times 10^{-9} \text{ m})}{2 \times 10^{-3} \text{ m}}$$

$$= 386 \times 10^{-6}$$

$$= 3.86 \times 10^{-4}.$$

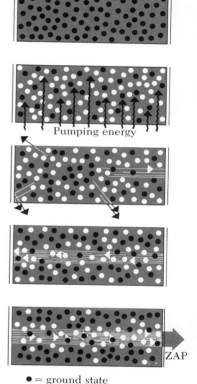

Pumping energy

ZAP

● = ground state

○ = excited atom

Figure 14-31
Steps in laser action represented symbolically.

In one kilometer (.62 mile) the beam will have spread to only 77.2 cm in diameter. Lasers which produce beams having a larger initial diameter are being used to measure the distance from the earth to the moon. A pulse of light is sent and the time required for the pulse to make a round trip is recorded electronically. The

(8) Schawlow, Arthur L., "Laser Light", *Scientific American 219*, Sept., 1968, pg. 120.

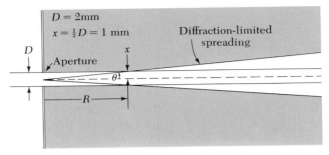

Figure 14-32
Diffraction-limited spreading of a coherent beam.

beams spread to a diameter of just 2 miles while traveling the 240,000 miles (397,000 km) to the moon. A reflector on the moon can easily gather enough light from such a beam to send a detectable signal back to earth (unless it's cloudy.)

Laser beams and radio beams spread in the same diffraction-limited manner. Neither of the beams spreads *at all* (except for scattering by air molecules) until they've gone far enough for the central diffraction fringe to widen out to the original beam diameter. For the red helium-neon beam on the previous page the required distance of travel, R, for the central fringe to widen to two millimeters can be found with the aid of Figure 14-32. For small angles θ, the approximation $\sin \theta = \tan \theta = \theta$ holds. Tan θ = x/R, so, by substitution, $\sin \theta = x/R$. Multiplying both sides by R gives $R \sin \theta = x$, and dividing both sides by $\sin \theta$ gives $R = x/\sin \theta$. Since x is equal to $D/2$, $x = 1$ mm for the laser we discussed previously. We saw that $\sin \theta = 3.86 \times 10^{-4}$, so:

$$R = \frac{x}{\sin \theta} = \frac{10^{-3} \text{ m}}{3.86 \times 10^{-4}}$$

$$= .259 \times 10 \text{ m}$$

$$= 2.59 \text{ m}.$$

The ratio λ/D is much larger for radio waves than for light waves because λ is about one million times larger, and there's a practical limit to the diameter, D, of a manageable antenna. However, the radio beam can be made more and more directional for a given antenna diameter, D, by decreasing λ.

Ordinary beams of incoherent light spread out much faster than laser light because each point in the ripply wavefront is able to spread independently of the other parts of the wave. The result is that the whole beam spreads out at an angle determined by the distance from the source to the aperture and the diameter of the source.

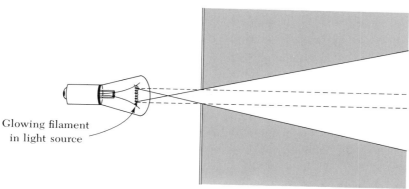

Figure 14-33
Shadow-limited spreading of an ordinary beam.

Coherence and Order

Chapter 13 began with the statement that much of physics was concerned with the movement of energy from one place to another and that this could happen without having to transfer mass as well. Toward the end of this chapter, we began to find that *coherence* was required in order to "steer" a wave in the direction that we wanted it to go; or, even just to test for the existence of waves. Coherence, we saw, implied some kind of control of the ripples in the wave, or at least some "cooperative" action (as in stimulated emission). In the next two chapters we shall look into the problem of thermal energy and getting *it* to move around. We will find that thermal energy is, by nature, *incoherent*, disordered, or chaotic and that this places limits on our ability to use it.

Summary

Standing wave patterns can develop in a medium surrounded by reflecting walls if the proper relation between wavelength, λ, and the distance, *L*, between reflecting walls exists:

$$L = \frac{n\lambda}{2}$$ Equation (14-1)

where $n = 1, 2, 3, 4 \ldots$.

Standing waves can be maintained with very little expenditure of energy thanks to the phenomenon of resonance.

When two or more similar waves pass through a medium at once they can add constructively, if they are in step (in phase); or, they can cancel out destructively if they are out of phase. In either case, the result is called interference.

When two waves of frequency f_1 and f_2 interfere, they produce beats of frequency

$$f_1 - f_2 \qquad\qquad \text{Equation (14-2)}$$

A point at a distance, r_1, from one wave source, and r_2, from a second will experience constructive interference, if:

$$r_1 - r_2 = n\lambda \qquad\qquad \text{Equation (14-3)}$$

and destructive interference, if:

$$r_1 - r_2 = (n + \tfrac{1}{2})\lambda \qquad\qquad \text{Equation (14-4)}$$

These conditions will occur *only if* the two wave sources are coherent, that is, if they oscillate "in phase" and emit similar waves.

Non-reflecting coatings produce a destructive interference by making the wave reflected from the coating's bottom surface be exactly out of phase with the wave reflected from the coating's top surface.

At a large distance from the two sources the approximation, $r_1 - r_2 = d \sin \theta$, holds, so the rule for constructive interference becomes:

$$n\lambda = d \sin \theta, \qquad\qquad \text{Equation (14-5)}$$

and for destructive interference;

$$(n + \tfrac{1}{2})\lambda = d \sin \theta. \qquad\qquad \text{Equation (14-6)}$$

Interference phenomena are important for separating out particular wavelengths from a mixture, and are even the operational test for the presence of waves.

When more than two coherent sources emit similar waves, the same direction rule for constructive interference, as given by Equation (14-5) still applies, but the interference pattern is sharper. Fringes become "lines". Thus two wavelengths differing by a small amount can be separated. Gratings, which are essentially a row of hundreds to thousands of parallel sources per centimeter become excellent devices for analyzing a mixture of wavelengths into their component parts. The science based on the examination of these spectra is called spectroscopy.

To study radio waves coming from stars, a grating on a much larger scale is used. The experiment is reversed, in that a row of antennae receives instead of emitting the waves, and θ, rather than λ, is the variable that is studied.

When a wave passes the edge of a barrier it spreads out by a process called diffraction. A single, coherent wave coming through an opening of width D spreads out to an angle θ given by

$$\sin \theta = \lambda/D \qquad\qquad \text{Equation (14-7)}$$

A tiny portion of the energy spreads out even further but for most purposes, this constitutes the limits of the beam's divergence.

In geometrical optics the assumption that light travels in straight lines is made. This applies when λ is much smaller than any objects or openings that the wave might hit. This is the realm considered in Chapter 13. In this chapter the subject of wave optics is considered. Here the wave encounters objects and openings of size similar to λ, and diffraction may occur.

Lasers provide, at last, a source of bright, coherent, monochromatic, electromagnetic radiation. Their beams do not spread in the manner of ordinary spotlight beams, but are limited in divergency only by Equation (14-7).

Further Reading

Connes, Pierre, "How Light is Analyzed", *Scientific American 219*, Sept. 1968, pg. 72. (The basics of spectroscopy.)

Gunther, Max, "Lasers—The Light Fantastic", *Playboy*, Feb. 1968, pg. 72. Reprinted in *A Universe of Physics*, Jerry B. Marion, Ed., John Wiley & Son, N.Y., 1970. (A short, readable history of the laser.)

Lasers and Light, A series of readings from *Scientific American*, W. H. Freeman & Co., San Francisco, 1969.

Pierce, John R., *Quantum Electronics*. Doubleday, Anchor Science Study Series, S44, Garden City, N.Y., 1966. (A lucid discussion of recent developments in electronics including a readable section on lasers.)

Van Heel, A. C. S. and Velzel, C. H. F., *What is Light?* World University Library, McGraw-Hill, N.Y., 1968. (Pictorial and concise coverage of refraction, interferences, diffraction, and other wave phenomena.)

Problems

14-1 What is the lowest resonant frequency of a toilet booth 33 inches wide?

14-2 A half-wave antenna can be thought of as a "resonant cavity" for picking up electromagnetic waves. Suppose an antenna five meters long vibrates electrically (not mechanically) at the fundamental wavelength. What will be the frequency of the electromagnetic wave it picks up best?

14-3 What is the minimum thickness of bluish tarnish on silverware?

14-4 The ionosphere is a layer of charged particles in the earth's atmosphere. It tends to reflect radio waves back toward the earth. In the late afternoon, as the sun sets, this layer tends to move up. At the same time, an AM radio tuned to a distant

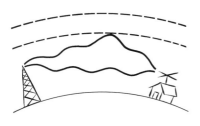

station will fade in and out. Explain this phenomenon on the basis of interference between two waves reaching the receiver: one direct, and one reflected, as shown in the margin.

14-5 A person stands near a pair of loudspeakers which are vibrating in phase, sending out a "middle C" tone (264 hertz). The temperature is 0°C. The person is 20 meters from one speaker, 15 meters from the other, and facing them. If the person moves his or her head to the left and then to the right, will a louder or a softer sound be heard?

14-6 Suppose that a grating has 6000 lines per cm. Find the angle θ, of constructive interference for $n = 1, 2$, and 3 with $\lambda = 4.5 \times 10^{-7}$ m (blue light). Show by means of a sketch the relative positions of constructive interference for red light ($\lambda = 6.5 \times 10^{-7}$ m) and for blue light.

14-7 Suppose that a double slit experiment is done with water ripples of two cm wavelength and with d equal to 8 cm. Find the distance between points of constructive interference at a distance, D, equal to 160 cm from the slits.

14-8 An apartment building shows this step-like construction when viewed from above:

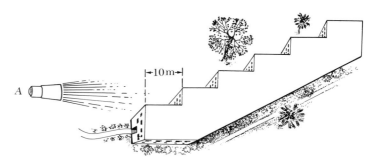

A person with a siren stands at point A, cranking it so that its frequency goes up and down. It is discovered that certain tones reflect back from the building louder than others. Figure out the frequency of two or three of these tones.

14-9 A scientist examines blue light (450 nm) by means of a grating and telescope and finds a first-order blue line at an angle of 16.3° from the "head-on" zeroth order line.

(a) At what angle will the next blue line be found?
(b) A new light source replaces the blue one, and produces a first order line at an angle of 19.25° from the head-on direction. What is the wavelength of the new source?

14-10 Suppose a factory whistle sends out a steady tone, with a frequency of 1000 hertz, on a day when the temperature is zero degrees C. A person hears the sound of the whistle coming through an opening between two large walls in a building under construction. The opening between the walls is one meter wide and the whistle is more than 100 meters behind the walls. The person walks along a line parallel to, and 50 meters from, the walls. Directly opposite the opening the sound is loud, but along the line in either direction, zones of silence are passed through. What is the distance between the zones?

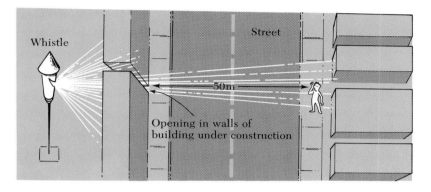

Whistle

Street

50m

Opening in walls of
building under construction

M. C. Escher, *Order-Chaos.* *(Courtesy of Escher Foundation, Haags Gemeentemuseum, The Hague, and Vorpal Gallery, San Francisco.)*

THERMAL PHYSICS

15

TEMPERATURE AND THERMAL ENERGY

Up to now thermal energy has been mentioned primarily as a bookkeeping item to keep the total energy of a closed system constant. If a certain amount of kinetic or potential energy disappeared, it was replaced with an equal amount of thermal energy. Thermal energy was also a way of accounting for work done against frictional forces. Recall:

Work done by net force $\rightarrow$ gain in *K.E.*

Work done by forces against force fields $\rightarrow$ gain in *P.E.*

Work done against frictional forces $\rightarrow$
gain in thermal energy, *U.*

The experimental fact that warming, and sometimes melting, accompany work done against frictional forces affirms the existence of thermal energy. A detailed discussion of thermal energy has been delayed until now because it is significantly different from, and more complex than, mechanical energy. Thermal energy is only indirectly visible. A warmed object expands, melts, glows, or burns our fingers when we touch it. But we cannot turn an object on its side and expect gravity to pull the thermal energy out of it. If we wish to use the thermal energy in an object, the best we can do is put the object near a cooler object and patiently wait for the thermal energy to spread from hot to cold object. If we do it right, we can cause some of the colder substance to boil and expand greatly, pushing against a piston and doing work for us. Aside from this, we have little control over the transfer of thermal energy. How do we measure it? Is there some way to picture it?

Temperature

Let us answer the first question first. Although measurements of thermal energy are less direct than measurement of the primary dimensions length, time, mass, and electric current, the same principles must be followed in order for the results of thermal measurements to be communicable. You can express the measurements in many cases in the form of *temperature* changes. If heat flows from one object to another, then we say that the object that loses the heat has a higher temperature than the one that gains it. If the two objects are at the same temperature, no heat moves from one to the other, on the average. Temperature is an *intensive* physical quantity, independent of how much matter is in the system. Thermal energy, on the other hand, is an *extensive* quantity, increasing with the size of the system. A burning match, for example, has a high temperature but contains little thermal energy. A bathtub full of warm water is much cooler, and is at a lower temperature, but it contains far more thermal energy. It is more capable of melting a big block of ice than the match is.

We are notoriously poor estimators of temperature. We know when something is warmer or cooler than our own bodies, but even this sense can be altered if our hands happen to be warmer or cooler than usual. It is a basic necessity to establish a temperature scale—a system for defining and comparing the temperatures of things in different parts of the world.

To be useful, a standard temperature scale should be defined in terms of certain temperature states of some fairly common substance. One of the first temperature scales was the Centigrade, or "hundred part scale", now called the Celsius temperature scale, after its inventor (the Swedish astronomer) Anders Celsius (1701–1744). It was based upon the *melting and boiling points of water*, a very common compound. By definition, the melting point of water was taken to be zero degrees (0°C), and the boiling point of water (at a certain pressure equal to the average barometric pressure at sea level) was taken to be 100 degrees (100°C). Celsius' temperature scale was then completed by dividing the temperature interval between zero and 100 into 100 parts, as well as extending it below zero and above 100.

But how can we use such an abstract scale for comparing the temperatures of objects not at zero or 100°C? How do we know when a sample of water is at 30°C, that is, 3/10 of the way up from freezing to boiling? Water has no easily measurable property which changes smoothly from freezing up to boiling, so by itself water does not make a good thermometer. We require some material parameter which can be measured easily and which increases without hitches over a long temperature interval. (Temperature intervals can't really be spoken about until we've decided how to measure them, but we can still tell whether a property is changing in a complex or simple way during the steady heating of an object.) Early in the nineteenth century, the French physicists Jacques Charles (1746–1823) and Joseph L. Gay-Lussac (1778–1850) found that if gases are maintained at a constant low pressure, they expand almost linearly (that is, the graph of volume *vs* temperature is nearly a straight line) with heating, over quite a wide range of temperatures. The rate of expansion was very nearly 1/273 per degree of temperature change on the Celsius scale. If the volume of the gas is defined as unity at 0°C, it would have increased from 273/273 to 373/273, at 100°C. A sample of gas kept at constant pressure then became a very good thermometer for measuring temperatures between zero and 100°C as well as below and above this interval. Insofar as various gas thermometers agreed with each other, their temperature readings were really communicable, and eventually they came to *define* temperature as well.

Since the gas appeared to shrink by 1/273 of its volume at 0°C with every degree of cooling, its volume should, in principle, decrease to zero at approximately −273° on the Celsius temperature scale[1] (Figure 15-1). This property of gases led to the creation of a new "ideal gas"[2] temperature scale, the *Kelvin temperature scale*, with degrees or temperature intervals matching those of the Celsius scale, but with the zero point adjusted to fall at −273.15°C. Temperatures measured on this scale are called Kelvin temperatures: 0°C becomes 273.15 K, and 100°C becomes 373.15 K, or 373.15 Kelvin.

The Kelvin, or Absolute temperature scale, has come to be widely used in scientific work because many physical and chemical processes and properties are expressible more simply in the Kelvin scale, or seem to approach zero, as zero on the Kelvin scale is approached.

In 1948 the International Committee on Weights and Measures agreed on a modification of the Celsius temperature scale to take

(1) To be exact, −273.15°C (Celsius). Actually gases condense to liquids and freeze to solids before −273.15°C is reached, so the volume-temperature plot is not a straight line down to zero.

(2) An ideal gas is one whose molecules are points, occupying no volume, and which exert no forces on each other except when they collide. Real gases approach the behavior of ideal gases as their pressure is lowered.

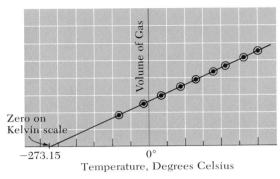

Figure 15-1
*Change in volume of an ideal gas as the temperature
is changed.*

advantage of a more accurately measurable temperature state.
Instead of depending on two states of water, it was redefined in
terms of a single temperature state of water, plus the known
thermal expansion properties of gas. The single state is called
the "triple point" of water: under a standard condition of pressure, liquid water, in a closed container, will exist with the solid,
and gaseous states in equilibrium at only one temperature. The
triple point temperature can be measured to within .0003°C, far
more accurately than either the ice point or the steam point. The
new Celsius scale is identical with the old scale with the exception that the tie between the Kelvin and Celsius scales is made
at the more accurately measurable triple point rather than the
ice or steam point.

A minor portion of the world still uses the *Fahrenheit* temperature scale, invented by Gabriel D. Fahrenheit in 1714. Orig-
iginally, he used the temperature of a mixture of ammonium
chloride, ice and water as the zero point and, on a suggestion
from Newton, chose the oral temperature of the human body as
the high-temperature reference point. He set this temperature
at 96, because that number is evenly divisible by 2, 3, 4, 6, 8, 12,
etc., and would be convenient for calculations. The freezing
point of water came out to be approximately 31.2°F and the
boiling point approximately 206.5°F on this first scale. Later,
Fahrenheit discovered that, by stretching the scale so that the
boiling point of water became exactly 212°F, the freezing point
became 32°F, and the temperature interval between the freezing
and boiling points came out to be 180 units. This was a number
which was evenly divisible by five and nine as well as all the
numbers mentioned earlier, so Fahrenheit was very pleased. By
making the boiling point of water the high-temperature reference
state, he also did not depend on the more variable human body
temperature. On the new scale, the "normal" human oral temperature rose from 96°F to 98.6°F, where it stands today.

Table 15-1 *Comparing the Three Main Temperature Scales*

	Kelvin Temperature	Celsius Temperature	Fahrenheit Temperature
Steam point	373.15	100	212
Triple point	273.16	.01	32.018
Ice point	273.15	0	32
°F & °C scales match		−40	−40
Absolute zero	0	−273.15	−459.67

To convert Fahrenheit temperatures to Celsius, use the relation

$$T°C = (T°F - 32)(\tfrac{5}{9})$$ Equation (15-1a)

To convert Celsius temperatures to Fahrenheit, use the relation

$$T°F = (T°C)(\tfrac{9}{5}) + 32.$$ Equation (15-1b)

Or, if you prefer, recall that $0°C = 32°F$ and that $100°C = 212°F$, then apply your own algebra.

Kelvin temperatures are obtained by adding 273.15 to the Celsius temperature.

But even the "ideal gas" Kelvin temperature scale is dependent upon a physical property of a real substance, and the change of the measured property (volume) is not exactly linear with temperature. The temperature scale could still be forced to fit the properties of the gas, but in repeated comparisons with various properties of various substances, we would find that certain degree intervals would be slightly larger than others. This leads to the suggestion that temperatures be defined in terms of gains and losses of heat under very closely controlled conditions. Instead of defining the temperature in terms of how much a certain sample of gas expanded, scientists instead spoke in terms of how much energy the gas had absorbed. This did not change temperature measurements themselves, but it made available a more "absolute" temperature scale whose definition was independent of the usually complex way in which real substances respond to changes in temperature.

Once temperature scales have been decided upon, the freezing and boiling points of other materials can be measured. They, in turn, become good reference temperatures because a material tends to stay at its boiling and freezing points, not warming up or cooling off during the addition or withdrawal of small amounts of energy. (Why this is so will be shown later.) In 1968 the International Committee accordingly established a "practical tem-

Table 15-2   *The International Practical Temperature
Scale*

Assigned Temperature		State
1337.58 K	1064.43° C	Freezing point of gold
1235.08 K	961.93° C	Freezing point of silver
692.73 K	419.58° C	Freezing point of zinc
373.15 K	100° C	Boiling point of water
273.16 K	.01° C	Triple point of water
90.188 K	−182.962° C	Boiling point of oxygen
54.361 K	−218.789° C	Triple point of oxygen
27.402 K	−246.048° C	Boiling point of neon
20.28 K	−252.87° C	Boiling point of hydrogen
17.042 K	−256.108° C	Equilibrium between liquid and gaseous states of hydrogen at 33×10^3 N/m² pressure
13.81 K	−259.34° C	Triple point of hydrogen

perature scale" defined in terms of 11 accurately measured boiling and freezing points.[3]

A lot of physics tends to be done at these temperature points because these boiling or freezing materials can provide a highly stable temperature "bath" for the apparatus being studied, protecting it from going up or down in temperature.

Having defined a temperature scale, and certain fixed temperature points, a convenient means is still needed for comparing the temperature of something with a gas system without having to carry the object to a gas thermometer. You need a portable device which has an easily measured physical property that increases almost linearly with temperature in the temperature range where you want to use it. The most commonly used property is expansion, or as in the mercury-glass thermometer, *differential* expansion. A quantity of mercury is put in a glass capsule which has a very thin neck, as shown in Figure 15-2. When the thermometer is put into something warm, the glass expands but the mercury expands even more, so it rises in the thin tube. If you want to make a Celsius thermometer, it must first be put into a bath of ice and water to make the mercury move to the zero level, which can now be marked on the tube. The thermometer can then be put into a steam bath at 100 degrees Celsius and

Figure 15-2
Mercury-glass thermometer.

(3) Gray, William T. and Finch, Donald I., "How Accurately Can Temperature be Measured?" *Physics Today 24*, Sept., 1971, pg. 32.

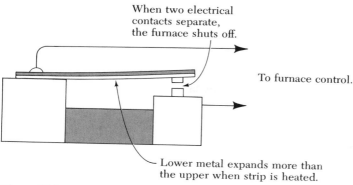

When two electrical
contacts separate,
the furnace shuts off.

To furnace control.

Lower metal expands more than
the upper when strip is heated.

Figure 15-3
Bimetallic switch used as a thermostat.

another mark made on the tube at the new mercury level. After this, the distance between the two reference marks can be divided into 100 equal parts, each corresponding to one degree temperature change. The marks can be extended below zero and above 100. If the thin glass tube (called a capillary tube) has any thick or thin places in it, the rise of mercury will be too little or too great per given change of temperature. Thus, a glass-mercury system would not be a good basis for *defining* temperature. Nevertheless, it is close enough to being correct to be a useful means for measuring temperature.

The differential expansion principle is also used in the bimetallic strips which are the heart of some thermostats that regulate temperatures of buildings. The two metals in the strip expand at different rates when the strip is heated, causing it to warp. This can be used to open or close a switch if the temperature of the room gets too high or too low.

Other physical properties which depend on temperature, such as electrical resistance or luminosity, can also be used for temperature measurement. For example, the temperature of the stars can be established by comparing the color of the light they emit with the color of objects (red-hot or white-hot filaments of metal) which have been heated to known temperatures.

Thermal Energy as Substance

The answer to the second question on page 392: — is there some way to picture thermal energy — has been stated outright or hinted at in several previous chapters. Nevertheless, it might be well to see, from a historical viewpoint, how the answer has changed. In so doing we will find still another field in which physics has withdrawn from a mechanistic, or at least a materialistic, point of view. The idea that thermal energy is a fluid or *substance* goes back to ancient Greece. Fire was one of the four material elements. Therefore, when something was heated, it was being mixed with some extra "fire". The concept of thermal energy as a substance was developed into the theory of *caloric* by Joseph Black (1728–1799), a Scots chemist. According to the theory, caloric had no weight and particles of caloric repelled each other. However, caloric was attracted to ordinary matter, and combined with it chemically to turn solids into liquids and liquids into gases. Count Rumford, who was mentioned in Chapter 7, made a scientific career of attacking the caloric theory. He was particularly bothered by the fact that in machining cannon barrels, he saw heat coming out of them, seemingly without limit. If rubbing a piece of metal with a tool squeezed the caloric out of it, you would think the caloric would come to an end. Tests of machined chips of metal showed them to be the same thermally as the unmachined metal, so "robbing them of their caloric" seemed to have no effect. Rumford eventually concluded that thermal energy is not a substance but *motion*, on a molecular scale. Rumford's ideas did not lead to the immediate demise of the caloric theory because they were not quantitative. However, the caloric theory was finally overthrown by the work of Julius Robert von Mayer (1814–1878) and James Prescott Joule (1818–1889). Mayer was a German physicist who showed in 1846 that work done, and heat produced, are equivalent. Joule was a Britisher who actually measured the mechanical equivalent of heat; that is, the amount of work required to produce a certain amount of thermal energy. Caloric became a thing of the past. As is stated in Chapter 7, the evidence that thermal energy is produced by work done against frictional forces, and is just another form of energy, put the conservation of energy principle on the strong footing that it still has.

Thermal Energy and Statistics

Modern physicists now visualize heat or thermal energy as motion of the molecules or atoms in a substance. It is kinetic energy on a fine scale. But there is an important further difference: it is chaotic, *disordered* kinetic energy. For these reasons an approach different from the mechanistic, deterministic Newtonian physics must be used. The concept of mechanism implies that you can identify and locate an object, keep tabs on its position, measure its velocity, and predict where the object will be at some time, Δt, later. If you can predict, you can also control; you can, in principle, reach in and speed it up, slow it down, or lift it. However, in the case of random molecular motion it is impossible to follow the motion of each object in the system. You have to take a statistical or probabilistic approach and content yourself with statements about the probable behavior of the whole assembly when certain things are done to it. Although the behavior of individual atoms is uncontrolled and unpredictable, out of the random fluctuations of their individual motions comes a smoothed-out large-scale predictability. You observe properties and phenomena which are the statistical result of many events occurring on an atomic scale.

The statistical point of view, whether it is used in science, or for calculating insurance premiums, does not mean that you are just interested in facts and figures. It means that predictions are made concerning *large numbers* of events or objects. When thinking statistically, you can never be 100% sure that something must or must not happen. You learn to live with "fluctuations" from the most probable event. An insurance company can never be sure which of its customers will live to be 130 years old. But the company survives by being able to predict quite accurately what fraction of 100,000 people now 21 years old will be alive in 50 years. In an analogous fashion, temperature is a statistical concept, that applies to an assembly of billions of atoms or molecules. At a certain temperature, the atoms in the assembly will have various random kinetic energies, but will possess a certain *average* kinetic energy. It has no meaning to speak of the temperature of a single particle. When physicists use the word temperature, they are implicitly speaking of a statistical average over events occurring on a finer scale than the object whose temperature they are referring to. We can now distinguish between temperature, thermal energy, heat, and work:

(1) Temperature, T—related to the average random kinetic energy of the atoms or molecules in a system.

(2) Thermal energy, U—the total random kinetic energy of all the molecules in a system, as measured in a frame of reference in which the total momentum of the system is zero.

(3) Heat exchanged, ΔQ—energy being transferred (crossing the boundary of a system) as a result of temperature differences. It can either be added to a system from its surroundings or removed from a system and given to its surroundings. The distinction between what is change in thermal energy, ΔU, and what is heat depends upon where you draw the boundary of your system.

(4) Work, ΔW—energy which is added to, or removed from, a system by means of forces.

One thing should be said about numerical values of the thermal energy of a system. Just as was the case for potential energy, there is always an arbitrary choice involved: you *define* the thermal energy, U, to be zero under such and such a set of conditions. What is more important than the numerical value of thermal energy is the *change* in thermal energy, ΔU, which a system undergoes when you do various things to it. All we really can measure is heat or work added to, or removed from, the system.

For historical reasons, thermal energy and heat are usually measured in calories. One calorie is defined to be the heat required to warm one gram of water from 14.5 to 15.5°C. The reason for the temperature specification is that various amounts of heat are required to warm water 1°C at various temperatures. The 15 degree calorie was taken as the standard for thermal measurements. (The unit of energy which weight watchers worry over is the Calorie, which equals 1000 calories.)

Joule's experiments measured the increase in thermal energy of many systems when mechanical work was done on them (by rubbing, stirring, or letting an electric current run through them). The modern value for this equivalence is

$$1 \text{ calorie} = 4.184 \text{ joules.} \qquad \text{Equation (15-2)}$$

Another unit of thermal energy commonly used in England and America is the British thermal unit, or "B.t.u.". It is the amount of heat required to increase the temperature of one pound of water one degree Fahrenheit. Since one pound = (1/2.205) kg and 1°F = (5/9)°C,

$$1 \text{ B.t.u.} = \left(\frac{1}{2.205}\right)\left(\frac{5}{9}\right)(1 \text{ calorie}) = .252 \text{ Cal.} = 1054 \text{ joules}$$

Equation (15-2), which states "the mechanical equivalent of heat", was arrived at experimentally and provided an important link between thermal and mechanical physics. Once the equivalence was known, it could be used as a *definition* of the calorie, and the basic units in thermal physics became the joule and the degree Kelvin (now called simply, the kelvin). We shall speak primarily in terms of joules from this point on.

The First Law of Thermodynamics

Imagine a closed chamber filled with steam. A certain amount of heat, which we shall symbolize as ΔQ, is added to the steam by building a fire under the chamber. The temperature of the steam will rise by an amount which depends on how much steam is in the chamber. All of the heat added to the system goes into increasing the thermal energy of the steam: $\Delta Q = \Delta U$. According to our molecular model of a gas the water molecules in the steam will be racing around with greater random speed than they did before (Figure 15-4).

But now suppose a wall of the chamber can move, as a piston in a cylinder. If we add the same amount of heat, ΔQ, to the steam it will now expand, doing some work, ΔW, against the piston, so that less of the heat added is left to increase the thermal energy of the steam. Its temperature will not increase as much as before. If ΔW is the work done *by* the expanding steam,

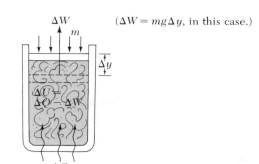
ΔW ($\Delta W = mg\Delta y$, in this case.)

Figure 15-4
When heat, ΔQ, is added to a system, the increase in thermal energy, ΔU, is greater if the system is not allowed to expand and do work, ΔW.

then by the principle of energy conservation the sum of ΔU and ΔW should be equal to the heat added to the system, ΔQ:

$$\Delta Q = \Delta U + \Delta W$$

or

$$\Delta U = \Delta Q - \Delta W \qquad \text{Equation (15-3)}$$

This relation is called the *first law of thermodynamics*. It is a statement of energy conservation, applying to situations in which heat and mechanical energy are entering or leaving a system. If a system is thermally insulated, $\Delta Q = 0$. If it is mechanically isolated, $\Delta W = 0$. If work is done *on* a system; for example, if the steam were compressed, then ΔW is negative, and Equation (15-3) becomes $\Delta U = \Delta Q + \Delta W$. Both the heat added and the work done on the system add to its thermal energy in this case.

If these ideas are considered in the light of Chapter 7, we might be tempted to call the first law of thermodynamics a statement of the non-conservation of energy of a system. It still insists that energy is not created or destroyed. But instead of focusing on systems which are isolated and whose energy therefore remains constant, the first law of thermodynamics determines the amount of non-isolation in systems by providing a determination of how much heat they take on, and how much work they do on (or receive from) their surroundings. Of course, we must still keep track of *all* the energy, and that's what counts. As we've seen before, conservation or non-conservation of energy is always dependent on where one draws the boundaries of the system. If we allow for all possible changes in potential, kinetic, nuclear, and thermal energy in a system, it is possible to write down one grand energy conservation equation:

$$\Delta K.E. + \Delta P.E. + \Delta U + \Delta mc^2 = \Delta Q - \Delta W \qquad \text{Equation (15-4)}$$

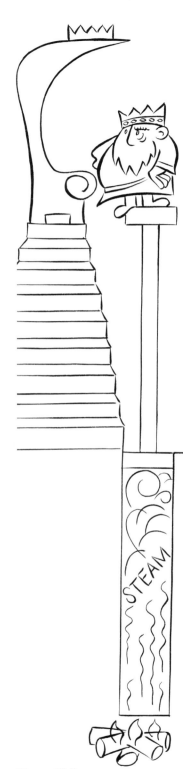

To illustrate these ideas, consider a fable. It concerns a fat and lazy king. His throne was mounted on a pedestal five meters above the floor of the throne room. It was reached by a flight of stairs, but over the years the king grew too fat to climb them. He therefore commissioned the court engineer to build a steam lift which would majestically raise his highness up to the level of the throne. The engineer sank a round, metal-lined cylinder about seven meters deep and with an area of 1 m² down through the floor of the throne room. Under it he put a fireplace which could be stoked from the cellar below. Water was put into the bottom of the cylinder, and as the fire blazed, heat was absorbed by the water. Some of it boiled, making steam. Inside the cylinder the engineer fitted a piston which supported a tree trunk with a platform on top of it. Things were arranged so that when the king entered the room the platform was even with the floor, with the piston about five meters below. When the king got onto the platform, his weight, that of the platform, the pole, and the piston totaled 565 pounds, or 2500 N, approximately. The air pressing down on the piston added another 101,000 N. As more heat entered the water, more steam formed, and since it was not confined, it expanded, lifting the king up to the throne. The work done by the steam was

$$\Delta W = F \Delta y = (2500 \text{ N} + 101,000 \text{ N})(5 \text{ m})$$
$$= 517,500 \text{ J}$$

Meanwhile enough extra water needed to be evaporated to provide steam at constant pressure to fill the extra five cubic meters of space under the rising piston. Water expands tremendously when it turns to steam. With 103,500 N force acting down on the piston, a cubic centimeter of liquid water becomes about 1635 cubic centimeters of steam. This meant that 5 m³/1635 which equals $(5 \times 10^6 \text{ cm}^3)/1635$ or 3058 cm³ of water had to be turned to steam. We shall see on page 407 that to turn one cm³, or one gram, of water to steam requires 2255 joules of heat. This meant that the heat that had to be added to the system was

$$\Delta Q = (3058 \text{ cm}^3)(2255 \text{ joules/cm}^3) = 6,899,000 \text{ J}.$$

The lift does about one joule of work for every 13 joules of heat added to it: $(6,899,000/517,500 \cong 13)$.

The increase in thermal energy of the water-steam system must then be:

$$\Delta U = \Delta Q - \Delta W = 6,899,000 \text{ J} - 517,500 \text{ J}$$
$$= 6,376,000 \text{ J}.$$

The king was most pleased by the steam lift and ordered the court engineer to begin design of a steam-powered rocket for trips to Mars.

Figure 15-5
The King's steam lift.

When we examine the response of an object to the addition or removal of some heat, we see some interesting things. First, the amount of heat required to increase the temperature of an object always is proportional to its mass, just as the weight of an object in a gravitational field is proportional to its mass. Second, the *material* comprising the object is important. For example, more heat is required to warm up a kilogram of water one degree Celsius than to warm a kilogram of lead the same amount. The relationship between the increase in thermal energy and the temperature change can be summarized by the relation

$$\Delta U = mc\Delta T \qquad \qquad \text{Equation (15-5)}$$

where m is the mass of the object in grams, c is the *specific heat*, or the amount of heat required to warm a gram of the substance one degree Celsius (measured in J/gm·°C) and ΔT is the temperature change, in °C. Table 15-3 (below) shows some typical values of specific heat at or near room temperature for a number of materials. Water at 15°C is taken as the standard, with a specific heat of 4.184 J/gm·°C.

The difference between the specific heats of liquid and solid forms of water hints at the complexity of thermal properties. Specific heat is not a constant, but depends on temperature. The

Table 15-3   *Specific Heat, c, For Various Materials*

Material	c, in J/gm·°C	Temperature
Liquid water	4.184	15°C
Ice	2.059	0°C
Aluminum	.923	100°C
Ammonia	4.720	20°C
Copper	.3815	20°C
Glass, pyrex	.839 (avg)	19–100°C
Glycerine	2.510	50°C
Iron, steel	.473 (avg)	18–100°C
Lead	.1296 (avg)	20–100°C
Mercury	.139	20°C
Silver	.2354 (avg)	15–100°C

explanation of this and of the differences in specific heats of various substances requires a knowledge of the molecular characteristics of matter which is beyond the scope of this text. It was, in fact, a problem which physicists were not able to solve satisfactorily until the late 1920's.

Table 15-3 shows that water in the liquid form has a relatively large specific heat. For this reason, large bodies of water are more difficult to heat up and cool down than the land which surrounds them. A body of water, then, can temper the climate along its shores, especially if breezes blow from the water over the land. During the early winter, the water cools slowly, and during the summer, it is slow to warm up during hot days.

Heat Exchanges

When objects having different temperatures are put together in a system, heat moves from the hot objects to the cooler ones. The movement can occur through *conduction, convection,* or *radiation,* as illustrated in Figure 15-6. In the first case the heat is transmitted directly by molecular contact, as when two bodies touch, or when thermal energy spreads across a single object. "Hot" molecules—or, more correctly, groups of hot molecules in the object—collide with molecules having less random kinetic energy and the thermal energy is shared. In the case of *convection,* a heat transfer medium is warmed and *it* moves (either spontaneously or by being pumped) carrying the heat from one place to another where it can warm up cool objects. Finally, hot objects can *radiate* heat which travels through space to reach cooler objects, warming them up. These heat waves have wavelengths in the infrared region of the electromagnetic spectrum (Table 13-2, page 339). Hot water radiators, as well as stars, emit heat waves.

In all three instances the net transfer of heat continues as long as there are differences in temperature between the objects. If a system is left to itself, all objects in it will come to the same final temperature, but not necessarily with the same internal thermal energy. The system will then have reached *thermal equilibrium.*

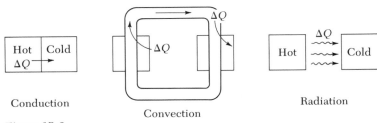

Conduction

Convection

Radiation

Figure 15-6

This process can take considerable time. Anyone who has bitten into a cold pea taken from a bunch of frozen peas that have just been brought to a boil in a pot knows this.

Even after thermal equilibrium is reached in a system, heat exchanges still continue; but the amount of heat leaving one object in a system will be balanced by energy received, on the average. The statistical nature of the process allows for energy fluctuations, but they are very unlikely to endure.

The human body is almost never in thermal equilibrium with its surroundings. We are "warm-blooded" animals, meaning that we have in us a form of thermostat which tries to keep our body temperature at 37°C (98.6°F). If our surroundings are cold, our metabolism increases; food is converted to thermal energy more rapidly, and the extra energy keeps our body temperature at the normal level. If our surroundings get too warm, we perspire. As the perspiration evaporates, the surface of our bodies is cooled, for reasons that will shortly be examined. In humans, fever results when the body's heat production exceeds its heat loss. For instance, when infection is being battled by the body the extra effort requires additional metabolism, and the body temperature goes above "normal". In contrast to this, cold-blooded animals, such as snakes, undergo increases and decreases of their body temperatures as their surroundings get warmer or cooler. In cold weather they literally cannot move.

A problem: suppose you drink a bottle of soda which is at the temperature of melting ice. How many calories must the soda contain in order to provide your body with just enough energy to raise the soda to body temperature?

First, let us suppose that there are 12 ounces (340 gm) of pop in the bottle and the specific heat of the pop is 4.184 joules/g·°C.

The pop must be heated from 0°C to 37°C. Therefore,

$$\Delta Q = \Delta U = mc\Delta T$$

$$= (340 \text{ g})\left(\frac{4.184 \text{ joule}}{g \cdot °C}\right)(37°C - 0°C)$$

$$= 52{,}635 \text{ joules}$$

$$= 12{,}580 \text{ cal.}$$

Another problem: a chunk of unknown metal having a mass of 100 grams and a temperature of 25°C is dropped into an insulated cup containing 290 grams of water at 90°C. The system is stirred and comes to a final temperature of 88°C. Question: which of the metals in table 15-3 is the chunk made of? First, the thermal energy given off by the water is equated to that absorbed by the metal:

$$-M_x c_x (\Delta T_x) = M_w c_w (\Delta T_w)$$

or

$$(100 \text{ g}) (c_x) (88 - 25)°C =$$

$$- (290 \text{ g}) \left(\frac{4.184 \text{ joules}}{g \cdot °C} \right) (88 - 90)°C$$

$$c_x (100 \text{ g}) (63°C) = +2426.7 \text{ joules}$$

$$c_x = \frac{2426.7 \text{ joules}}{6300 \text{ g} \cdot °C}$$

$$c_x = .385 \frac{\text{joules}}{g \cdot °C}$$

The metal is probably copper.

Change of Phase

Consider an experiment in which 100 gm of ice at $-10°C$ are placed in a closed container on a hotplate. You turn on the hotplate and measure the temperature of the ice or water continuously. The hotplate gives off 1000 watts or 1000 joules per second. Assume that 500 joules of this goes into the ice or water and 500 joules into the container. A graph of temperature *vs* time would look like the one shown in Fig. 15-7. After a quick climb to 0°C, the temperature lingers there for a minute or so. Then the temperature rises steadily up to 100°C, where it remains for quite a long time. After all the water has boiled away, the tempera-

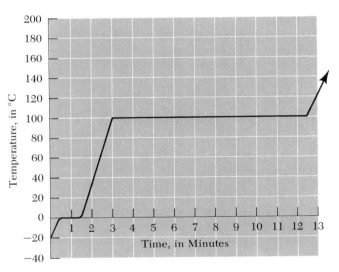

Figure 15-7

Time record of the temperature of a specimen of ice which receives heat at a uniform rate.

ture of the steam enclosed in the container begins to rise again. Behavior of this sort led physicists to suggest that energy is required to melt ice to water and to boil water to steam, *without increasing the temperature* of the water. When a substance melts, freezes, vaporizes, or condenses (changes back from gas to liquid) it is said to undergo a change of phase. The amount of thermal energy required to melt one kilogram of a substance is called its *heat of fusion*. (Not to be confused with the nuclear fusion processes discussed in Chapter 9.) For water, this amounts to an energy of 3.35×10^2 joules. When a kilogram of a substance freezes, the heat of fusion is given off. The amount of thermal energy required to turn one gram of a substance from liquid to vapor or gas is called its *heat of vaporization*. For water this amounts to 2.255×10^6 joules per kilogram. It takes five times as much heat to evaporate one kilogram of water as it does to increase the temperature of that water from zero to 100 degrees Celsius! From this you can see why so much heat was required to make the extra steam for the king's lift and how the evaporation of sweat cools us off: whenever a liquid changes to a gas, it absorbs a large amount of heat. But when the water condenses from gas or vapor back to liquid again, the heat is given off.

The amount of energy released when large quantities of water vapor condense into liquid water, as in a rainstorm, can be stupendous. It becomes a major factor in the large turbulent movements of air which accompany storms. James E. McDonald[4] estimates that in an ordinary rainstorm (with about a one kilometer radius, and producing an average rainfall of one centimeter) the energy released by condensation of the water amounts to more than 8×10^{13} joules—about as much as was released by the nuclear bombs dropped on Japan in 1945! This energy helps drive the updrafts of air which separate positive from negative charge, producing the huge voltages in thunderstorms that were spoken of on page 294.

By now we can conclude that water is a common but peculiar substance. It expands instead of contracting when it is cooled to freezing; it has a relatively large specific heat, making it difficult to warm up or cool, and it absorbs or gives off large amounts of heat when it changes phase between gas, liquid, or solid.

Because of the heat of fusion of water, ice cubes are more effective for cooling drinks than plain cold water. A large amount of energy is absorbed from the rest of the liquid just to turn the cubes from *ice* at 0°C to *water* at 0°C. To illustrate this, suppose we first pour 50 grams of water at 0°C into 200 grams of coffee at 70°C. The simplifying assumption is made that the specific heat of coffee is the same as for water, that is, 4.184 joules/g·°C. Assume also that the mixing occurs so rapidly that no heat is lost to

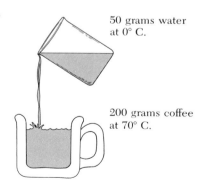

50 grams water
at 0° C.

200 grams coffee
at 70° C.

(4) *Advances in Geophysics*, Vol. 5, Academic Press, New York, 1958.

Figure 15-8

the surrounding air. The coffee will give off heat equal to that absorbed by the water until the mixture all comes to a single final temperature.

$$-\Delta Q_{cof} = -(m_{cof})(c_{cof})(\Delta T_{cof}) = \Delta Q_w = (m_w c_w \Delta T_w)$$

But

$$c_{cof} = c_w = 4.184 \frac{\text{joules}}{\text{g}°\text{C}},$$

so

$$m_{cof}\Delta T_{cof} = -m_w \Delta T_w; \quad \frac{\Delta T_{cof}}{\Delta T_w} = -\frac{m_w}{m_{cof}} = -\frac{50 \text{ g}}{200 \text{ g}} = -0.25$$

$$\Delta T_{cof} = -0.25 \, \Delta T_w.$$

But how can we find the final temperature (T_f)? We know that the final temperature of both coffee and water is the same. So

$$\Delta T_{cof} = T_f - 70°\text{C};$$

$$\Delta T_w = T_f - 0°\text{C} = T_f$$

So then we have:

$$T_f - 70°\text{C} = -0.25(T_f);$$

adding 70°C and 0.25(T_f) to both sides, we get

$$T_f + 0.25 \, T_f = +70°\text{C}$$

or

$$1.25 \, T_f = +70°\text{C}$$

Dividing by 1.25, we get:

$$T_f = \frac{70°\text{C}}{1.25} = 56°\text{C}$$

The coffee cools from 70°C to 56°C, and the cold water is warmed from 0°C to 56°C. Note that the coffee, with four times the mass, undergoes one-fourth of the temperature change.

Now suppose we repeat the experiment, but use 50 g of ice at 0°C instead of 50 g of water at 0°C. The following relation now holds:

$$\begin{array}{c} \text{Heat given off} \\ \text{by coffee} \end{array} = \begin{array}{c} \text{Heat absorbed by} \\ \text{melting ice} \end{array} + \begin{array}{c} \text{Heat absorbed by} \\ \text{water warming to} \\ \text{final temperature} \end{array}$$

$$-m_{cof}c_{cof}\Delta T_{cof} = (\text{Heat of fusion})m_w + m_w c_w \Delta T_w$$

$$(-200 \text{ g})\left(4.184 \ \frac{\text{J}}{1 \text{ kg} \cdot {}^{\circ}\text{C}}\right)(T_f - 70{}^{\circ}\text{C}) =$$

$$\left(3.35 \times 10^2 \ \frac{\text{J}}{\text{g}}\right)(50 \text{ g}) + (50 \text{ g})\left(4.184 \ \frac{\text{J}}{\text{g} \cdot {}^{\circ}\text{C}}\right)(T_f - 0)$$

$$\left(-837 \ \frac{\text{J}}{{}^{\circ}\text{C}}\right)(T_f - 70{}^{\circ}\text{C}) = (1.675 \times 10^4 \text{ J}) + \left(209 \ \frac{\text{J}}{{}^{\circ}\text{C}}\right)(T_f)$$

$$(-837 - 209) \ \frac{\text{J}}{{}^{\circ}\text{C}} \ (T_f) = (1.675 \times 10^4 \text{ J}) - (5.86 \times 10^4 \text{ J})$$

$$\left(-1046 \ \frac{\text{J}}{{}^{\circ}\text{C}}\right)(T_f) = -4.185 \times 10^4 \text{ J}$$

$$T_f = \frac{+4.185 \times 10^4}{1046} \ {}^{\circ}\text{C}$$

$$= 40{}^{\circ}\text{C}$$

The ice cools the coffee 30°C, more than double the effect of just the cold water!

Molecular theory provides answers as to why energy is absorbed or released during phase changes. According to the model the molecules or atoms in a crystalline *solid* are closely packed and arranged in neat rows and layers. They vibrate randomly, but the vibrations are usually about a particular site in the solid. As a solid is warmed the random motion of the molecules increases and, on the average, each molecule requires more space. Thus, most solids expand as they are heated. Eventually the random energy of the molecules is such that they are able to wander away from their "home sites" in large numbers. The atoms are still closely packed but their arrangement is less orderly. The solid melts.

Some rigid materials, such as glass, do not have as much order or structure in their atomic arrangements. They are better classified as very viscous liquids. They do not have a definite melting point, as do crystalline solids, but soften gradually as they are heated. A crystalline solid such as ice, on the other hand, melts just on the surface. It does not soften but maintains a core

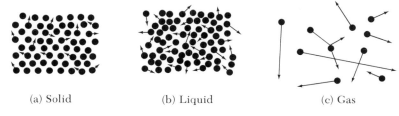

(a) Solid   (b) Liquid   (c) Gas

Figure 15-9

of solidity until the last of it turns to liquid. As we have mentioned, ice is unusual in that it contracts when it melts. For reasons beyond the scope of this discussion, the molecules in ice are not spaced as closely as the molecules in liquid water at 0°C.

If one continues to heat a sample of liquid, the atoms or molecules in it finally gain enough energy to tear themselves away from each other. The liquid boils or evaporates, depending on the rate of heating, and it becomes a gas. A lot of work has to be done to separate these atoms, which are held together by forces of attraction, increasing their potential energy as well as their kinetic energy. For this reason, considerable heat must be applied to a liquid to make it vaporize. (Some materials, such as carbon dioxide) change directly from the solid form (dry ice) to gas, by-passing the liquid phase.

The Conversion of Mechanical to Thermal Energy

Because of the prevalence of friction forces in the real physical world, mechanical energy is continually being converted to thermal energy. Sometimes we *want* this to happen, as when we rub our hands together to warm them up, or, as on page 180, a truck driver steps on the brakes as the vehicle comes down a long hill. But most of the time we generate thermal energy without wanting to.

Suppose a 500 gram projectile made of lead is traveling at a speed of 400 m/sec. Suppose further that it smashes into a thick steel wall and that half of its original energy is given to the wall, warming it. Half of the energy is kept by the projectile and, we claim, causes it to heat up. Question: does any of the lead melt? How much?

If the projectile has a temperature of 60°C before it hits the wall, then a certain amount of heat is required to warm it up to 327°C, the melting point of lead. This energy,

$$\Delta Q = \Delta U = mc\Delta T = (500 \text{ g}) (1296 \text{ joules/g°C}) (267°C)$$
$$= 17{,}302 \text{ joules}$$

The initial kinetic energy of the projectile is

$$\tfrac{1}{2}mv^2 = \tfrac{1}{2}(500 \text{ g}) (400 \text{ m/sec})^2 = 40{,}000 \text{ joules}.$$

Half of this is available for heating the projectile. Therefore, it should heat up to 327°C and there should be $20{,}000 - 17{,}302 = 2698$ joules left over for melting.

The heat of fusion of lead is 24.7 joules/g. Then the amount of lead melting must be

$$\frac{2698 \text{ joules}}{24.7 \text{ joules/g}} = 109 \text{ grams}.$$

About one-fifth of the projectile melts.

As far as energy requirements or conservation are concerned, the melted lead could rejoin the shell, the warm wall could cool off, giving its thermal energy back to the shell, which could rebound, heading back toward the cannon and its amazed gunners at 400 meters per second. But this has never happened—at least in the recorded annals of artillery. The next chapter will look into the interesting question of why it doesn't happen.

Summary

The existence of thermal energy is affirmed by the warming, melting, and vaporization of objects. Until the mid-1800's, heat was thought to be a substance, with which objects combined. Now it is visualized as fine-scale, molecular kinetic energy, chaotic and uncontrollable.

To measure temperature we need a means as well as some well-defined scale of temperature. The Celsius temperature scale, based upon water, is widely used. The freezing temperature of water is given a value of $0°C$ and the boiling temperature of water at normal atmospheric pressure is taken to be $100°C$. The Kelvin temperature scale is based upon the expansion of gases, and takes a zero point $0\ K = -273.15°C$.

Celsius temperature is $T°C = (T°F - 32)\left(\dfrac{5}{9}\right)$

Equation (15-1a)

Fahrenheit temperature is $(T°F = T°C)\left(\dfrac{9}{5}\right) + 32$

Equation (15-1b)

Portable thermometers actually used for measuring temperatures must have some property associated with them which increases almost linearly with increasing temperature. Differential expansion, electrical resistance, and luminosity have all been used as thermometric variables.

Thermal physics requires a shift from mechanical to statistical thinking. Probabilities, not certainties, are obtained. Temperature itself is a statistical concept, related to the average random kinetic energy of an assemblage of many objects.

Thermal energy is seen as the total random kinetic energy of the molecules in a system. Heat is energy which is added to or removed from a system as a result of temperature differences. Work is energy which is added to or removed from a system by means of forces.

The first law of thermodynamics states that $\Delta Q = \Delta U + \Delta W$, or:

$$\Delta U = \Delta Q - \Delta W, \hspace{2cm} \text{Equation (15-3)}$$

Where: ΔU is the change in internal thermal energy of the system; ΔQ equals the heat given to the system; and, ΔW is the work done by the system.

For a non-isolated system, the principle of energy conservation can be expressed as;

$$\Delta(K.E.) + \Delta(P.E.) + \Delta U + \Delta mc^2 = \Delta Q - \Delta W$$
$$\text{Equation (15-4)}$$

Thermal energies are generally measured in calories, the amount of heat required to raise the temperature of one gram water 1° Celsius.

$$1 \text{ calorie} = 4.184 \text{ joules} \hspace{2cm} \text{Equation (15-2)}$$

Increases in thermal energy are measurable from increases in temperature:

$$\Delta U = mc\Delta T \hspace{2cm} \text{Equation (15-5)}$$

In this equation, m is the mass of object in grams; c is the specific heat of system, (the thermal energy required to raise the temperature of one gram, one degree Celsius, measured in joules/g·°C); and, ΔT the temperature change in degrees Celsius.

Water has a relatively large specific heat. It is difficult to warm up and cool off, which enables it to affect local climate.

Heat may be exchanged by conduction, convection or radiation. In any case, temperature equilibrium is reached when all objects in a system come to the same final temperature. To do this, objects in a system exchange a certain amount of heat: one losing, the other gaining, but their temperature changes may not be the same.

Large amounts of energy (3.35×10^2 joules/g for water) are required to melt a substance from solid to liquid without changing its temperature. This energy is called "heat of fusion". An even larger amount of energy, called the "heat of vaporization" is required to change a substance from liquid to vapor or gas $(2.255 \times 10^3$ joules/g for water).

Mechanical energy is continually being converted to thermal energy, but the opposite conversion is more difficult to accomplish.

Further Reading

Brown, T. M., "Resource Letter EEC-1 on the Evolution of Energy Concepts from Galileo to Helmholtz," *American Journal of Physics* 33, pg. 759, 1965.

Holton, Gerald and Roller, Duane H. D., *Foundations of Modern Physical Science*, Addison-Wesley, Reading, Mass., 1958. (Chapters 19 and 20 give interesting history of the development of thermal concepts.)

MacDonald, D. K. C., *Near Zero: The Physics of Low Temperatures*, Doubleday Anchor Book, Garden City, N.Y., 1969. (About the properties of materials at very low temperatures.)

Problems

15-1 On the basis of heat being molecular motion, how do you explain the flow of heat to the handle of a silver teapot?

15-2 Where do liquids go when they evaporate?

15-3 Suppose you have a vacuum or thermos bottle, good for keeping hot liquids hot. Would it also be good for keeping cold liquids cold? Why or why not?

15-4 What would a thermometer read if you were to take it to outer space, where there is no air. Does it make any difference if the sun shines on it?

15-5 Using the molecular motion model, can you explain why a gas heats up when it's compressed?

15-6 Several objects lying side by side on a table will not all feel equally cool to the touch even though they are at the same temperature. One reason for this is that they do not conduct heat equally well. A piece of plastic which does not conduct heat away will warm more quickly when you touch it. What is another factor to consider in explaining how cool an object feels?

15-7 How much energy in joules must be added to a system if the internal energy is increased by 500 joules and 1000 joules of work are done *on* the system?

15-8 Suppose you put two identical clocks, one unwound and the other fully wound, into separate beakers of acid. After the metal parts have dissolved, how can you tell which beaker held the wound-up clock? Trying out a few numbers, indicate if it would be easy, or whether it would require a very careful experimental set up to identify which beaker is which.

15-9 Starting from basic principles, such as the electric force acting on an electron in an electric field, explain why a toaster gets warm when you plug it in.

15-10 A 500-gram block of copper is pushed 10 m at constant speed along a horizontal floor by a horizontal push of 6 N. Find (a) the thermal energy developed, and (b) the temperature rise of the copper if 2/3 of the thermal energy developed goes into the copper.

15-11 A suspicious person wants to find out if a cup received as a gift is really made of silver or of silver-plated brass (brass and copper have about the same specific heat.) She weighs the cup and finds its mass to be 200 g. Then she heats the cup to $100°C$ in boiling water. After that she drops the hot cup into an insulated bucket containing 500 g of water at $20°C$. After stirring things a bit, she measures the final temperature of the mixture and finds it to be $34.7°C$. What is the cup made of?

15-12 The water coming from a cold faucet has a temperature of $15°C$ while that coming from the hot faucet has a temperature of $70°C$. How much water must be drawn from each faucet to obtain a bath of 75 kg of water at $45°C$?

15-13 Suppose you are in a Paris hotel. You look out the window and see a large thermometer on the wall of a building across the street. It reads $36°$. Do you put on a coat before you go out?

15-14 When 5 grams of steam at $100°C$ are condensed in 200 grams of water contained in a 100 gram aluminum cup, the final temperature of the system is $40°C$. Assuming no loss of heat to the surroundings, calculate the temperature rise of the water in the cup. Calculate the heat released to the water and the cup. Explain why heat is given off when the steam condenses to water.

15-15 125 grams of iron pellets are put into a cardboard tube 1.64 m long. The ends are closed with tape. The tube is then rotated like a large drum major's baton. The shot is lifted from one end of the tube to the other, as the tube rotates. How many times would you have to rotate the tube to heat up the iron $1°C$? A "rotation" occurs when you turn the tube 180 degrees so that the end which was down is now up.

16

USING THERMAL ENERGY; DISORDER AND THE DIRECTION OF TIME

The Second Law of Thermodynamics

Chapter 15 said that thermal energy is visualizable as fine scale, chaotic energy and that statistical, rather than detailed mechanical analysis is needed to understand it. It also said that in one instance after another, thermal energy seems to grow at the expense of large-scale mechanical energy. This one-way arrow pointing from mechanical toward thermal energy is one example of a general principle which seems to be at work in the physical world. It is called *the second law of thermodynamics*. The purpose of this chapter is to present various statements of the second law, to examine examples of the law at work, and to seek a general explanation of the principle.

The second law of thermodynamics is a different sort of law from those we have discussed thus far. It merely indicates directions or tendencies. It is not quantitative, in the sense that it does not tell how *rapidly* a system will follow a certain tendency. Though it has profound implications, it is based on probability, not mechanistic certainty. In whatever manner it is stated, the words "tend to" or "will most probably" are implicit. A first statement of the law might take two equivalent forms:

(1a) Heat tends to flow from hot to cooler objects.

(1b) A warm and a cool object placed together will eventually come to the same temperature.

416

*Using
Thermal Energy;
Disorder and
Direction of
Time*

These statements all fit common sense. One does not expect to see a glass of cool water spontaneously turn into ice cubes and warm water. If a barrier between a portion of hot gas and one of cool gas is opened, the two gases will mix; eventually coming to the same temperature on both sides of the wall. Each portion will contain approximately the same number of more energetic and less energetic molecules. Maxwell, who contributed fundamental ideas to thermodynamics as well as to electricity and magnetism, thought carefully about this mixing of molecules on either side of a wall between two gases. He proposed an experiment (in principle) which has led to a great deal of discussion since.[1,2] He pointed out that if a tiny "demon" were stationed at a door in a barrier separating two groups of gas molecules that were at the same temperature, the demon could, for instance, identify the more energetic molecules on the right side of the wall. Then, as they approached the door, "Maxwell's demon" would quickly open it and let the molecule through to the left chamber. In the same way, the demon could pick out less energetic molecules and let them through the door into the right chamber. Pretty soon the gas on the left would be hot and the gas on the right would be cool. The second law of thermodynamics would be violated. That is, two portions of gas, initially at the same temperature, would have developed a difference in temperature with no apparent work being done on either one.

Since then other writers have shown that such a miniature intelligence could see and identify the molecules only if it received a signal from them. As we've seen before, signals represent energy. The separation of hot from cold would NOT be obtained for nothing. However, Maxwell's purpose was not to criticize the second law or find a way around it, but to show that its assertions are probable, not certain. Even without a demon, fluctuations might cause temporary temperature differences between two bodies in contact. For our purpose, the demon can best be seen as an illustration of the basic difference between mechanical and thermal energy, and the availability of either type. To quote from Maxwell:

> Suppose that a body contains energy in the form of heat. What are the conditions under which this energy or any part of it may be removed from the body? If the heat in a body consists in a motion of its parts, and if we are able to distinguish these parts, and to guide and control their motion by any kind of mechanism, then by arranging our apparatus so as to lay hold of every moving part of the body, we could, by a suitable train of mechanism transfer the energy of the moving parts of the heated body to any other body in the form of

(1) Ehrenberg, W., "Maxwell's Demon," *Scientific American 217*, Nov, 1967, pg. 103. Offprint 317.
(2) Klein, M., "Maxwell, His Demon and the Second Law of Thermodynamics," *American Scientist 53*, pg. 89–91, Jan-Feb, 1970.

Figure 16-1
Maxwell's demon.

ordinary motion. The heated body could thus be rendered perfectly cold and all its thermal energy could be converted into visible motion of the other body. Now this supposition involves a direct contradiction of the second law of thermodynamics, but is consistent with the first law. The second law is therefore equivalent to a denial of our power to perform the operation just described, either by a train of mechanism or by any other method yet discovered. Hence, if the heat consists in the motion of its parts, the separate parts which move must be so small or so impalpable that we cannot in any way lay hold of them to stop them.[3]

Maxwell is essentially saying that if molecular motion *could* be controlled, then the distinction between work or mechanical energy, and thermal energy would disappear.

Getting Work Out of Thermal Energy

Maxwell's statement, along with the assertion made earlier, that thermal energy is increasing at the expense of mechanical energy, helps to formulate a second set of expressions of the second law:

(2a) Heat cannot be transformed completely into work or mechanical energy.

(2b) When heat is withdrawn from an object for purposes of doing work, some heat must always be rejected to a cool reservoir.

In this form, the second law can be seen as a restriction on the first law of thermodynamics: conversions from mechanical to

(3) *Theory of Heat*, 1871, 1908, Longmans, Green and Co. pp. 153–154.

418

*Using
Thermal Energy;
Disorder and
Direction of
Time*

thermal energy can be 100 percent efficient, with no energy lost to other processes but converting thermal energy to mechanical energy is an uphill battle against energy losses to other processes.

Any device which uses thermal energy for doing work is called a *heat engine.* Let us consider what is involved in the operation of such an engine. Since we cannot connect an apparatus directly onto the molecules whose energy we would like to tap, we must bring the assembly of energetic molecules, that is, the hot object, in contact with another assembly of less energetic molecules—a cooler body—which will most probably accept some of the molecular energy. In so doing, the cooler body will expand: or, if we arrange conditions properly, it will vaporize and expand a great deal. We can take advantage of the tendency to expand by making the hot vapor or gas push against a piston or rotate a turbine. Thus, some of the energy withdrawn from the hotter body can be made to do useful work. But we get into trouble very soon. The hotter and the cooler objects will eventually come to the same temperature, and we can no longer expect energy to pass from one object to the other on any large scale. To get a heat engine which will keep on withdrawing the energy from a hot body, we must either keep furnishing new cool material or else we will have to re-cycle the heat-absorbing medium, cooling it periodically and bringing it back into contact with the hotter body. Use of a closed loop permits the hot gas—steam, for example—to be expanded to less than atmospheric pressure. This allows more work to be accomplished by it and less heat lost to the cooling water. Furthermore, the water used in high-temperature boilers must be quite pure, or else deposits will form that interfere with their proper operation. Use of the closed loop avoids the trouble of re-purifying water all the time. Nevertheless, one must throw out thermal energy in the form of wasted exhaust during each cycle.

To illustrate some of these ideas, consider the schematic diagram of an industrial steam turbine engine in Figure 16-2.

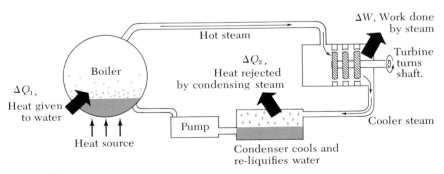

Figure 16-2
Steam turbine engine.

Oil or coal is burned and used to heat water in a boiler. (The water could also be heated by passing it through a hot nuclear reactor.) A portion, ΔQ_1, of the heat produced is absorbed by the water. Some of the water boils under pressure and the hot steam is passed through a pipe to a cylinder where it can expand while pushing against a piston, or the blades of a turbine. The more the steam can be made to expand, do work, and cool down in the process, the larger will be the fraction of thermal energy ΔQ_1 converted into work, ΔW. At this point the energy can be kept in mechanical form, directed into gear systems which will run other machines, or connected to propellers, as in a modern ocean-going ship. More often the mechanical energy is used to drive generators which convert it into electrical energy that can be sent via power lines to where it is needed.

The cooled steam must then be cooled further until it turns back to liquid water. If this is not done, steam pressure builds up behind the turbine, and it is more difficult for the incoming hot steam to turn the turbine. In the process of cooling and liquifying the steam, some thermal energy, ΔQ_2, is given off to the surroundings of the engine. If cooling water is used, the runoff water is warmer than when it entered the condenser. To save energy, some of this rejected heat can be used to keep the steam hot on its way to the turbine. But to get the water down to a temperature at which it can accept more energy ΔQ_1, some thermal exhaust energy, ΔQ_2, must be thrown away.

Such a steam turbine is an example of an *external* combustion heat engine. On the other hand, gasoline engines, where the fuel is burned inside the cylinder or expansion chamber, are called *internal* combustion heat engines.

Heat engines, with all their boilers, expansion chambers, condensers, and exhaust systems can be quite complex. However, they can *all* be visualized as having a *hot reservoir*, or source of thermal energy; a *working fluid*, such as gas, water, or molten metal; and a *cold reservoir*, or place into which the exhaust energy goes as shown in Figure 16-3.

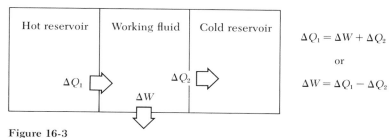

$$\Delta Q_1 = \Delta W + \Delta Q_2$$

or

$$\Delta W = \Delta Q_1 - \Delta Q_2$$

Figure 16-3
The principle behind heat engines. ΔQ_1 is the heat absorbed by the working fluid, ΔW is the work done by the fluid, and ΔQ_2 is the heat exhausted to the cold reservoir.

The first law of thermodynamics, $\Delta Q = \Delta U + \Delta W$, still applies, but the working fluid is continually being re-cycled, and brought back to its original temperature and pressure, so that $\Delta U = 0$. (The steam lift on page 402 was not carried through a complete cycle, so ΔU was not zero.) The second law also applies in this situation, at the point where the energy ΔQ_2 must be rejected during each cycle so that new thermal energy ΔQ_1 can be absorbed in the next cycle.

The effectiveness with which a heat engine can withdraw mechanical energy from heat is defined in terms of its thermal efficiency, abbreviated here as Eff.

$$\text{Eff.} = \frac{\Delta W}{\Delta Q_1} = \frac{\Delta Q_1 - \Delta Q_2}{\Delta Q_1}$$

$$= \left(1 - \frac{\Delta Q_2}{\Delta Q_1}\right)$$

Equation (16-1)

A fractional number results. If this number is multiplied by 100, the thermal efficiency is expressible as a percentage. Anything which will reduce the exhaust heat ΔQ_2 for a given heat input ΔQ_1, will increase the thermal efficiency of an engine.

The efficiency of the king's steam lift was

$$\text{Eff.} = \frac{\Delta W}{\Delta Q_1} = \frac{517{,}500}{6{,}899{,}000} = .075 = 7.5\%.$$

To operate it as a continuing machine, one would have to let out the expanded steam or else cool it, so that the platform could drop down, ready to lift another load. As heat engines go, the lift would not be very efficient. Even so, the early steam engine, designed, in the mid-1700's, by James Watt to lift water out of English coal mines, operated on much the same principle.

Furthermore, it can be shown[4] that in an ideal engine, the ratio of the heat rejected to the heat absorbed is simply equal to the ratio of the Kelvin temperatures of the cold reservoir, T_2, and the hot reservoir, T_1:

$$\frac{\Delta Q_2}{\Delta Q_1} = \frac{T_2}{T_1}$$

Therefore,

$$\text{Eff.} = \left(1 - \frac{T_2}{T_1}\right).$$

Equation (16-2)

This constitutes an overall theoretical limit to the efficiency with which any engine can get work out of heat. It applies to the case where the temperature differences between the hot reservoir

(4) Resnick, R. and Halliday, D., *Physics*, John Wiley & Sons, Inc., 1960, pg. 626.

and the working fluid, and between the working fluid and the
cold reservoir later in the cycle are small. Real heat engines have
an efficiency somewhat less than this when friction and heat
losses are accounted for.

To illustrate this, suppose the boiler described on page 418
heats the steam to 577°C (850 K) under very high pressure, and
that the condensed, cooled water has a temperature of 38°C
(311 K) as it leaves the condenser and enters the pump. If the
turbine is frictionless, if no heat is lost from any of the pipes, and
if the heat which is given off by the steam when it condenses
aids in reheating water in the boiler; then the limiting efficiency
of the steam turbine system would be

$$\text{Eff.} = 1 - \frac{311}{850}$$

$$= 1 - .366$$

$$= .634 \quad \text{or} \quad 63.4\%$$

Losses in a real system would probably bring the efficiency down
to about 40 percent.

The Heat Death of the Universe

Regardless of the efficiencies involved, the main idea to keep in
mind is that, in order for heat to be utilized, (or to be moved) it
must be able to flow from a hot place to a cool place. Energy
needs to be concentrated, as was shown toward the end of Chapter 7, but it has a strong tendency to spread out evenly. We on
earth are lucky to be located close to a large hot body or reservoir,
the sun. It pours a continual rain of energy toward us and all the
space around it. Eventually, it will cool down, reaching the same
temperature as the earth. (Only after it has flared up to a temperature which will make life on earth impossible anyway, according
to some theories of stellar evolution, e.g. pages 323-324.)

Although vast amounts of energy will still be contained in the
lukewarm sun, the planets, and in the form of radiation in the
space between them, the energy will not be usable. There will
be no cool objects around to make possible the operation of heat
engines. This is the "heat death" of the universe which the second law of thermodynamics predicts. It caught the imagination of
the Western world in a dismal sort of way after it was proposed
in the nineteenth century, as illustrated by Figure 16-4. However, this fate is a long way off in the future. The best attitude is
to worry over more immediate thermodynamic factors which are
equally capable of destroying, or at least spoiling, life on earth.

422

*Using
Thermal Energy;
Disorder and
Direction of
Time*

Figure 16-4
Camille Flammarion, "The Heat Death." (From *La Fin du Monde*, Paris, 1894, pg. 115.)

Thermal Pollution

We have seen that the systems we use to produce energy and to do work for us reject exhaust heat to their surroundings. This means that in a country like the United States, where (as stated in Chapter 7) large amounts of energy are consumed, large amounts of unwanted heat are also produced. Until recently, little notice was paid to this waste heat. But now ecologists are showing that large power plants can upset the life balance in the rivers which are usually used for cooling the working fluid during the cyclical process of removing mechanical work from thermal sources. The metabolism, lifetime, fertility and other characteristics of animal life in river water are very sensitive to temperature changes.[5] This means that when cooling water is dumped

(5) Clark, John R., "Thermal Pollution and Aquatic Life", *Scientific American* 220, March, 1969, pg. 19. Offprint 1135.

Figure 16-5
Air cooling towers at the Fort Martin Power Station, Fairmont, West Virginia.
(Courtesy of the Marley Co., Mission, Kansas.)

back into a lake or river at a higher temperature than it had before intake, it can destroy fish which live in the water near the power plant. Thermal pollution is thus one more dimension to humanity's disturbance of its surroundings. Power plants now under design are including air cooling towers or cooling basins so that cooling water has an opportunity to dissipate its heat to the air before it re-enters rivers or lakes where it might upset the ecological balance. But even these measures can upset the local climate by increasing moisture in the air, causing fogs. Furthermore, they will prove insufficient if consumption of energy continues to increase at its present rate.

Nuclear power plants have been criticized because they operate at lower peak temperatures, T_1, thus are less efficient and reject more waste heat per joule of work done. This situation can be analyzed as follows. If we let ΔW equal the work done, or useful energy withdrawn, ΔQ_1, equal the heat absorbed from the hot reservoir, and ΔQ_2 equal the heat given off to exhaust, then from Equation 16-1:

$$\text{Eff.} = \frac{\Delta W}{\Delta Q_1} = 1 - \frac{\Delta Q_2}{\Delta Q_1}.$$

Subtracting 1 from both sides gives

$$\text{Eff.} - 1 = -\frac{\Delta Q_2}{\Delta Q_1},$$

or

$$\frac{\Delta Q_2}{\Delta Q_1} = 1 - \text{Eff.}$$

Multiplying both sides by ΔQ_1 we get the waste heat,

$$\Delta Q_2 = \Delta Q_1 (1 - \text{Eff.}). \qquad \text{Equation (16-3)}$$

We can re-arrange Equation 16-1 to obtain $\Delta Q_1 = \Delta W/\text{Eff.}$ If this is then substituted into Equation 16-3, we can obtain an expression for the waste heat which involves only ΔW and the efficiency:

$$\Delta Q_2 = \frac{\Delta W}{\text{Eff.}} (1 - \text{Eff.}).$$

Performing the indicated division by "Eff." gives:

$$\Delta Q_2 = \Delta W \left(\frac{1}{\text{Eff.}} - 1 \right). \qquad \text{Equation (16-4)}$$

From Equation 16-2,

$$\text{Eff.} = 1 - \frac{T_2}{T_1} = \frac{T_1 - T_2}{T_1}.$$

By substituting this latter value for Eff. in Equation 16-4 and dividing it into 1 we obtain:

$$\Delta Q_2 = \Delta W \left(\frac{T_1}{T_1 - T_2} - 1 \right)$$

In order to perform the subtraction indicated in the parentheses, the 1 must be replaced by an expression equal to one and having the same denominator $T_1 - T_2$ as the other fraction. Therefore the expression becomes:

$$\Delta Q_2 = \Delta W \left(\frac{T_1}{T_1 - T_2} - \frac{T_1 - T_2}{T_1 - T_2} \right) = \Delta W \left(\frac{T_2}{T_1 - T_2} \right), \quad \text{Equation (16-5)}$$

in the theoretical limit. A lower peak temperature, T_1, for a given exhaust temperature, T_2, results in larger exhaust heat, ΔQ_2, for a given amount of useful work done, ΔW. However, when the efficiency of nuclear power plants is compared with fossil fuel plants *presently in use*, there's little distinction.

   Table 16-1 and 16-2 compare the efficiency of several types of stationary power plants as well as some modes of transportation. The efficiencies of fossil fuel and nuclear systems, and the steam and gas-powered automobile, are limited by Equation (16-2), which applies to cyclic heat engines. The other devices have efficiencies which are limited by other considerations. However, in every case, efficiency denotes the amount of work done per unit of energy fed into the system.

Table 16-1 *Comparative Efficiency of Various Stationary Heat Engines*

Type of Power Plant	Actual Efficiency in Percent	Joules of Waste Heat Produced Per Joule of Work Done or Electrical Energy Produced. Equation (16-3)
Modern Fossil Fuel System	40	1.5
Average of Existent Fossil Fuel Systems	32	2.1
Nuclear Power Systems	32–39	1.6–2.1
Magnethohydrodynamic systems[6]	20–30	2.33
Solar Cells	14	6.1

Solar cells produce less thermal pollution than indicated because they withdraw energy from sunlight which would otherwise warm up the earth anyway. They *do* redistribute the energy, concentrating it in certain places. The "waste heat" from the other devices has also been used to melt snow, to remove salt from sea water, or to operate additional cyclic systems which convert part of the heat into useful work.

Table 16-2 *Comparative Efficiency of Modes of Transportation*

Mode of Transportation	Actual Efficiency in Percent	Joules of Waste Heat Produced per Joule of Work Done (Equation (16-4))
Diesel-electric rail car	30	2.3
Bicycle	20–30	2.3–4
Electric passenger rail car	21	3.8
Passenger jet in level flight	21	3.8
Pedestrian	18	4.5
Steam powered automobile	15	5.7
Gasoline-powered automobile	8.5	10.7

(6) An electrically conducting "fluid" (a plasma of ionized particles) is pumped through a magnetic field, producing electric voltage and energy by induction (see page 313).

426

*Using
Thermal Energy;
Disorder and
Direction of
Time*

Whether we generate our power using fossil or nuclear fuel, we are close to the point where we must either limit world population or our energy consumption per capita, or we will begin to irrevocably alter the balance of life around us — just by the heat our machines give off.[7] The overall average thermal energy wasted on earth is about 2.5 joules per joule of electric energy generated. Of course, the energy we count as not being wasted also (eventually) ends up as heat.

As demonstrated in Chapter 7, the bicycle looks good as a means for getting about.[8] Aside from this, Table 16-2 ranks the various modes differently than does Figure 7-19. This is because Table 16-2 looks at the work involved in just moving the vehicle around, rather than into the number of passengers being transported. On this basis, a passenger jet ranks rather well in efficiency because so much measurable work is done to keep it in the air and moving, whether it carries passengers or not.

Mixedupness

Why should thermal energy prevail? Physicists answer this by saying that thermal energy is more disorderly and add a third, more general statement of the second law of thermodynamics:

(3a) The physical world tends toward disorder.

Why then does disorder prevail? Simply because it can occur in more ways than order can! A large group of objects can be sorted out to give a limited number of specific arrangements, but the objects can be mixed together in almost limitless ways. If each "way" is equally likely, then the mixed-up state is more likely to occur.

Let us illustrate by means of ten balls (or molecules), to be placed into two halves of a box. Let someone who is blindfolded drop the balls in. There is only one way for all ten balls to end up in either the left, or the right side of the box: they must *all* fall there. But we can get one ball on the left and nine on the right in ten ways: by having either ball number 1, or 2, or 3, or 4, etc. — fall on the left. The same is true for having nine on the left and one on the right. An arrangement of two and eight can happen 45 ways. (Any of ten balls could be the first on the left — or the right, and for each of the ten, any of nine could be the second one. This would give a total of 90 possibilities. However, we don't

(7) Weinberg, Alvin M. and Hammond, R. Philip, "Limit to the Use of Energy", *American Scientist* 58, July-Aug., 1970, pp. 412–418.

(8) (The efficiency here is calculated on the assumption that it takes 1/6–1/8 horsepower or 90–125 J/sec to pedal a bike 12 miles/hour, so the amount of work done in a five hour ride is $(90 \text{ to } 125) (3600) (5) = (1.62 \text{ to } 2.25) \times 10^6$ J. The energy consumed (ΔQ_1) is about 1,800,000 cal, or 7.5×10^6 J. The Efficiency, $\Delta W / \Delta Q_1 = (1.62 \text{ to } 2.25)/7.5 = .2 \text{ to } .3$).

distinguish between various sequences of dropping. If ball number 5 and later ball number 7, for example, falls on the left (or right) side, we call the arrangement the same as if ball number 7 and later ball number 5 fell on the left (or right). Thus, there are only 90/2 or 45 different two-eight arrangements.) The two-eight arrangement can then be considered 45 times as likely as the zero-ten arrangement. The three-seven arrangement can happen 120 ways, the four-six arrangement 210 ways, and the five-five arrangement 252 ways. From this it can be seen that the spreadout arrangement of balls is the more likely one.

If there are 10^{23} molecules in the boxes instead of ten balls, then the statistical probability of the even distribution becomes overwhelming. If, instead of positions of molecules in boxes, we now speak of molecules having high and low energies, the same principles apply: the two sorts of molecules will most likely mix together, with equal numbers of high energy molecules throughout the container, mixed evenly with the less energetic molecules. Two objects thus come to the same final temperature (have the same proportions of high- and low-energy molecules) because this is the state which can occur in the most ways and is thus most probable.

Measuring Disorder; Entropy

"Disorder" and "mixed-up" are terms that cry for closer definition. First, the physicist's *measure* of disorder is *entropy*. In the most general sense, entropy represents lack of information in the face of a great many information possibilities. If a system composed of many particles has enough energy to exist in many states (that is, if its molecules can have many locations and energies) but you do not know for sure which state the system is in, then the system is said to have a high entropy. A liquid, for example, has higher entropy than a solid because the individual molecules are able to migrate around whereas in a solid they are largely tied to certain sites in the material. The liquid is more disordered, or, at the same time, can exist in a richer variety of states.[9]

The entropy of a system will also increase if we add particles to it, increase its volume, or decompose the molecules in it into separate atoms. Physicists have ways of adding up all these additional possible states to arrive at the relative probabilities involved, so that a numerical expression of the relative entropy of a system can be obtained.

(9) If we know or become familiar with a particular state, then we could decide to call that state "ordered" if we wished. Certain combinations of cards are no more rare than other specific combinations, but according to certain rules of the game, an ace, king, queen, jack, and 10 of one suit strikes us as very rare and orderly. (With a liquid, on the other hand, you are denied that detailed knowledge.)

Another, more large-scale approach defines the *change* in entropy, ΔS, of a system which undergoes a change of thermal energy, ΔQ, as

$$\Delta S = \frac{\Delta Q}{T}$$

Equation (16-6)

where T is the Kelvin temperature of the system. This assumes that the system is large enough so that it remains at the same temperature T in spite of the addition or removal of the thermal energy. If entropy still represents disorder, then this relation states that the relative change in disorder of a system for a given ΔQ is greater at lower temperatures. If the system is at a lower temperature, then its possible states are fewer and a given amount of thermal energy will increase those possibilities more than when the same thermal energy is given to a system which already has considerable energy and many possible states or arrangements. *Since ΔS and ΔQ can be measured or calculated, the relation above provides a direct, non-material definition for temperature.* This was anticipated when "absolute" temperature was spoken of on page 395.

Suppose a certain amount of thermal energy, ΔQ_1 leaves a large hot reservoir at temperature T_1 and enters directly into a cooler reservoir at temperature T_2. The entropy of the hot reservoir is reduced by the amount: $\Delta S_1 = -\Delta Q_1/T_1$. The entropy of the cold reservoir is *increased* by the amount: $\Delta S_2 = +\Delta Q_1/T_2$. The overall change in entropy for the system is

$$\Delta S_1 + \Delta S_2 = -\frac{\Delta Q_1}{T_1} + \frac{\Delta Q_1}{T_2}$$

Equation (16-7)

Since T_2 is smaller than T_1, the net change in entropy of the system is *positive*. If the two bodies continue to exchange thermal energy, they will come to the same final temperature. But, the body which was cooler initially will undergo an increase in entropy which is larger than the decrease in entropy undergone by the body that was initially hotter. (The ΔQ's are equal and opposite, but the body which was warmer originally has a higher average temperature. Thus $\Delta S = -\Delta Q/T$ is smaller than for the initially cool object.) The overall entropy increases. The molecular energy which was concentrated in the hot object at the start ends up dispersed between the two. If great care is taken so that thermal energy does not cross any large temperature differences as it goes from one object to the other, it is possible in principle to keep the entropy of the system from increasing. But for entropy to decrease in an isolated system is very, *very* improbable. This gives rise to the final statement of the second law of thermodynamics which will be made in this text:

(3b) The entropy of an isolated system increases or remains the same as time passes.

The alert reader might be objecting that life, intelligence, and civilization ought to be able to counteract the thrust toward physical disorder that has been described. Life itself represents a marvelous means for maintaining genetic order from one individual to the next. But, according to the second law, every living being leaves in its wake more physical disorder than it can match with its own powers of organization. We are not talking about refuse, sewage, air pollutants or the other waste products that spoil our surroundings, and which hopefully can be controlled. Even if everyone were careful to clean up after themselves, each person would be increasing the entropy of the universe just by breathing out warm air. George Gamow and W. Britten[10] have shown that by absorbing radiant energy from a hot sun, and using that energy to perform photosynthesis at much lower earth temperatures, a growing plant increases the total entropy of the universe. The entropy of the sun *decreases* but, by Equation 16-7, the entropy of the plant *increases* even more.

But is this all there is to it? Before we decide whether order or disorder is going to win out, we need to be sure what the words mean. When physicists say *order*, they mean *low probability*. A particular arrangement or sequence is orderly if there are not many ways to obtain it. Physicists can also show that the arrangements that can be obtained in the most ways will be most probable, and that these arrangements will indeed be the ones which will occur. In aesthetics, on the other hand, order comes out of the whole, implying pattern, structure or form, qualities which have no definable rank on the probability scale. Wholes *are* incorporated into physics. Symmetry, conservation, and wave interference patterns are all examples of this. Nevertheless, the methods of physics make it very difficult to see patterns or evaluate their entropy. A good example of our inability to codify and measure structure is the music and poetry "written" by computers.

As humans, we recognize structure even though we don't understand our knowledge of it enough to design machines that can produce it. We learn to *encode* certain sequences of signals with meaning. The poker hand referred to in footnote (9) was one example. The sequence of chords leading to the resolution of a musical theme is another. This supra-statistical feature is the basis of *symbol*, whereby a set of letters or other signals can convey more than can be expressed in the statistical probability of the particular sequence of signals. In other words, we can question the way in which physics keeps score as it adds up the total order in the universe.

Another factor which thermodynamics has difficulty accounting for is that some arrangements seem to be *preferred* in nature

(10) *Proceedings of the National Academy of Science*, 47, pp. 724–7, May, 1961.

430

*Using
Thermal Energy;
Disorder and
Direction of
Time*

for ("organic") properties they possess *as a whole*. (Recall that the basic assumption of the statistical arguments we have just gone through is that all arrangements are equally likely.) A large number of atoms will settle into a coherent, crystalline, orderly array if they don't have too much individual energy. In so doing, the whole assembly can get to a more stable arrangement having lower over-all energy. A heat of fusion is given off. The crystal will maintain this order as time passes unless it is fed too much energy, in which case it melts. Another arrangement of atoms, supposedly no more improbable than any other particular arrangement of the same atoms, forms the DNA molecule, the building block of biological genes. It is not only energetically stable, but has the capacity to form more complex patterns of atoms, eventually becoming a living human being who is conscious, recognizes patterns and is able to organize them. The pattern and order are maintained. A number of thinkers have tried to examine this tendency of organic evolution to combat physical disorder or entropy. Charles S. Pierce called it "agapism".[11] Rudolf Arnheim[12] speaks of an "anabolic tendency" for orderly structure to develop. R. Buckminster Fuller, perhaps the most optimistic of all, uses the word "syntropy" in his writings to represent the powers of human intelligence to marshall and organize the natural resources of the earth.[13]

To maintain life, movement of energy is required. The energy won't move unless it has cool places to move to; that is, unless it is concentrated in hotter places. The second law shows quite clearly that energy tends to spread out evenly; so life does seem to be in jeopardy. However, we may be far from knowing all there is to know about the efficiency of focused intellect in keeping itself going. When we recall that mass itself is something like concentrated energy, then the doom forecast by the second law might be very distant indeed.

The Direction of Time

Conservation principles are the mainstay of mechanical physics. They are useful for predicting how systems will behave, but there is one thing they cannot do. They cannot tell you what is past and what is future. As we've seen, symmetry principles always hide something. In the case of momentum and energy conservation, what's hidden is the difference between "before" and "after". For example, if momentum and energy are conserved in

(11) *Collected Papers of C. S. Pierce, Volume I*, paragraph 411 and *Volume VIII*, paragraph 317.

(12) *Entropy and Art*, University of California Press, Berkeley, 1971.

(13) See, for example, the poem, "Now and When", from *Intuition*, Doubleday and Co., 1972. Reprinted in *Harper's*, April-May, 1972, pp. 58–66.

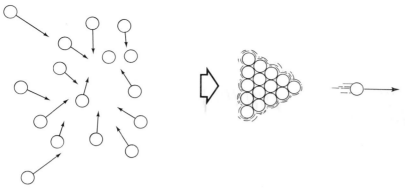

Figure 16-6
It's improbable, but not impossible.

the collision of two billiard balls, one could show a motion picture of the collision backwards or forwards and no one could be sure which was which.

We reached the same conclusion (from another direction) toward the end of Chapter 4. In the section "Mechanism and Determinism" it was stated that insofar as we can predict exactly and correctly the behavior of a system we can also "retrodict" it, seeing what the object had to be doing in the past in order to get where it is and travel with the velocity it has now. There really would be no future and past; just an unwinding of mechanism.

Yet common sense does tell us when certain things are running backwards. We laugh when a TV commercial shows a mound of shave cream disappear into the dispenser. We disbelieve when shown a film in which a broken egg gathers itself together, the cracks disappear, and the egg leaps up from the floor onto a kitchen table.

None of these phenomena is energetically *impossible*. They are merely *statistically improbable*. They go against the second law of thermodynamics, which lacks the seeming certainty of Newton's laws, but which (as Maxwell says):

> . . . has the same degree of truth as the statement that if you throw a tumblerful of water into the sea, you cannot get the same tumblerful of water back out again.[14]

Suppose we consider a system which is more complex than two billiard balls but less complex than a broken egg. A motion picture of a cue ball hitting a neatly stacked triangle of 15 billiard balls gives a strong sense of before and after. If the film is shown backwards, everyone laughs because "absurd" things happen. A lot of balls converge on each other, jostle into a triangle and

(14) Letter to Lord Rayleigh, December 6, 1870.

give their kinetic energy to one white ball which rolls away from them with great speed. In this case, the "backwards" motion *could* be achieved. We could very carefully roll 16 balls toward each other with proper velocities, and after sufficient tries, we might get them to nestle into a triangle, etc. But the chances of this happening spontaneously are very rare, and we know it. The difference between before and after, then, is not mechanical and certain. It is tied up with probabilities and order.

When a system consists of just two balls the concept of "order" gets lost. We could say, "White on left, and red on the right constitutes order", but this would be too arbitrary to be useful or communicable. The idea of order-disorder and thus the direction of time becomes more and more convincing *as the system becomes more complex,* or as we look carefully enough to see the complexities. Sixteen billiard balls are quite convincing; a large step toward separating past from future in a collision.

Now suppose we examine the "simple" two-ball system a little further. In any real collision a little of the initial mechanical energy is converted to thermal energy. In other words, collisions between large-scale objects are always inelastic to some extent. If we calculate the initial and final mechanical energies and find one lower than the other, we can determine whether we were shown the movie backwards, or in the right sequence. The lower mechanical energy comes *after.* But how can we account for the change in energy? We say some thermal energy has been formed. And what is thermal energy? Molecular motion! In noticing that the collision is inelastic, we are no longer seeing the balls as just simple spheres. They become highly complex assemblies of many molecules. The distinction between past and future resolves again to one of recognizing a change in a complexity.

A person could claim that he or she certainly knows "before" from "after" just by feeling it, without worrying over losses of mechanical energy. But in so doing, individuals include themselves in the system, and it becomes very complex indeed; with whiskers lengthening, or hunger and fatigue increasing, not decreasing, as the clock ticks.

Recent studies of the decay of certain sub-atomic particles (neutral K mesons) have suggested that the decay process is not invariant with respect to time reversal. That is, a reaction made to go in the opposite direction does not produce the same result as a film of the original reaction shown backwards.[15] The distinctions between past and future are very faint and can be arrived at only after a very subtle and indirect interpretation of the data. Nevertheless, the phenomenon has great significance in physics because it furnishes the first evidence that the physical world "recognizes" the direction of time in simple, non-statistical systems.

(15) Sachs, Robert G., "Time Reversal", *Science 176*, pp. 587–597, May 12, 1972.

Does the unsolid statistical explanation of the behavior of physical systems that has been discussed constitute a defeat in our
quest for simple, certain explanations for physical phenomena?
Probably so. But at the same time it can be seen as a step toward
describing the world as it really exists, physically. Newtonian
physics is large-scale physics predicting the behavior of things
large enough to see and massive enough not to be blown about
in the breeze. It predicts by idealizing, which is the same as
simplifying. We can know the net force acting on a baseball
having a mass, m, and predict accurately where the (middle of
the) ball will be at time, Δt, later: $\Delta x = \frac{1}{2}a(\Delta t)^2 = \frac{1}{2}F_{net}/\text{mass }(\Delta t)^2$,
(from Chapters 3 and 4). Classical physics treats baseballs as
simple spheres having no other describable details. But look at
a baseball on a fine scale. While the ball is sliding in a simple
fashion along a table top a particular molecule in that baseball
may be undergoing a far more complex motion; combining the
simple large-scale motion with its own pulsations. A given molecule will very probably travel anything *but* $\frac{1}{2}a(\Delta t)^2$. Newtonian
physics ignores the detailed structure of the objects whose mo-
tion it describes. Statistical physics tries to contend with the
details of matter, and the complexity of the job requires speaking
in probabilities, not in certainties.

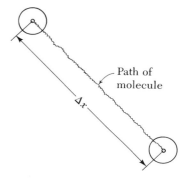

Path of
molecule

Newtonian determinism is really a statistical determinism
which, because of the large size of the bodies involved, becomes
so probable as to appear certain. *Cause and effect* have always
been a troublesome set of words, implying that a given set of
conditions will lead inevitably to another set of conditions. The
idea has been attacked by philosophers from David Hume
(1711–1766) on down to this day and is not used by most contemporary philosophers of science. *Determinism* is another word
which gained great popularity in the generations after Newton
and came to be applied to everything from economics to sociology. It also connotes consequences which *must* follow from a
given set of circumstances. Contemporary physicists are more
concerned that the laws of physics be *causal*. R. Bruce Lindsay
says:

> Whenever patterns or order has been established within a given range
> of experience, leading to physical laws descriptive of this order,
> causality is preserved.[16]

Causality thus implies consistent patterns of experience, whether
strictly deterministic or merely statistical. The laws of physics
are certainly causal by this definition.

It should be pointed out that indeterminism and a resultant
uncertainty arise from the *practical* impossibility of knowing the

(16) *American Scientist 56*, Summer, 1968, pp. 93–111.

434

*Using
Thermal Energy;
Disorder and
Direction of
Time*

instantaneous state of a complex system in detail. This could be called *classical* indeterminism. It is not the same as the more basic *quantum* indeterminism which forms an integral part of the observing and measuring of atomic-sized physical systems. (This will be encountered in the next Chapter.) Both are so small as to be negligible when describing the motion of something as large as a baseball, but they represent a modesty which has been forced on physicists in the last century. There are limits to the predictive powers of physics, even when applied to the physical world. These limits were overlooked during the golden days of Newtonian mechanics but they loomed larger and larger as physicists tried to understand thermal, and then atomic, phenomena.

Summary

The second law of thermodynamics is a principle based on statistics or probabilities and can be stated in several ways.

(1) Thermal energy tends to flow from warm to cooler objects. A warm and a cool object placed together will come to the same final temperature.
(2) There is a trend from mechanical to thermal energy which is difficult to reverse. Heat cannot be transformed into work without rejecting some waste heat to a cool reservoir. Heat engines convert thermal energy into work. They are limited in their ultimate thermal efficiency.

The thermal efficiency, Eff, of a device is equal to the work it can do per unit heat given to it:

$$\text{Eff} = \frac{\Delta W}{\Delta Q} \qquad \qquad \text{Equation (16-1)}$$

The theoretical limit to the efficiency of a cyclic heat engine is given by

$$\text{Eff} = 1 - \frac{T_2}{T_1} \qquad \qquad \text{Equation (16-2)}$$

where T_2 is the Kelvin temperature of the coldest heat absorber (reservoir) available and T_1 is the Kelvin temperature at which heat is supplied to the engine from a hot source or reservoir.
Since exhaust heat is always rejected by heat engines, we are in danger of upsetting many balances in nature by thermal pollution.

(3) The physical world is tending toward mixedupness, or increasing entropy. Change in entropy, ΔS, equals $\Delta Q/T$.

Mixedupness will prevail eventually because it can happen more ways and is thus more probable. Eventually the energy of the universe will be evenly spread around, and will no longer be useful for doing work.

It is difficult to apply physics to the recognition and evaluation of large-scale patterns, as in paintings or symphonies, or life itself. Therefore, it is difficult to say how final is the claim of the second law of thermodynamics that disorder will ultimately prevail.

There are many phenomena which could be reversed without violating the principles of mechanism; that is, they conserve energy. The observed fact that these phenomena have never run backwards seems to require probability for an explanation. Statistical physics thus gives a direction to time which mechanics cannot identify. It separates "before" from "after", if the system involved is sufficiently complex—has enough objects in it.

In the step from mechanics to thermal physics, we gave up a great deal of determinism. The future of specific gas molecules cannot be predicted or controlled because of the impossibility of knowing their positions and velocities at a certain initial time. However, the statistical approach retains a measure of causal understanding by being able to predict what a large number of gas molecules will on the average (most probably) be doing in the future.

Further Reading

Ayer, A. J., "Chance", *Scientific American 213*, Oct. 1965, pg. 44. (The word "chance" has many definitions, not all of which apply to the word probability. The vagueness of the concept persistently contributes to errors in its application.)

Bork, Alfred M., "Randomness and the Twentieth Century", *Antioch Review XXVII*, No. 1, pp. 40–61. (Suggests that randomness is a basic theme in contemporary culture, and tries to explain why.)

Bronowski, Jacob, "The Idea of Chance", Chap. 6 in *The Common Sense of Science*. Random House Vintage Book, New York, 1953. (Probability still allows for predictability.)

Bronowski, Jacob, "The Idea of Statistics and the Arrow of Time", Chaps. 8 and 10 in *Insight*, MacDonald, London, 1964. (A very visual presentation of some probability ideas. Gives an interesting example of decoding a message.)

Dyson, Freeman, "What is Heat?", *Scientific American 191*, Sept. 1954, p. 58. (Discusses the relations between heat flow, entropy, and order.)

Eddington, A. S., "The Decline of Determinism", Chap. IV in *New Pathways in Science*, Cambridge University Press, 1935. (Discusses the form which physical law takes when a system is no longer mechanically determined.)

Gardner, Martin, "Can Time Go Backward?", *Scientific American 216*, Jan. 1967, pp. 98–108. Offprint 309. (A short, popular article now somewhat dated. However, it provides an amusing introduction to ideas on time symmetry.)

Gold, Thomas, "The Arrow of Time", *American Journal of Physics 30*, June 1962, pg. 403. (A more technical but non-mathematical discussion of how entropy increase separates past from future.)

Harleman, Donald R. F., "Heat, the Ultimate Waste", *Technology Review 73*, Dec. 1971, pg. 44. (A complex problem. Best solution may *not* be through rigid water temperature controls.)

Krutch, J. W., "The Colloid and the Crystal", from *The Best of Two Worlds*, Wm. Sloane Assoc., N.Y., 1950. Reprinted in *Great Essays in Science*, Martin Gardner, Ed., Washington Square Press, N.Y., 1957, pg. 108. (A thoughtful essay on order in living and non-living things.)

Kuntz, Paul, Ed., *The Concept of Order*, University of Washington Press, Seattle, 1968. (A symposium which discussed order in history, science, art, religion, language, and society.)

Lindsay, R. B., "Entropy Consumption and Values in Physical Science", *American Scientist 47*, Sept. 1959, pg. 376. (Talks of a "thermodynamic imperative", whereby all humans should struggle to increase the order in their environment so as to combat the natural tendency to transform order into disorder.)

MacDonald, D. K. C., *Faraday, Maxwell, and Kelvin*, Doubleday & Co., Anchor Book Science Study Series, 1964. (A popular account of the lives of three great physicists of the nineteenth century.)

Morrison, Philip, "The Mind of the Machine," *Technology Review 75*, Jan. 1973, pg. 13. (With each age of machines that simulate life, the question of why life differs from non-life returns on a new level.)

Mueller, Robert E., *The Science of Art, The Cybernetics of Creative Communication*, John Day, N.Y., 1967. (Looks at art and science as ways of encoding information and communicating it.)

Pierce, J. R., "Information Theory and Art", Chap. 13 in *Symbols, Signals, and Noise*, Harper Torchbook, N.Y., 1965. (A primer on information theory, with applications.)

Sandfort, John F., *Heat Engines*, Doubleday Anchor Book, Garden City, 1969. (Theory and practice of getting work out of thermal energy. An introductory paperback.)

Summers, Claude M., "The Conversion of Energy", *Scientific American 225*, Sept. 1971, pg. 149. Offprint 668. (Most energy is converted into electricity. Whenever conversion from one form of energy occurs, some energy is always lost.)

Whyte, Lancelot L., "Atomism, Structure and Form", pp. 20–28 in *Structure In Art and Science*, G. Kepes, Ed., G. Braziller, New York, 1965. (Distinguishes between structure and form.)

16-1 Suppose a person tossed 49 straight heads without cheating. What is the probability that the 50th toss will be a head?

16-2 A 100 kilowatt power plant operates at an overall efficiency of 30 percent. How much thermal energy does it contribute to its surroundings every hour? (Assume that it is always producing at peak power.)

16-3 Why do the molecules of a gas rush to fill up a vacuum when they are released into it?

16-4 Is the efficiency of a power plant greater in the winter or the summer? What factors enter in?

16-5 Do persons increase or decrease the entropy of the universe when they write music? When they play it? Listen to it?

16-6 Does the entropy of a heat engine increase with time?

16-7 Suppose a fossil fuel power plant is able to run at an efficiency which is equal to two-thirds of its ideal efficiency. If the maximum temperature of the high-pressure steam in the boiler is 1220°F, (660°C) and the temperature of the heat sink is 212°F (100°C), show that the real efficiency should be 40 percent.

16-8 A cook stirs cake batter with an electric mixer driven by a one-eighth horsepower motor. (The motor is controlled so as to use 93.3 watts.) Suppose the mixer does 83.3 joules of work on the cake batter every second. What is the efficiency of the motor? If the cake batter has an effective mass times specific heat of 837 joules/°C, how much will its temperature increase in 10 minutes, assuming that no heat is lost to the surroundings?

16-9 Calculate the efficiency of a skier who burns 42,000 joules per minute and who travels 20 miles cross country in four hours. Assume that the skier overcomes an average friction force of 16N and that air resistance amounts to 2.2N. If cake contains 18×10^3 joules (2000 calories) per gram how much cake does the skier need to eat to supply just the energy needed in the run?

16-10 In Chapter 9 it was stated that in order for fusion reactions to start, the gases involved must be heated to a temperature of more than 50 million degrees. Why couldn't the requisite temperatures be produced by focusing the sun's rays into a very small region by means of huge mirrors: a sort of super-solar furnace?

Katsushika Hokusai, *A Group of Blind Men Examining an Elephant.* (From the *Hokusai Sketchbooks*, by James A. Michener, Charles E. Tuttle Co., Tokyo and Rutland, Vermont.

QUANTUM PHENOMENA AND KNOWLEDGE

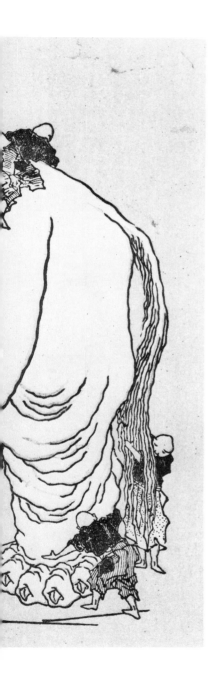

17

PHYSICS, KNOWLEDGE, AND SYMBOL

Looking Backwards—and Around

We have swum through a portion of physics in much the same way as a breaststroker gets from one end of the pool to the other: by plunging under water, coming up for breath, plunging under, coming up, etc. Now that we're near the end of the pool, let's take a good look back at what we've plunged into. In particular, let's try to make ourselves aware again of the particular tools physicists use and the way they use them in building their models of the world. After that we can look sideways and forward to see how this method of working compares with other means for describing the world.

First, consider observation itself. We saw in Chapter 2 that physicists try to make measurement an important part of their observations. In measuring, they determine a ratio between the magnitude of the thing being measured and the magnitude of some standard object or event. Out of this process come numbers. The numbers constitute objective knowledge simply because they are easily communicated. The measurement process also provides a good backdrop for pointing out that the *meaning* of terms (whether it is something simple like "length" or is a more complex quantity or concept) always is based on an *operation* or experience which can be shared.

Chapter 3 led us into the problem of locating objects and describing how they move in time. Knowing an object's acceleration, along with its position and velocity at a certain instant, we found out where the object *had* to be at all instants before and after. We also had an opportunity here to manipulate algebraic symbols representing physical quantities so as to uncover new relationships—perhaps not yet observed—between the physical quantities such as acceleration, velocity and distance moved. In Chapter 4 we added Newton's explanation of acceleration as a result of interactions between the object and its surroundings. Then we surfaced, looked around and realized that we had said something about determinism. An object is a complete victim of its past history and the forces acting on it. We also saw that the Newtonian world view, the view closest to what seems "natural", implies certain concepts of space (Euclidean, three-dimensional), and time (separate and marching on).

In Chapter 5 we studied the growth of a theory, with gravity as the example. The process, flowing from measurement through summarization in law to hypothesis and back to measurement again, involved elements of logic. However, creative fabrications were probably more important than logic in the growth of theory. Aesthetic ideas, such as simplicity, were important in selecting the "best" among alternative theories. The theory, usually expressed in terms of masses, momenta, molecules, and other concepts not directly observable, was symbolic. As Max Born said,

> "Man seems willing—eager—to put his trust in concepts more enduring than sensations".[1]

The interesting thing about those first five chapters was that a thoughtful discussion of how and why simple objects move brought us directly up against most of the time-worn problems in epistemology (see page 4): problems such as perception, meaning, objectivity, determinism and idealization. Physics is fun because it deals with observing, describing, and understanding the simplest possible systems in nature. This means that problems which might be hidden or ignored during the study of complex biological or political phenomena stand a better chance of exposure in physics.

With Chapters 6 and 7 we became more aware of aesthetics; style, if you will. Physics, given many possible words or concepts for describing the behavior of matter, will choose concepts which stay the same in the midst of over-all change. *Energy* and *momentum* call attention to themselves by remaining the same within a closed system. We should recall that individual objects within a system do not usually have constant energy or momentum. Only when we look at the whole system do we be-

---

(1) "Symbol and Reality", published as appendix to *Natural Philosophy of Cause and Chance*, Dover Publications, New York, 1964.

come aware of a "form" that is carried along smoothly by a wildly varying bunch of elements.

Einstein's special relativity then carried the principle of invariance further by stating that all laws of physics must be invariant between inertial frames; no matter what this does to our comparisons of time, position, and mass between those two frames. The particular datum which does approach absoluteness is the speed of light. Einstein puts a time lapse into all our communications so that "now" becomes just a local concept, limited to single places in space. Time and space are intermixed.

Although the concept of energy arises at a bookkeeping, or aesthetic level, it provides a direct route toward the dollars and cents, pollutive, warlike, political, and technological dimension of scientific knowledge. By focusing on energy, physicists found out what enables machines to do what they do, and how to make them do more. Since the beginning of the twentieth century physicists have even shown how to turn inert matter into energy by means of nuclear reactions. Thus the process of physics, visible as harmless applied philosophy, suddenly flips over and becomes a genie, ready to do either good or bad things.

Chapters 11 and 12, on fields and waves, mark a departure from the study of massive objects and their motions. The concept of force is lifted from things banging or pulling on each other and allowed to run free in space. How do you live philosophically with the idea that an object can reach out through space and pull or push on another object? We traced some of the history of the problem and saw that some of the twentieth century answers describe force as a form of space "warpage", or as sub-microscopic games of catch. The principle of simplicity requires that we throw out the concept of a mechanical aether to transmit forces because it's not necessary. The discussion of waves and their interactions retained the energy concept, but it is seen as being a characteristic of the whole. Once again, the form is carried by cooperative elements which vibrate in turn. They make the wave but they do not go where it goes. Furthermore, the elements can either be objects such as air molecules, or immaterial entities such as an electric field. The whole picture becomes quite nonmaterial but the punch delivered by the wave—light, sound or whatever—is just as capable of moving matter around as a stick or a piece of rope. The study of waves also unearths—gives rise to—the even more general cooperative phenomenon of coherence. Waves which beat "in time" possess a degree of potency which independently vibrating waves lack.

Chapters 15 and 16 depart even further from Newtonian determinism: solid, certain predictability is replaced by probabilities. This occurs because the systems being studied are made up of billions of elements. Once heat is seen as molecular motion the concept of energy is expanded. When friction saps away mechanical energy, thermal energy is produced and the sum of mechanical plus thermal energy is conserved. But there's an important change: thermal energy is chaotic and it is made up of motions too small to control or keep track of. The fact that chaos can occur in more ways than order gives it a statistical advantage. As time moves on, more and more of the universe's energy goes into thermal form, until the universe will at last be made up of many lukewarm masses incapable of generating power or doing work. Life, the will to order, must yield, it seems. But we are well advised to remember that the observations on which physics is based and the arena in which it is played, are limited in space and time. It is quite limited in its ability to provide measures for pattern or to define order. Physicists have to determine, by *induction* what the other end of the universe is like, and they know little about where life and intelligence came from, in the first place. We ought to realize also that the chances of political chaos destroying culture and even human life are eminently closer than the doom forecast by the second law of thermodynamics.

So this has been our swim through part of physics. We have had to learn details and formulae and certain thought sequences in order to see what physics leads to. We have seen that physics is intimately involved with many of the basic problems of philosophy. But our approach has always been, "How does physics best fit our experiences together", rather than "Does this help build a sound philosophical system?" It would be much better if we could quit seeing science as holy scripture to be memorized and start seeing it as one kind of reflective human response to outside experiences. Then the aesthetic, satisfying, extralogical dimensions of science appear, quite naturally, as being natural. It shows humans in tension with nature, trying to be separate and objective, yet being involved and participating. The "philosophy of physics", if one can be spelled out at all, has to be flexible and precarious.

Philosophizing about Physics

Then how good is scientific knowledge? Its *completeness* has been questioned from the very first pages of this book, since abstraction and analysis are so important in it. It has been *successful* in terms of verified hypotheses and making rational that which was puzzling initially. Therefore, most scientists do not lose sleep over the coherence and meaning of the science they fabricate. Nevertheless, philosophers of science (and sometimes, scientists who are approaching retirement age) have a great deal to say in the realm of interpreting and evaluating the "truth" of science.

The two branches of philosophy which bear most directly on science are *ontology,* which is concerned with being and existence, and *epistemology*, which has already been mentioned on pages 4 and 442. The rapid growth of empirical science in the last 250 years has generated great interest in epistemology. Although problems in ontology still occupy some philosophers, the emphasis has been more on understanding the world. At least four philosophical "schools" can be named, by which scientific knowledge can be assessed.[2]

(1) *Idealism*. The real world lies in mental images. The knower contributes a great deal. Physical theory is mental construction; empirical evidence is secondary.

(2) *Realism*. Being is prior to knowing. Mental concepts represent the structure of events in an external world which continues on even when we're not examining it. The scientist *discovers* and *explores*.

(3) *Instrumentalism*. Theories are useful tools; invented, not discovered. These constructions are intertwined with the language and thought patterns of the times which produce them. But experience is also important.

(4) *Positivism*. Modern Empiricism. Operationalism could be considered a subdivision, with its emphasis on operation as the basis for meaning of terms. Here theory is simply a summary of data. The observer is neutral. A concept has meaning only if it is observable.

In the previous chapters the latter two approaches have been emphasized for three reasons. First, we wanted to underline the operational basis for definitions, which Bridgman preached. (page 17.) Second, we wanted to undo the naive realistic attitude which sees scientists as merely uncovering a reality which is independent of their mental structuring. Third, we wanted to emphasize the creative role of scientists in order to give their activities more humanity and to show the similarities between science and other modes of knowing.

Positivism, with its emphasis on word meanings, *has* been a healthy influence on all modern thought, including science. The philosopher Ludwig Wittgenstein (1889–1951) clarified the distinction between mathematical terms which are either true or false as a result of word definitions (they are called tautologies if true)—and facts, which are empirically testable. The latter refer to perceived things, but are neither true nor false. Wittgenstein thus helped to show why proofs are not of primary importance in science. The job is to check against the empirical world, not arrive at the proof of theorems.

However, not many scientists are ready to buy the Positivists' idea that theory is *only* a summary of experience. Einstein's statement that no information can travel faster than light and that a universal "now" does not exist is a positivistic idea. Yet Einstein refused to be claimed by the Positivists. In fact, he was surprisingly Realistic, believing that there was an external reality

(2) Barbour, Ian, Chapter 6 in *Issues in Science and Religion*, Prentice-Hall, Englewood Cliffs, 1966.

waiting to be understood.[3] In general scientists' minds are less tidy than positivism calls for. They like to keep a few spare concepts in their mental attics just in case they might want to use them to interpret some phenomenon in the future. The positivistic neutral observer is largely a myth. We saw on page 117 that the "raw data" which the positivist begins with are in truth loaded with preconception and theory. Leon Rosenfeld, a coworker with Niels Bohr, has said:

> "No scientist would accept the extreme positivist convention that there is nothing more in statements about phenomena than the expression of relations between sensations".

Einstein said further:

> "Without the belief in the inner harmony of the world there would be no science. This belief is fundamental to all scientific creation."

The next section discusses some of the philosophical consequences of atomic physics, where the connection between theory and experiment is quite indirect: the object being studied is never really seen. One effect has been to encourage interpretations at the extremes of the scale: idealism and positivism. Positivists point out the invisibility of the particles which atomic theory calls for, and end up saying that only the experiment is real. Idealists, on the other hand, praise theoretical physics as a mathematically coherent system and they look for reality at that abstract mental level. Still, the average physicist probably spends time in all four schools that were mentioned. Physicists act as if they were convinced realists. (Who ever heard of applying for a government grant to study something that really doesn't exist?) Nevertheless, if a physicist is the least bit reflective he or she will recognize the importance of his or her own mind in shaping what is seen and how it is interpreted.

The Problem of Objectivity in Microphysics: Uncertainty

Up to this point we have avoided getting bogged down in discussions of objectivity. We have merely emphasized the importance of unambiguous communication—the establishment of mutual understanding. However, we did implicitly assume some things about the basic measurement process which need to be brought out and examined. First, we took for granted that the observer and the thing being observed were separate and dis-

(3) Cline, Barbara L., "The Debate Between Niels Bohr and Albert Einstein", Chapter 13 in *The Questioners*, Thomas Y. Crowell, N.Y., 1965. Reprinted as *Men Who Made a New Physics*, New American Library Signet Book, N.Y., 1969.

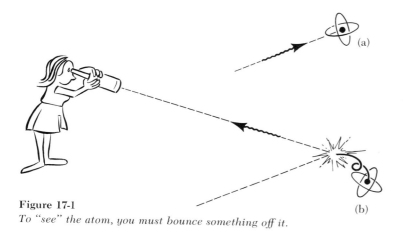

Figure 17-1
To "see" the atom, you must bounce something off it.

tinct. This is, of course a prime prerequisite for gathering objective knowledge. We also assumed that the thing being measured was not affected by the measuring process. Certainly laying a meter stick along side a curtain to measure its width does not affect the curtain! Or does it?

Observing something requires the exchange of a signal to detect where the object is. With man-sized, well-illuminated things like curtains it is easy to forget that the object is the source of reflected light waves which permit you to see where it is.

If, instead of a curtain, you want to measure up an atomic or subatomic particle, the process is different in several ways. First, the observation is less direct, with electronic counters, recorders, and cameras interposed between you and the particle since it is not visible to the unaided eye. Second, when you *look* at such a particle you are bombarding it and "asking" it to reflect back a signal — a quantum of light whose energy is not small in comparison with the energy of the particle itself.[4] The particle is different afterwards. You are no longer sure where it was and what its momentum was before you examined it.

The problems we've been encountering here can be expressed quantitatively as the *uncertainty principle*. It states that for a given single measurement made on an object, the uncertainty in its position (Δx, let's say) times the uncertainty in its momentum (Δp_x, where p_x is the component of momentum in the same direction as the coordinate x) must be equal to or

---

(4) George Gamow gave a good macroscopic analogy to this disturbance-by-measurement phenomenon: if you want to measure the temperature of a tub full of water, the initial temperature and the specific heat of the thermometer don't make much difference. But if you measure the temperature of a test tube full of warm liquid with the same thermometer, it's going to make a big difference if the thermometer is initially cold, or hot.

greater than an energy-time constant, $h/2\pi$, where h is equal to 6.6×10^{-34} joule·seconds (*not* joules per sec) and is called Planck's constant, after the pioneer German atomic physicist, Max Planck (1858–1947).

$$(\Delta p_x \Delta x) \geq \frac{h}{2\pi} \qquad\qquad \text{Equation (17-1)}$$

Planck's constant is so small that we can forget this kind of uncertainty when the system is man-sized. In fact, almost *all* quantum effects — graininess — smooth into insignificance when the system under observation contains trillions of atoms, as do all man-sized systems. For example, if we were to measure the position of a 0.15 kg baseball to the nearest tenth of a millimeter (10^{-4} m), the uncertainty or inaccuracy in its momentum would be:

$$\Delta p_x = \frac{h}{2\pi \Delta x};$$

$$\Delta v_x = \frac{\Delta p_x}{m} = \frac{h}{2\pi m \Delta x}$$

$$\Delta v_x = \frac{6.6 \times 10^{-34} \text{ J·sec}}{(2\pi)(1.5 \times 10^{-1} \text{ kg})(10^{-4} \text{ m})} = 7.0 \times 10^{-30} \text{ m/sec}$$

This inaccuracy is far less than any instrumental inaccuracy we could imagine. The uncertainty principle does not spoil macroscopic measurements. On the other hand, suppose an electron, whose rest mass is 9.1×10^{-31} kg, were confined in the nucleus of an atom. That is, suppose we wanted to be able to say that the electron existed somewhere within a space only 10^{-15} of a meter in diameter. The uncertainty in its momentum, which can be interpreted as the amount of momentum we have to allow for the electron to have, will be:

$$\Delta p_x = \frac{h}{2\pi \Delta x} = \frac{6.6 \times 10^{-34} \text{ J·sec}}{(2\pi)(10^{-15} \text{ m})}$$

$$= 1.05 \times 10^{-19} \text{ kg} \cdot \text{m/sec}.$$

Momentum can also be expressed as $p = \gamma m_0 v$. Suppose we figure out what γ, the relativistic mass increase factor, would be, assuming that the speed, v, is almost equal to 3×10^8 m/sec.

$$\gamma = \frac{\Delta p_x}{m_0 v} = \frac{1.05 \times 10^{-19} \text{ kg·m/sec}}{(9.1 \times 10^{-31} \text{ kg})(3 \times 10^8 \text{ m/sec})}$$

$$= .0385 \times 10^4 = 385.$$

Very roughly, we would have to allow for the possibility of the electron having energy about 400 times its rest energy if we were to try to pin down its location to within 10^{-15} meters. An electron with such gigantic relativistic energy could not be contained in so small a place. For this reason, nuclear physicists feel certain that electrons do not exist in the nucleus. This is just one of many strange results which the uncertainty principle leads to.

One immediate epistemological consequence of the uncertainty principle is that the observer, the measuring apparatus and the object being observed form a system, none of which can be omitted from a full description of the observation made. The line separating the observer and the observed is not definite. Classical Newtonian objectivity is in question.

Signals carry energy which may, in some cases, not be negligible in comparison with the energy of the system under observation. This raises fundamental problems, just as did the finite velocity of signals in special relativity theory. Widespread simultaneity went out the window. Now, it seems, precise measurement must go.

Up to this point the conclusion that uncertainty results from *disturbance* during the measurement process has perhaps been encouraged. However, uncertainty can be interpreted in other ways. For one thing, it arises directly from the theoretical physics which is used to explain the behavior of atom-sized systems. We call it *quantum mechanics*, and its basic approach uses probability. For example, instead of saying that a particle moves at constant speed from point a to point b, it is said that the maximum in the particle's probability (or wave) function is found first near point a, then later near point b. The uncertainty in the position and momentum of a particle then becomes related to the width of the probability curves representing the position and momentum. When mathematical operations are performed on these probability functions, one finds that the product of their widths has to be at least as large as $h/2\pi$!

Figure 17-2
Motion of probability function.

This probability-function way of describing motion might appear less strange if we considered the following example: suppose you had to describe the motion of a rapidly oscillating pendulum. It vibrates so rapidly that it cannot be located even with a high speed camera. How would you study it? You could put a dark screen behind the pendulum, shine a light on the pendulum, then open the camera shutter and take a time exposure. If the pendulum travels more slowly in some parts of its swing it will produce a greater exposure of the film there. When you look at the time-exposure photograph you can say that the chances of the pendulum being at a particular part of its swing is related to the illumination there. In places where the pendulum swings faster the probability of the pendulum being there would be less. Your description of the motion will thus not be very good for locating the pendulum on a one-try basis. But it will tell what the chances are of finding the pendulum at that spot if you reach in a large number of times. On this basis, uncertainty can be interpreted as a statement of the statistical spread (lack of precision, page 15) which results when you make a series of very accurate measurements of some variable (momentum, position, etc.) on a series of identical physical systems.

On the other hand, some physicists, including Einstein and Planck, interpret uncertainty as a result of *inadequate theory*.[5] They had the faith (but since about 1930, little else) that quantum mechanics would eventually be replaced by a more deterministic, less probabilistic scheme. Other scientists, such as Henry Margenau and Werner Heisenberg, insist that the adequacy of quantum mechanics has been verified so often that uncertainty must be seen as a *property of nature*. Particles just don't have the sharp position and velocity of macroscopic objects. They are, so far as we can tell, fuzzy. These men say that the act of observation does not disturb so much as it narrows down the potentialities of the system being observed. Still another interpretation, favored by Niels Bohr, sees uncertainty as *conceptual*. The terms which have proved to be useful for describing large objects are stretched beyond their limit when we apply them to atom-sized objects.

Complementarity

Thus far we have seen the uncertainty principle as limiting our powers to measure and describe atomic systems. Niels Bohr (1885–1963) has expanded it into a wide-ranging epistemological principle called *complementarity*. It claims that whenever we concentrate on knowing one dimension of a situation, we always exclude ourselves from knowing much about some other complementary dimension. If we concentrate on the trees we don't see the woods.[6] Let us start with atomic observations as an example and then expand.

Bohr would say that when you study a particle, the complementarity principle takes on definite, quantitative form; the uncertainty principle. There is no way to get around uncertainty. If you build an apparatus which is rigid and yields precise position measurements it won't be good for measuring the velocity or momentum of the particle. On the other hand, if you use a

(5) Although Einstein was happy to accept and use statistical methods in thermal physics, where the systems involve many elements, he could never accept the idea that the behavior of a single particle could be predicted only with probability. It was at the International Solvay conference in Brussels in 1926 that he made his famous statement: "I simply cannot believe in a God who plays with dice".

(6) The picture used to introduce this last section of the book illustrates an old Oriental fable. It tells how each blind man fails to understand what an elephant is like because he investigated just one part of the beast. Although this is a better illustration of incomplete knowledge than of complementarity, it can get us thinking about the problem. With complementarity another idea has been added: that your method of investigation actually prevents your seeing the rest of the thing that is being studied. The trees and the woods analogy comes closer.

flexible, recoil-type of apparatus it will tell you the momentum of the particle which bounces off it but will give very uncertain information about the particle's position. No single experiment —no single apparatus—will give you sharp information about position and momentum at the same time.[7]

Another way of saying the same thing would be to state that some apparatus localizes an object well and makes it appear to be a *particle*. Another will measure energy better than position, so that the object appears fuzzed out like a *wave*. A lot of unnecessary nonsense has been written about electrons, light energy, and other atomic-sized objects being waves one day and particles the next. The dilemma is completely unjustified and results from too literal an interpretation of experimental results. An electron is not *really* either a particle or a wave or even a wiggly combination of the two. (Recall the comments on page 368 regarding light.) We simply get different perspectives on it depending on the apparatus that is used. The type of apparatus used always plays a part in the information that is gathered.

Bohr extends the principle of complementarity, a sort of principle of fuzzy knowledge, into other realms. It removes the apparent contradictions between material and spiritual elements in human experience. The process of observing material parameters precludes being aware of spiritual parameters. And vice versa. One can draw up lists of complementary concepts. In each case attention to one of the pair leads us to think the other is irrelevant or non-existent: mechanism and organism, compassion and justice, thought and sentiment, individual and society, free will and determinism, mind and brain, etc. However justified or unjustified these extrapolations of complementarity from microphysics to other fields of experience may be, they express a humbling realization about human knowledge: all concepts, all words, can be used only within a limited domain of validity. When we try to extend the use of these terms beyond their range of validity, they conflict and lead to contradictions.

The following quotation from Leon Rosenfeld[8] could form a good summary to what we have been trying to say:

> "The view of physics, and in fact of science, which results from the development of quantum theory is very different from the scientific ideal of the nineteenth century; but it is by no means, or so it seems to me, inferior to it. . . . the supremacy of human reason over all forms of irrationality was based on the deterministic view of a Nature governed by inflexible laws, which was the task of science to discover and formulate. These laws, by their very nature, would

(7) Bohr, Niels, "Discussion with Einstein on Epistemological Problems in Atomic Physics", pp. 199–242 in *Albert Einstein: Philosopher—Scientist*, Paul A. Schlipp, ed., Open Court, 1949.

(8) "Foundations of Quantum Theory and Complementarity", *Nature 190*, April 29, 1961, pg. 384.

have absolute validity, and they would give a picture of the universe . . . which could be contemplated but from which the spectator, the human spectator, would be excluded, or at least reduced himself to an object of contemplation. The present conception, on the contrary, stresses the fact that at every stage the human observer intervenes in a very fundamental way by determining the very language in which the laws of Nature are formulated. The significance of the laws of Nature, from this point of view, is not that of drawing a picture of the world independent of who is looking at it, but rather of giving an analysis of the interaction between ourselves and the external world. At first sight it would seem that one thus introduces in science a subjective element. But the fear that the objectivity and rationality of the old view would be abandoned is entirely unfounded. True objectivity is simply the possibility of guaranteeing that the account of the phenomena will convey equivalent information to all observers, that it will consist of statements intelligible to all human beings. . . . It is just the invariance of the laws of physics . . . that guarantees their objectivity.

The Language Problem

Terminology is a serious problem in microphysics. According to Bohr, language is *the* main problem. (Certainly the number of times he rewrote his own papers before publication, and worked to clarify the descriptions of others would attest to this.) To quote him directly:[9]

> "However far quantum effects transcend the scope of classical physical analysis, the account of the experimental arrangement and the record of the observations must always be expressed in common language supplemented with the terminology of classical physics. . . . The word "experiment" can be used only in reference to a situation where we can tell others what we have done and what we have learned."

At the same time, the use of classical words leads to contradictions such as an object being both particle and wave, as we've seen. Classical words such as "diameter" may impose an unjustified structure on the atomic systems they are applied to. Bohr suggests that much of the conflict will disappear if people are careful to include details of the method of measurement with the report of what was measured.

In the atomic realm physics is a search not so much for reality as for rigorous use of language. Bohr said,

> "In our description of nature the purpose is not to disclose the real essence of phenomena but only to track down as far as possible relations between the multifold of our experience".[10]

(9) "On the Notions of Causality and Complementarity", *Science 111*, January 20, 1950, pg. 52.

(10) *Atomic Theory and the Description of Nature*, Cambridge University Press, 1934, p. 18.

Newtonian physics tried to mirror the macroscopic world. But instead of trying to mirror the atomic world, quantum physics describes the experimenter's interaction with the atomic world.

We are raising here questions which reach far beyond the limits of quantum physics. Many but not all linguists feel that language precedes and determines thought in the first place. Benjamin L. Whorf (1897–1941) has said:[11]

> "It is not possible to define event, thing, object, or relationship from nature. To define them always involves a return to the grammatical categories of the definer's language."

It has been shown[12] that certain languages, called "nominal", tend to use collective terms. Speakers of such languages see themselves as part of a whole, not a separate, self-conscious observer of the world. The idea of number and of tense is vague. On the other hand, "verbal" languages such as English make the actor—the subject—distinct and important. Objects are enumerated. Past, present, and future are separate concepts. In certain languages the concept of animate and inanimate are nearly the reverse of Western ideas. For example, a tree sitting in the middle of a plain is seen as inanimate because it does not affect the life of the speaker in any dynamic way. However, a rock perched at the top of a cliff is seen as animate because it may come tumbling down at any moment. The structure of languages reflects (determines?) the way humans look at the world.

On the other hand, the work of Piaget with children which was referred to in the early chapters of this book suggests that operations or motions carried out with the body are equally important in concept formation. Ideas of object, permanence, conservation, and motion itself are built up with the help of repeated reachings, manipulations, and pourings out and back in again. The concept of invariance which has played such an important role in this book could have been born when a child first looked at a stone on the ground from different angles, picked it up and, on rotating it, realized that some invariant thing kept sending out signals! In an ingenious article, Jerome Rothstein[13] describes how a civilization of intelligent worms buried in the ooze at the bottom of the ocean might build up a description of their surroundings—a physics—starting from a sense of touch, temperature, taste, and nothing else. (As with the ants in Chapter 2, one wonders how they would record, and thus store up, or accumulate their knowledge.) Repeated bodily experiences such as hunger, pulse, and events happening outside in the mud can be grouped to give a sense of time passage. Repeated slitherings against stones leads to the idea of permanent objects. The

(11) "Science and Linguistics", in *Language, Thought and Reality*, John B. Carroll, ed., MIT Press, Cambridge, 1957.

(12) Werkmeister, W. H., *A Philosophy of Science*, Chap. 5, Harper & Row, New York, 1940.

(13) "Wiggleworm Physics", *Physics Today 15*, Sept., 1962, pg. 28.

fact that stones can either be hit or missed gives a sense of dimension. This all builds up until finally some brave worm captures a bubble and rises above the ooze to perform the first space voyage. Changes in body temperature with day and night can then introduce a whole new set of time variables, etc.

Whether based on language or on body motions, thinking is a symbolic activity. Whether they are used for science, persuasion, or worship, our words are a condensation or abstraction from a complex of experiences. The new phenomena which appear in the atomic realm should be seen not so much as revealing weaknesses in scientific description as forcing us to recognize the symbolic nature of *all* thought. The divisions between natural and social sciences and the humanities would seem less important if we could see them all as alternative ways of manipulating symbols, or trying to express something about ourselves or our surroundings.

The Unity of Knowledge—A Final Step
Toward Invariance

None of this overcomes the fact that communication of symbols, with some understanding achieved, requires shared experiences. (The most rudimentary example of this would be the international standards discussed at length in Chapter 2.) Recognition of the basically symbolic nature of all knowledge sets the stage for tolerance and trust, but perhaps not for much added understanding. Maybe this is enough. If Lawrence Durrell is inspired by Einstein's special theory of relativity to write his *Alexandrian Quartet* need we worry over whether he understands the theory? However, this text has tried to make clear that the connections between physics and art, music, psychology, and philosophy go far beyond mere mutual respect for each other. Even if fact is as important in science, as feeling is in art, we have tried to show that fact is not so hard and is involved with subjective choices. If the application of logic seems to separate science from more emotional or creative human activities, we have tried to emphasize the uncharted creativity which is vital to the construction of new scientific theories. We have also shown that science is shot through with aesthetic principles, such as symmetry and simplicity, which are also important in art.

The connections between physics and art become even more marked if we recall some of the similar ways in which the two of them have changed during certain periods. To Western peoples during medieval times, both art and science were an expression of the world's unity under God. Artists were usually anonymous and their painting taught or pictured humanity as an integral part of God's concern and mastery. The science of

455

*The Unity
of Knowledge—
A Final Step
Toward
Invariance*

the times, as described in Chapter 4, concerned itself with visible manifestations of God's masterpiece, nature. With the Renaissance, Western art became more self-conscious. Rules of perspective were worked out. The artists are known, and their art becomes their view of the world. Over on the science side, reason rules. People were confident of their own thought, and the explanation of planetary orbits is in terms of physical law which humans have discovered. These laws are reality itself. Soon after the oval becomes acceptable in Western art and architecture, it becomes acceptable in the description of planetary orbits. Newtonian physics partakes of Baroque grandeur. God is still in evidence, but more in a role of cheering humanity on to greater rational exploits. With self-awareness and self-confidence, people eventually separated themselves from the world.

Near the end of the nineteenth century drastic changes occurred. With the appearance of Impressionistic and then Cubistic painting, the rules of representation which were perfected during the Renaissance were thrown out. Artists no longer sought to imitate nature and then step aside, objectively. They became more involved in the painting and the way in which paint is put on the canvas became more important than any object which the painting might be depicting. As was mentioned at the end of Chapter 8, the painter might become relativistic, showing several perspectives of a face on the same canvas. But above all, they were bringing themselves and the viewers back into the creation. At these same times, physics and the role of physicists were also changing. When they began to study matter on a fine scale, the solid, unchanging Newtonian masses began to evaporate, and were replaced by empty space and "cloudy" electrons rotating around concentrated nuclear centers of force. From special relativity, the scientist learned that the mass of an object even depended upon its speed, so that inert matter was unsolid, and had to be visualized as energy in a different form. In order to appreciate modern physics, as well as modern art, one must think at a more symbolic, less literal level. And as has just been described, quantum theory did away with the idea of neutral, objective observation of nature on a fine scale. In many ways, science and art both gave up literal representation at the same time, and at the same time the scientist and the artist recognized that they are more involved in their own work—less objective—than they had thought before. Hopefully, the realization of this similarity of history in two modes of knowing will encourage us to look for other unifying concepts. As they are found they will help overcome the great differences in vocabulary which have fragmented human knowledge.

In 1955, John Wheeler, a physicist at Princeton University, presented a paper before the Symposium on Communication held by the 5th Annual Conference of Graduate Alumni at

Princeton. It was called "A Septet of Sibyls: Aids in the Search For Truth". In it, he tried to answer the question of whether a unifying concept from one field can be applied to another. His concluding statement might well serve to conclude this text:[14]

> "As we go about the world, we are exploring a marvelous jungle, full of strange birds, sights, and sounds. We find that there are a few paths that allow us to go about from one part of the jungle to another. It is perhaps too early to look for a law of these paths. Let us only be happy that there we indeed have for paths some great unifying concepts and that we can communicate with one another."

Summary

Philosophy underlies physics, if it is not at the core. Idealism, realism, instrumentalism, and positivism represent just four efforts to systematize attitudes that scientists might hold as they try to order, and communicate, knowledge of the world around them.

The relation between the knower and the known becomes most delicate in the realm of microphysics. The two seem interconnected, not independent.

The uncertainty principle states that when you try to investigate simultaneously certain pairs of variables, (x component of momentum, p_x, and the position, x, for example) the precision of your measurement of them is limited. If Δp_x is the uncertainty in measurement of p_x and Δx is the uncertainty in measurement of x, then

$$\Delta p_x \Delta x \geq h/2\pi, \qquad \text{Equation (17-1)}$$

where $h = 6.6 \times 10^{-34}$ joule-sec. The product of the two uncertainties cannot be less than $h/2\pi$.

Quantum mechanics, the branch of physics with which physicists try to predict the behavior of atomic-sized systems, works in terms of probabilities. The probabilities arise not because of complexity, as in thermal physics, but because of the more basic uncertainty principle. The uncertainty principle is at work in all physical investigation, but its effects are so small that they are noticeable only in the atomic realm.

Uncertainty can be interpreted as arising from

(1) disturbances during measurement;
(2) the statistics of fine measurement;
(3) inadequate theory;
(4) the inherent fuzziness of the atomic world; or,
(5) inadequate terminology, or concepts.

(14) *A Septet of Sibyls: Aids in the Search for Truth, American Scientist 44,* 1961, pp. 360–377, 1956.

The concept of complementarity expands the uncertainty discovered in quantum physics into a philosophical principle of knowledge. It points out that our specific approach to problems prevents our recognizing the validity of concepts which arise when different approaches are used.

The problem of science is one of communicating unambiguously, rather than uncovering absolute reality. Our use of symbols — our language — has tended to make "discoveries" in various realms of thought more disconnected than they really are.

Further Reading

Bridgman, Percy W., "Quo Vadis", *Daedalus* 87, Winter, 1958, pp. 85–93. (The father of operationalism writes about the importance of word meanings and how they limit knowledge.)

Bronowski, J., "The Discovery of Form", pp. 55–60 in *Structure in Art and Science*, G. Kepes, Ed., G. Braziller, N.Y., 1965. (A short but excellent account of the re-emergence of form and structure in science and art in the 20th century.)

Bronowski, J., "The Vision of Our Age", Chapter 15 in *Insight*, McDonald, London, 1964. (The author interviews a sculptor, an architect, a physicist, and an author; and finds them all to be searching for a pattern of meaning.)

Bronowski, J., *The Identity of Man*, American Museum Science Book, Natural History Press, Garden City, 1965. (Examines consciousness, thought, and the similarities between the activities of poets and scientists.)

Cantore, Enrico, "Humanistic Significance of Science, Some Methodological Considerations", *Philosophy of Science 38*, Sept. 1971, pg. 395. (Pictures science as a humanistic reaction to one's surroundings.)

Einstein, Albert and Infeld, Leopold, *The Evolution of Physics*, Simon and Schuster, N.Y., 1938. (Still one of the clearest discussions of how classical Newtonian physics gave way to quantum and relativistic physics.)

Gabo, Naum, "The Constructivistic Idea in Art", pg. 104 in *Modern Artists on Art*, Prentice-Hall, N.Y., Englewood Cliffs, N.J., 1964. (Says that the same spiritual state propels artistic and scientific activity at the same time and in the same direction.)

Hafner, E. M., "The New Reality in Art and Science", *Comparative Studies in Society and History 11*, Oct. 1969, pg. 385. (Finds many similarities between modern science and art.)

Heisenberg, Werner, "Language and Reality in Modern Physics", Chap. X in *Physics and Philosophy*, Harper and Row, N.Y., 1958. (Source of many ideas in this chapter.)

Heisenberg, Werner, "Discussions About Language", Chap. 11 in *Physics and Beyond*, Harper and Row, N.Y., 1971. (The author tries to recreate conversations between himself and Niels Bohr on meaning, complementarity, and language.)

Langer, Suzanne, *Philosophy in a New Key*, New American Library Mentor Book, 1942, 1951. (A good discussion of the symbolic nature of all knowledge.)

Morick, Harold (ed.), *Wittgenstein and the Problem of other Minds*, McGraw-Hill, New York, 1967. (Ten essays on Wittgenstein's philosophy, several of which deal with language and operationalism.)

Seeger, Raymond J., "Scientist and Poet", *American Scientist 47*, Sept. 1959, pg. 350–360. (A physicist discovers revealing similarities between poetry and science.)

Tuve, Merle A., "Physics and the Humanities", The Verification of Complementarity", from *The Search For Understanding*, Carlyle P. Haskins, Ed., Carnegie Institution of Washington, 1967. (A leading American statesman of science suggests complementarity as being the basic contribution of physics to modern thought.)

Young, John A., *Doubt and Certainty in Science*, Oxford University Press, London, 1956, especially pp. 111–112. (Uses the transition from medieval to Newtonian physics to illustrate how the form we give the external world is a construct of our minds.)

APPENDIXES

APPENDIX A

SOME CONSTANTS AND PROPERTIES

Speed of light in vacuum, c	3×10^8 m/sec $= 186{,}000$ mile/sec.
Speed of sound in air (0°C)	331 m/sec $= 1090$ ft/sec.
Acceleration due to gravity at the earth's surface, g	9.8 m/sec$^2 = 32.2$ ft/sec^2.
Average radius of the earth, r_e	6.4×10^6 m $= 4000$ miles.
Mass of the earth, M_e	6×10^{24} kg.
Average speed of the earth in its orbit around the sun	30 km/sec $= 18.5$ miles/sec.
Average earth-moon distance	3.86×10^5 km $= 2.4 \times 10^5$ mile.
Average earth-sun distance	1.5×10^8 km $= 93 \times 10^6$ mile.
Density of water	1.000×10^3 kg/m^3.
Specific heat of water	4.184 J/g·°C.

Triple point of water	273.16 K = .01°C.
Heat of fusion of water	3.34×10^5 J/kg.
Heat of vaporization of water	2.255×10^3 J/g.
Index of refraction of water, η	1.333.
Range of visible electro-magnetic radiation	430 to 690×10^{-9} m.
Mass-energy relation: $E = mc^2$	9×10^{16} J/kg.
Universal gravitational constant, G	6.67×10^{-11} N·m²/kg².
Electrical constant, K	9×10^9 N·m²/C².
Elementary charge, e	1.6×10^{-19} C.
Electron rest mass	9.1×10^{-31} kg.
Proton rest mass	1.67×10^{-27} kg = 1.007825 AMU
Alpha particle rest mass	4.02600 AMU.
Planck's constant, h	6.6×10^{-34} J·sec.
Circumference of a circle $= 2\pi r = \pi d$	$\pi \cong 3.1416$.

Area of a circle $= \pi r^2$.

$\sin 0° = 0 = \cos 90°$

$\sin 30° = 0.500 = \cos 60°$

$\sin 45° = 0.707 = \cos 45°$

$\sin 60° = 0.866 = \cos 30°$

$\sin 90° = 1.000 = \cos 0°$

APPENDIX B

CONVERSION FACTORS

Mass:
$$1 \text{ kg} = 2.205 \text{ lbs.}$$
$$1 \text{ ounce} = 28.3 \text{ gm.}$$
$$1 \text{ AMU} = 1.66 \times 10^{-27} \text{ kg.}$$

Length:
$$1 \text{ m} = 39.37 \text{ inches.}$$
$$1 \text{ inch} = 2.54 \text{ cm.}$$
$$1 \text{ mile} = 5280 \text{ feet} = 1.61 \text{ km.}$$

Time:
$$1 \text{ day} = 86,400 \text{ sec.}$$
$$1 \text{ year} = 365.25 \text{ days} = 3.16 \times 10^7 \text{ sec.}$$

Velocity: $1 \text{ mile/hour} = 1.47 \text{ ft/sec} = 0.447 \text{ m/sec.}$

Energy:
$$1 \text{ J} = 10^7 \text{ dyne.}$$
$$1 \text{ B.t.u.} = .252 \text{ cal} = 1054 \text{ J.}$$
$$1 \text{ cal} = 4.184 \text{ J.}$$
$$1 \text{ kWh} = 3.6 \times 10^6 \text{ J.}$$
$$1 \text{ electron-volt} = 1.6 \times 10^{-19} \text{ J.}$$

Power: $1 \text{ horsepower} = 746 \text{ watts.}$

APPENDIX C

COMMON METRIC PREFIXES

nano $= 10^{-9}$

micro $= 10^{-6}$

milli $= 10^{-3}$

centi $= 10^{-2}$

kilo $= 10^{3}$

mega $= 10^{6}$

APPENDIX D

SYMBOLS USED

a	vector.
a	acceleration.
A	amperes.
B	magnetic field, or flux density, in weber/m².
c	velocity of light, 3×10^8 m/sec.
c	specific heat, J/g·°C.
C	coulombs.
°C	degrees Celsius.
d	distance between sources in interference phenomena.
D	width of a single source in diffraction phenomena.
e	charge on the electron, 1.6×10^{-19} C.
E	energy, in J.
$\mathscr{E}$	electric field intensity, in N/C or V/m.
f	frequency, in hertz.
F	force, in newtons or dynes.
g	gravitational field intensity, or acceleration due to gravity.
G	universal gravitational constant, 6.67×10^{-11} n·m²/kg².
h	Planck's constant, 6.6×10^{-34} J·sec.
$^1_1$H	proton, or hydrogen nucleus.
I	current, in amperes.
J	joules.
K	electric constant, 9×10^9 n·m²/C².
K	Kelvin temperature.

K.E.	kinetic energy, in J.
kWh	kilowatt hour.
m	mass, in grams or kg.
m_0	rest mass.
m	meters.
n	a whole number: 1, 2, 3, 0, -1, -2, -3, etc.
$_0^1n$	neutron.
N	newtons.
p	momentum, kg·m/sec.
P.E.	potential energy, in J.
q	electric charge, in C.
Q	heat, in J.
r	distance between two points, or radius.
R	electrical resistance, in ohms.
R	vector resultant (sum of two or more vectors).
S	entropy, in J/°C.
t	time, in sec.
t_o	initial time.
T	temperature, in °C or kelvins, K.
u, v	velocity.
v_0	initial velocity.
U	thermal energy, in J.
V	electrical voltage, in volts.
W	watts.
W	work, in J.
x	position.
x_o	initial position.
y	vertical position, or position in a field.
α (alpha)	an angle; also a particle.
β (beta)	an angle; also a particle.
γ (gamma)	$1/\sqrt{1 - v^2/c^2}$; also high-energy electromagnetic radiation.
Δ (delta)	"change in".
η (eta)	index of refraction.
θ (theta)	an angle.
λ (lambda)	wavelength.
μ (mu)	coefficient of friction.
ν (nu)	neutrino.
Φ (phi)	gravitational potential, in J/kg.
τ (tau)	period of a motion, in sec.

APPENDIX E

POWERS OF 10

Physicists have to manipulate some very large and small numbers in their attempt to describe nature. To make these operations easier, they use the mathematical "powers of ten" notation. According to this scheme, increasing the exponent of the number 10 by one is the same as multiplying by 10 or moving the decimal point one place to the right. Decreasing the exponent by one is the same as dividing by 10 or moving the decimal point one place to the left (Table E-1).

To handle other numbers, they are best written as a digit between 1 and 10 (easiest to conceptualize) times the appropriate powers of 10:

$$365 = 3.65 \times 10^2;$$
$$3{,}650{,}000 = 3.65 \times 10^6$$
$$.0365 = 3.65 \times 10^{-2};$$
$$.365 = 3.65 \times 10^{-1}$$

Suppose you want to divide 36,500 by .0051, making use of powers of 10 notation.

First, you re-express each number as a number between one and ten times the appropriate power of 10:

$$\frac{36{,}500}{.0051} = \frac{3.65 \times 10^4}{5.1 \times 10^{-3}}$$

Table E-1. *Powers of 10 notation.*

$$\frac{1}{1,000,000} = \frac{1}{10^6} = 10^{-6} = .000001$$

$$\frac{1}{100,000} = \frac{1}{10^5} = 10^{-5} = .00001$$

$$\frac{1}{10,000} = \frac{1}{10^4} = 10^{-4} = .0001$$

$$\frac{1}{1000} = \frac{1}{10^3} = 10^{-3} = .001$$

$$\frac{1}{100} = \frac{1}{10^2} = 10^{-2} = .01$$

$$\frac{1}{10} = 10^{-1} = 0.1$$

$$1 = 10^0$$

$$10 = 10^1$$

$$100 = 10^2$$

$$1000 = 10^3$$

$$10,000 = 10^4$$

$$100,000 = 10^5$$

$$1,000,000 = 10^6$$

To divide 10^4 by 10^{-3} you simply subtract exponents:

$$\frac{10^4}{10^{-3}} = 10^{(4-(-3))} = 10^7$$

Thus

$$\frac{3.65 \times 10^4}{5.1 \times 10^{-3}} = \frac{3.65}{5.1} \times 10^7$$

Alternatively, you could multiply the numerator and denominator by 10^3:

$$\frac{(3.65 \times 10^4)(10^3)}{(5.1 \times 10^{-3})(10^3)} = \frac{3.65 \times 10(4+3)}{5.1 \times 10(-3+3)} = \frac{3.65}{5.1} \times 10^7.$$

Carrying out the long division,

$$\frac{3.65}{5.1} = .716$$

(If you do this on a slide rule, you know in your head that the answer is near 0.7, not 70, or .007.)

The answer is thus $.716 \times 10^7$, or 7.16×10^6. If .716 is multiplied by 10, then the exponent on the 10 must be reduced from 7 to 6, thus dividing by a factor of 10 so that the magnitude of the number is kept the same.

When doing problems in physics it is always more important to keep an eye on the rough magnitude of the numbers rather than get every digit right, but miss by a factor of 1000!

APPENDIX F

TRIGONOMETRIC FUNCTIONS

Consider a line of length, L, which is pivoted at the origin and which rotates counter-clockwise, as shown. Suppose that at a particular instant it makes an angle θ (theta) with the x axis. At that instant, the line of length L will have *projections* or *components* X on the x axis, and Y on the y axis. As the line rotates, these projections will change. We define certain *trigonometric* relations or functions between the line, its components, and the angle θ. They are:

$$\frac{X}{L} = \text{cosine } \theta$$

$$\frac{Y}{L} = \text{sine } \theta$$

$$\frac{Y}{X} = \text{tangent } \theta$$

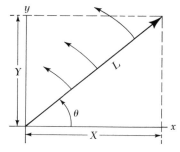

Figure F-1

Table F-1 shows how these functions depend on θ, for θ varying between 0 and 90 degrees (or $\pi/2$ radians.) (When the line rotates through one whole turn, θ increases by 360°, or 2π *radians.*

Table F-1.   *Trigonometric Functions.*

$\theta°$	$\sin\theta$	$\cos\theta$	$\tan\theta$	$\theta°$	$\sin\theta$	$\cos\theta$	$\tan\theta$
0	.0000	1.000	.0000	50	.7660	.6428	1.192
5	.0872	.9962	.0875	55	.8192	.5736	1.428
10	.1736	.9848	.1763	60	.8660	.5000	1.732
15	.2588	.9659	.2679	65	.9063	.4226	2.145
20	.3420	.9397	.3640	70	.9397	.3420	2.747
25	.4226	.9063	.4663	75	.9659	.2588	3.732
30	.5000	.8660	.5774	80	.9848	.1736	5.671
35	.5736	.8192	.7002	85	.9962	.0872	11.43
40	.6428	.7660	.8391	90	1.000	.0000	∞
45	.7071	.7071	1.000				

Note that $(\cos\theta^2) = X^2/L^2$ and that $(\sin\theta)^2 = Y^2/L^2$. Since, for a right triangle, the square of the hypotenuse is equal to the sum of the squares of the other two sides, $L^2 = X^2 + Y^2$. If we add the two;

$$(\sin\theta)^2 + (\cos\theta)^2 = \left(\frac{X^2}{L^2}\right) + \left(\frac{Y^2}{L^2}\right)$$

$$= \frac{X^2 + Y^2}{L^2} = \frac{L^2}{L^2} = 1.$$

Thus $(\sin^2\theta) = 1 - (\cos^2\theta)$.

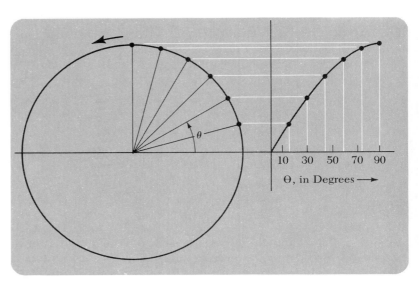

Figure F-2
Graphical representation of sine θ, for θ between zero and 90°.

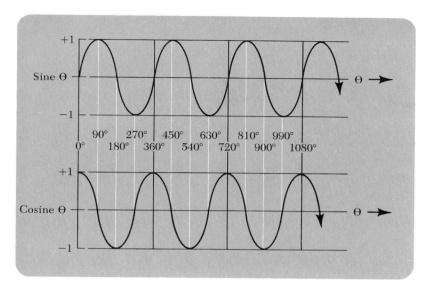

Figure F-3
(a) Graph of sine θ for θ ranging beyond 90°.
(b) Graph of cosine θ for θ ranging beyond 90°.

Another way to picture the sine (or cosine) of an angle is in terms of a point moving around a circle in a counterclockwise direction. The radius of the circle is taken to be one. If we make a graph of the vertical position of the point versus the angle θ, we obtain a plot of the sine! (In Figure F-2 positions are marked for every 15 degrees.) What makes this interesting is that the point can keep going right on around the circle. It need not stop at θ equal to 90 degrees. If we let this happen, we get a "sine wave", which can be used to represent oscillations of a swing (Chap. 7 (page 162)), of an electrical AC voltage (Chap. 11 (page 290)), or of waves in a string (page 333), or of an electromagnetic field (page 338).

APPENDIX G

ON VECTORS

Throughout physics quantities such as force, velocity, and momentum are encountered which have a *direction* as well as numerical value (magnitude). Putting it another way, they have *components* when they are referred to a frame of reference. Such quantities can be represented by the mathematical concept, *vectors*. They can be pictured nicely by means of arrows. If you want to be graphical about it, you can make the length of the arrow proportional to the magnitude of the vector quantity. The shaft and head will show the direction. As we saw on page 51, the arrow symbolism shows the various ways in which a vector quantity can change. When writing a vector as an algebraic letter, a small arrow can be drawn over the letter: $\vec{a}$. Or, as is done in this text, boldface letters (i.e. a) can be used to denote vector quantities when necessary.

When a vector slants between two coordinate axes, we can easily find its *components* in the direction of the axes. That is, we can find out how much "it's worth" in the chosen directions. The cartesian components of a vector are simply two (in two dimensions) or three (in three dimensions) mutually perpendicular vectors whose sum is the original vector itself. From Appendix F, it's apparent that the vector L will have an x component, L_x

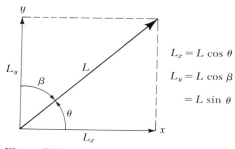

Figure G-1
Components of the vector **L**.

$= L \cos \theta$, and a y component, $L_y = L \sin \theta$. The y component equals $L \cos \beta$ (β is pronounced beta), since $\cos \beta = \cos(90 - \theta)$ $= \sin \theta$. (See the table in Appendix F.)

By Pythagoras' theorem, the magnitude of L, usually symbolized by $|L|$, is equal to $\sqrt{L_x{}^2 + L_y{}^2}$. But from what we have just seen, it is also equal to $\sqrt{(L \sin \theta)^2 + (L \cos \theta)^2}$, which is equivalent to $L \sqrt{(\sin \theta)^2 + (\cos \theta)^2}$

In a mathematical discussion, vectors are often described as *invariant*. Some have even suggested that this is why vectors find such wide application in physics. In what sense are vectors invariant? Simply this: when a frame of reference rotates, the apparent components of the vector will change but the magnitude (length) of the vector will not (Figure G-2).

$L_y{}^2 + L_x{}^2$ in one system is equal to $L_y'^2 + L_x'^2$ in the other system. This should come as no surprise. We just saw that $|L| = L \sqrt{\cos \theta^2 + \sin \theta^2}$, and from Appendix F, $(\sin \theta)^2 + (\cos \theta)^2 = 1$ no matter what θ is.

Vectors cannot be added using the rules of ordinary algebra. For example, one plus two might not be equal to three. Suppose a man walks one mile south and then two miles west. What is his overall vector change of position?

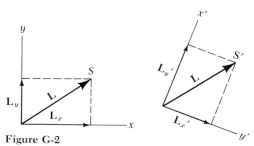

Figure G-2
The components of **L** *are different when measured in the rotated coordinate system,* s; *but* **L**$'$ *remains equal to* **L**.

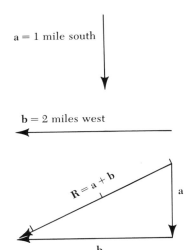

a = 1 mile south

b = 2 miles west

$R = a + b$

a

b

Figure G-3
Adding vectors.

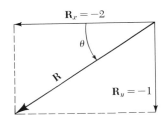

$R_x = -2$

θ

R

$R_y = -1$

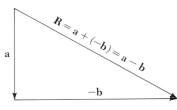

a

$R = a + (-b) = a - b$

−b

Figure G-4

This can be solved by means of a vector diagram in which the length of the arrows is proportional to distance. We could let 1/2 inch represent one mile. Then to add the vectors, we merely string them together, tail to head. Only the length and direction of the vector arrows matter. Their starting point is not fixed, so we can slide them around: The sum of several vectors is called their *resultant,* **R**. It is represented by an arrow drawn from the tail of the first arrow to the head of the last. By measuring the length and direction of the resultant, we find that the man's effective movement is around 2.2 miles in a direction about 25 degrees south of west. This method of adding vectors is called *graphical addition.* Its accuracy is limited.

A more accurate and often quicker way to add several vectors is simply to add their components. From these you can calculate the magnitude and direction of the resultant. In the example just cited, vector a and b have components

(East-West) $a_x = 0$ $\qquad\qquad b_x = -2$ miles

(North-South) $a_y = -1$ mile $\qquad b_y = 0$

(Vector quantities are usually taken to be positive to the right and up, and negative to the left and down.) Then the components of **R** are

$$R_x = a_x + b_x = -2 \text{ miles}$$

$$R_y = a_y + b_y = -1 \text{ mile}$$

The magnitude of $\mathbf{R} = \sqrt{(R_x)^2 + (R_y)^2} = \sqrt{4 + 1} = 2.236$ miles
To find the direction of **R**, recall that $\tan \theta = R_y/R_x$

$$\frac{R_y}{R_x} = -\frac{1}{-2}$$

$$= +0.5; \ \theta = 26.6° \text{ south of west,}$$

which we obtained only approximately using the graphical method.

To *subtract* one vector from another, you *add* a reversed vector; one of the same magnitude but which points in the opposite direction. We just added a and **b**. **a** − **b** would be represented graphically as shown in Figure G-4.

The magnitude of **R** is the same but the direction is 26.6° south of *east.*

For another example of the addition of vectors, consider a small airplane flying eastward at 150 miles per hour relative to the air. At the same time, a wind is blowing at 50 miles per hour in a northwesterly direction. What is the velocity of the plane relative to the ground?

First, let's try a rough graphical addition. If, in the diagram, we let one-half inch represent 50 miles per hour, then we should draw an arrow pointing east which is one and one-half inches long:

a = 150 miles per hour

To this we add an arrow representing the wind:

b = 50 miles per hour

When the two vectors are strung together they give a resultant of about 120 miles per hour in a direction 15 to 20° north of east.

To add the vectors analytically, we first find their components:

$a_x = 150$ miles/hour $b_x = (-50)(\cos 45°) = (-50)(.707)$

$\qquad\qquad\qquad\qquad\qquad\quad = -35.35$ miles/hour

$a_y = 0$ miles/hour $b_y = (+50)(\cos 45°) = (+50)(.707)$

$\qquad\qquad\qquad\qquad\qquad\quad = +35.35$ miles/hour

Then the components of **R** are:

$R_x = a_x + b_x = (150 - 35.35) = 114.65$ miles/hour

$R_y = a_y + b_y = (0 + 35.35) = 35.35$ miles/hour

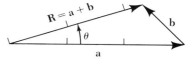

Figure G-5

The magnitude of the resultant, that is, the speed of the plane will be

$\mathbf{R} = \sqrt{(114.65)^2 + (35.35)^2}$

$\quad = \sqrt{14,394}$

$\quad = 119.97$ miles per hour

The direction of **R** is given by the equation:

$\tan \theta = \dfrac{R_y}{R_x} = \dfrac{35.35}{114.65} = .308$

$\theta = 17°8'$ north of east.

APPENDIX H

DEDUCTION OF THE LORENTZ TRANSFORMATIONS FROM EINSTEIN'S POSTULATE

Starting with $x^2 - c^2t^2 = (x')^2 - c^2(t')^2$, [Equation (8-5a)], we would like to find a set of equations relating x' and x, t' and t. Suppose that instead of $x' = x - vt$, and $t' = t$, we try using the more general expressions:

$$x' = k(x - vt)$$

and, for the fun of it,

$$t' = k(t - bx).$$

By substituting these new transformations for x' and t' into Equation (8-5a), we can find out what the constants k and b must be:

$$x^2 - c^2t^2 = (x')^2 - c^2(t')^2$$
$$= [k(x - vt)]^2 - c^2[k(t - bx)]^2$$
$$= k^2(x - vt)^2 - c^2k^2(t - bx)^2.$$

Performing the indicated multiplications gives:

$$x^2 - c^2 t^2 = k^2(x^2 - 2xvt + v^2 t^2) - c^2 k^2(t^2 - 2tbx + b^2 x^2)$$
$$= [k^2 x^2 - 2k^2 xvt + k^2 v^2 t^2]$$
$$-[c^2 k^2 t^2 - 2c^2 k^2 tbx + c^2 k^2 b^2 x^2]:$$

Now by collecting together terms involving x^2, t^2, and xt, we separate out x^2, t^2, and xt as factors:

$$x^2(1) + t^2(-c^2) = x^2[k^2 - c^2 b^2 k^2] + t^2[k^2 v^2 - c^2 k^2]$$
$$+ xt[-2k^2 v + 2c^2 k^2 b].$$

Now we bring all the terms over to the left side of the equation, so that the right side is equal to zero:

$$x^2[1 - k^2 + c^2 b^2 k^2] + t^2[-c^2 - k^2 v^2 + c^2 k^2]$$
$$+ xt[+ 2k^2 v - 2c^2 k^2 b] = 0.$$

Dividing the left side of this equation up as follows, we can work more conveniently with the individual bracketed expressions:

$$\underbrace{x^2[1 - k^2 + c^2 b^2 k^2]}_{(a)} + \underbrace{t^2[-c^2 - k^2 v^2 + c^2 k^2]}_{(b)}$$

$$\underbrace{+ xt[+2k^2 v - 2c^2 k^2 b]}_{(c)} = 0$$

The most general way for this equation to hold is for all the expressions in brackets [] to be equal to zero. By setting each of the bracketed quantities equal to zero, we obtain for (a),

$$1 - k^2 + c^2 b^2 k^2 = 0$$

for (b),

$$-c^2 - k^2 v^2 + c^2 k^2 = 0;$$

and, for (c),

$$+2k^2 v - 2c^2 k^2 b = 0.$$

Subtracting the quantity $(+2k^2 v)$ from both sides of expression (c) yields:

$$-2c^2 k^2 b = -2k^2 v.$$

Cancelling out the quantity $(-2k^2)$ from each side leaves the expression $c^2 b = v$, or,

$$b = \frac{v}{c^2}$$

Turning to expression (a) next, by setting it equal to zero, and subtracting 1 from both sides, we obtain the expression:

$$-k^2 + c^2 b^2 k^2 = -1.$$

Multiplying this by -1 yields:

$$k^2 - c^2 b^2 k^2 = 1.$$

Substituting the quantity (v/c^2) for b:

$$k^2 - c^2 \left(\frac{v}{c^2}\right)^2 k^2 = 1$$

$$k^2 - c^2 \left(\frac{v^2}{c^4}\right) k^2 = 1$$

$$k^2 - \left(\frac{v^2}{c^2}\right) k^2 = 1$$

Factoring k^2 from the left side

$$k^2 \left(1 - \frac{v^2}{c^2}\right) = 1$$

Divide both sides by $1 - (v^2/c^2)$. This gives:

$$k^2 = \frac{1}{1 - \dfrac{v^2}{c^2}}.$$

Solve for k by extracting the square root.

$$k = \sqrt{\frac{1}{1 - \dfrac{v^2}{c^2}}} = \frac{1}{\sqrt{1 - \dfrac{v^2}{c^2}}}.$$

Substitute these values for the constants b and k into the expressions for x' and t'.

$$x' = k(x - vt) \qquad\qquad t' = k(t - bx)$$

$$x' = \left(\frac{1}{\sqrt{1 - \dfrac{v^2}{c^2}}}\right)(x - vt) \qquad t' = \left(\frac{1}{\sqrt{1 - \dfrac{v^2}{c^2}}}\right)\left[t - \left(\frac{v}{c^2}\right)x\right]$$

$$x' = \frac{x - vt}{\sqrt{1 - \dfrac{v^2}{c^2}}}. \qquad\qquad t' = \frac{t - \dfrac{xv}{c^2}}{\sqrt{1 - \dfrac{v^2}{c^2}}}.$$

APPENDIX I

DERIVATION OF THE MASS-ENERGY EQUIVALENCE

Let's begin by expressing the relativistic mass dependence upon velocity in a slightly different way, to see what it suggests:

$$m = \frac{m_o}{\sqrt{1 - \dfrac{v^2}{c^2}}} = m_o\left[\left(1 - \frac{v^2}{c^2}\right)^{-1/2}\right]$$

$$= m_o\left[\left(1 + \left[-\frac{v^2}{c^2}\right]\right)^{+(-1/2)}\right]$$

We will treat the expression in rounded brackets first. By the Binomial Theorem, any quantity $(1 + x)$ raised to the nth power is expressible, mathematically, in the form of a series:

$$(1 + x)^n = 1 + nx + \frac{n(n-1)x^2}{(1)(2)} + \frac{n(n-1)(n-2)x^3}{(1)(2)(3)}$$
$$+ \cdots + \text{etc.}$$

Now, in this case,

$$(1 + x)^n = \left[\left(1 + \left[-\frac{v^2}{c^2}\right]\right)^{+(-1/2)}\right],$$

where

$$n = -\tfrac{1}{2} \text{ and } x = \left[-\frac{v^2}{c^2} \right].$$

Expanding this expression into the series form gives:

$$\left[\left(1 - \frac{v^2}{c^2}\right)^{-1/2} \right] = 1 + \left(-\frac{1}{2}\right)\left(-\frac{v^2}{c^2}\right) + \frac{\left(-\frac{1}{2}\right)\left(-\frac{3}{2}\right)\left(\frac{v^4}{c^4}\right)}{2}$$
$$+ \frac{\left(-\frac{1}{2}\right)\left(-\frac{3}{2}\right)\left(-\frac{5}{2}\right)\left(-\frac{v^6}{c^6}\right)}{6} + \cdots + \text{etc.}$$

Continuing to simplify by multiplication gives:

$$\left[\left(1 - \frac{v^2}{c^2}\right)^{-1/2} \right] = 1 + \frac{1}{2}\frac{v^2}{c^2} + \frac{\left(\frac{3}{4}\right)\left(\frac{v^4}{c^4}\right)}{2} + \frac{\left(\frac{15}{48}\right)\left(-\frac{v^6}{c^6}\right)}{6}$$
$$+ \cdots + \text{etc.}$$

$$= 1 + \frac{1}{2}\frac{v^2}{c^2} + \frac{3}{8}\frac{v^4}{c^4} + \frac{15}{48}\frac{v^6}{c^6} + \cdots + \text{etc.}$$

Then:

$$m = m_o\left[1 - \frac{v^2}{c^2}\right]^{-1/2} = m_o\left[1 + \frac{1}{2}\frac{v^2}{c^2} + \frac{3}{8}\frac{v^4}{c^4} + \frac{15}{48}\frac{v^6}{c^6}\right.$$
$$\left. + \cdots + \text{etc.}\right]$$

If both sides of the expression are multiplied by (c^2), each term will have the dimensions of energy; mass/(length²/time²).

$$mc^2 = m_oc^2 + \frac{1}{2}m_oc^2\left(\frac{v^2}{c^2}\right) + \frac{3}{8}m_oc^2\left(\frac{v^4}{c^4}\right) + \frac{15}{48}m_oc^2\left(\frac{v^6}{c^6}\right)$$
$$+ \cdots + \text{etc.}$$

$$= m_oc^2 + \frac{1}{2}m_ov^2 + \frac{3}{8}m_o\left(\frac{v^4}{c^2}\right) + \frac{15}{48}m_o\left(\frac{v^6}{c^4}\right) + \cdots + \text{etc.}$$

Or, by subtracting m_oc^2 from both sides of the equation, we obtain:

$$mc^2 - m_oc^2 = \frac{1}{2}m_ov^2 + \frac{3}{8}m_o\left(\frac{v^4}{c^2}\right) + \frac{15}{48}m_o\left(\frac{v^6}{c^4}\right) + \cdots + \text{etc.}$$

If v is very much smaller than c, the velocity of light, then all the terms on the right side of the equation except the first one will be extremely small. (If v/c is small $v^4/c^2 = v^2(v^2/c^2)$ will be very small.) Thus,

$$mc^2 - m_o c^2 = \tfrac{1}{2} m_o v^2,$$

which is simply the classical expression for kinetic energy. This therefore suggests calling the expression

$$mc^2 - m_o c^2 = \gamma m_o c^2 - m_o c^2$$

$$= (\gamma - 1) m_o c^2$$

the *relativistic kinetic energy*. When v is small compared to c, the expression $mc^2 - m_o c^2$ reduces to $\tfrac{1}{2} mv^2$, the classical expression.

APPENDIX J

ANSWERS TO
EVEN–NUMBERED PROBLEMS

Chapter 2

2. 60 years $= 1.89 \times 10^9$ sec.

6. c.

8. The rate at which a warm object loses heat to its surroundings depends, among other things, on its surface area. For a given volume, a sphere has the least surface area.

Chapter 3

2. 40 miles/hour; 45 miles/hour.

4. 10 miles.

6. 509 miles/hour.

8. 90 miles/hour.

10. (a) 1. x, (+); v is equal to zero and becoming positive; a, (+).
       2. x, (+); v, (−); a equals 0.
       3. x, (+); v, (−); a(−).
   (b) 1. x, (+); v, (+); a equals 0.
       2. x, (+); v, (−); a equals 0.
       3. x, (+); v equals zero and is becoming (+); a, (+).
   (c) 1. x, (+); v, (+); a, (−).
       2. x, (+); v, (−); a equals 0.
       3. x, (+); v, (+); a equals 0.

12. 4 minutes and 30 seconds.

14. 5 m/(sec)²; 10 m/(sec)².

16. 6.33 miles/hour.

18. 3.47 seconds.

Chapter 4

4. If the net force is zero, then the object will not accelerate, but will continue moving at a constant speed of 10 m/sec.

6. 0; 0; 2 m/(sec)²; 6 m/(sec)².

8. 0.5 m/sec.

10. 750 N.

12. When an object is at rest, rising (or falling) at a constant speed, the tension on the string from which it hangs is just equal to the object's weight. However, to accelerate the object upward, the string must exert an upward force which exceeds the weight. Otherwise no net force acts on the object. The increased tension might be enough to break the string.

14. (a) 686 N;   (b) 823 N;   (c) 549 N.

16. (a) 9.8 m/(sec)² down;   (b) *mg*;   (c) 2 g.

18. 16.66 N.

Chapter 5

4. In a certain kind of detective story, sometimes called closed room cases, assurance is given that only a limited number of people could have committed the crime. Occasionally a detective story writer will cheat and reveal late in the game that the room (or castle, island, or ship) was not really isolated from the rest of the world. But insofar as a limited number of people could have done it, the detective can eliminate all the suspects save one, and by deduction prove that *that* person did it. In a normal crime, the possible suspects are many, and it is rarely proven by deduction that a certain person did it.

6. To use a frame of reference larger than the earth in naming the stairs would cause problems. You could try naming them "solar" or "anti-solar", depending on whether they took you toward or away from the sun, but the names would be inappropriate half a day later.

8. The Material of which Neptune is made has an average density (mass/volume) only one-fourth that of the earth's.

12. g would be doubled, pressure would be quadrupled.

14. 31.3 m/sec.

16. .615 earth years; 247 earth years.

Chapter 6

4. The momentum of a system is conserved during a collision only if the system experiences no outside forces. However, it is always possible to include as part of the system the outside agent which is exerting the force. Then the force becomes internal and the momentum of the new, enlarged system will be conserved. Newton's first law says that a system, be it a single object or a more complex system, will maintain constant momentum if it experiences no net forces from its surroundings. If the mass of the system does not change, then the object or the "center of mass" of the system (if it contains several objects) will move along at constant speed.

6. 1 kg; -40 m/(sec)^2 for a, $+20$ m/(sec)^2 for b.

8. 0.31 m/sec; 2.5 kg·m/sec; to calculate the average force exerted, you must make assumptions about the time over which the impulse is applied.

10. 230 N.

12. 0.8 m/sec; -0.8 m/sec; 0.291 m/sec; -0.509 m/sec. If the man walks toward the right, he moves 1.45 m and the canoe moves -2.55 m (to the left) beneath him.

Chapter 7

2. 196 J. The work done does not depend on the direction in which you push. However, a larger force (40 N) must be applied if the push is horizontal than if the push (or pull) is upward parallel to the slope (39.2 N).

4. 14 m/sec; We know only that the final speed is the same in each case. The average speed and distance of travel are not the same, so the time required to go down the two hills is not necessarily the same. There is a particular shape for a hill, following a curve called the *brachistochrone* which will get an object to the bottom in shortest time. If friction acts, then the straight slope, for which distance moved is minimum, will result in minimum friction losses so that the kinetic energy and speed will be greater. (This ignores air drag, which depends on speed and is more complicated.)

6. Four times as far.

8. $Q\left(\dfrac{M}{M + m}\right)$

10. 1.02×10^5 kg/sec.

12. 2.5 m/sec to the left; 234,400 J.

14. .52 kg.

Chapter 8

2. 83.3 sec.

4. 11.11 m/sec; 22.22 m/sec.

6. If we never see or measure what is "really" happening, then that reality is not part of physical reality. Relativity—and science—are empirically based.

8. Suppose that for the ground observer, the lightning on the left occurs at $t_L = 0$ and $x_L = -L$ and that the lightning on the right occurs at $t_R = 0$, $x_R = +L$. Then, for the observer on the train, $t_L' = [t_L - (x_L v/c^2)]\gamma = t(Lv\ \gamma/c^2)$; $t_R' = [t_R - (x_R v/c^2)]\gamma = -(Lv/c^2)\gamma$. t_R', being negative, is before and t_L', being positive, is after $t' = 0$.

10. 1.66×10^8 m/sec.

12. 2.53×10^8 m/sec; 1.846×10^8 m/sec.

14. 2.94×10^8 m/sec.

16. 12×10^8 m.

Chapter 9

2. $v = c/2$, so $\gamma = 1.155$. Mass is increased by a factor γ and length is reduced by a factor γ, so density must be increased by γ^2, or 1.333.

4. 6.3×10^9 kg/sec.

6. 2.34×10^{-13} J are released for each silicon atom produced.

10. When a positron is emitted by a nucleus, it can be assumed that one of the positively charged protons in the nucleus has been transformed into a neutron. Heavy nuclei already have a preponderance of neutrons. Upon alpha decay, a set of two neutrons plus two protons is emitted, resulting in an even larger percentage excess of neutrons. To emit a positron would increase the number of neutrons relative to the number of protons even further, so it does not occur.

Chapter 11

2. The cork particles are attracted to the rod and to each other initially by induction. In time, however, charge is shared by the rod, moving gradually onto the cork particles. The more moist the air, the more easily charge moves over the surface. When a cork particle picks up charge, it is repelled from its neighbors, and from the rod.

4. 6.25×10^8 electrons/sec; 90 V.

6. 1800 V.

8. (a) b and c are at higher potential; (b) $E = 500$ V/m toward the left; (c) potential energy will be higher at (b) or (c); (d) -50×10^{-9} J.

10. 5060 V.

12. Electrical potential energy of the charges is used up. The charges enter the heating element with high potential energy and leave with low potential energy. The kinetic energy of the charges however, remains about the same.

14. (a) 2.31 Watt·hr, 8340 J, \$258; (b) \$142.44; (c) 1.65 cents; (d) at $t = 0$, $R = 5.45$ ohms. At $t = 10$ hours, $R = .667$ ohms. The resistance of the material in the filament increases with temperature. During the test, current decreased, and with it the temperature of the filament. Therefore, the resistance decreased.

16. 6 ohms; 8×10^{-17} N.

18. \$750,000.

Chapter 12

2. It will point southeast toward the magnetic pole. The geographic north pole is the point where the earth's axis of rotation meets the earth's surface. A compass will point vertically up and *down* at the earth's magnetic poles, if it is free to.

4. 5×10^{-4} N/m, east.

6. Uranium 238.

8. The force acting on the magnet will be upward, reaching maxima at the time when the field produced inside the ring, by the magnet, is changing most rapidly.

Chapter 13

2. 3.26 cm lies in the microwave range.

4. 495 m/sec.

6. (a) 4.62×10^{14} hertz; (b) 15×10^{9} or 15 gigahertz; (c) 331 hertz, at 0°C.

8. About 3 miles.

10. 28.8°

12. 23 m from a point directly below the boat.

14. They should follow a path made up of two straight lines. The path on land makes an angle θ_1 with a line perpendicular to the riverbank. The path in water makes an angle θ_2 with a line perpendicular to the riverbank. Sin θ_1 should be four times Sin θ_2.

Chapter 14

2. 30 megahertz.

4. If the path followed by the wave which bounces is $n\lambda$ longer than the path followed by the direct wave, then constructive interference will result. The signal received will be stronger. However, as the ionosphere moves up, the path difference will increase. Periodically it will equal $n\lambda$, and the signal will be strong. In between, the received signal will be weaker.

6. For $\lambda = 4.5 \times 10^{-7}$ m, $\theta = 15.7°, 32.7°, 54.1°$.
For $\lambda = 6.5 \times 10^{-7}$ m, $\theta = 23°, 51.25°$. No third order.

8. (Sound velocity) $n/6$.

10. 35.1 m.

2. They become gases or vapor, and if the system is closed they are still around. They can be retrieved by cooling the system. In a more open system, movements of air carry the evaporated material elsewhere.

4. Normally, a thermometer reaches thermal equilibrium with its surroundings by a process of conduction, wherein molecular bombardments enable energy to be exchanged until the thermometer reaches the same temperature as its surroundings. In space, where neighboring molecules are rare, energy can be exchanged only by radiation. If the thermometer is shaded from the sun, it radiates energy to its surroundings, and will reach very low temperature readings. On the other hand, if it is in the sunshine, it will absorb energy by radiation from the warmer sun, and will reach quite a high temperature. The actual final temperature would depend on whether the thermometer is largely transparent, letting much of the radiation pass through it, or whether it is black, in which case it absorbs all the radiation which reaches it. In the latter case, it should approach very high temperatures.

6. An object with a large specific heat should feel cooler because the heat flowing out from your warm fingers increases its temperature less.

8. The extra energy in the wound-up spring should warm up the acid in which it is dissolved. However, if you calculate the work done in winding a watch spring (remember: force times distance) it is only sufficient to heat the acid *very* slightly.

10. (a) 14.3 cal; (b) 0.21°C.

12. 40.9 kg hot, 34.1 kg cold.

14. 13.5°C; 2.995×10^{-3} cal.

Chapter 16

2. 8.4×10^8 J.

4. In theory, the availability of colder heat sinks in the winter would indicate higher efficiency. This is also true in practice. Less energy is required to pump cooling water through the condensers of a power plant during winter. Since the cooling water is colder, less has to be pumped through. Heating the building which houses the power plant is in winter not a problem.

6. No. Insofar as the material in the engine is brought back to the same state, cycle after cycle, its entropy does not increase. The entropy of the surroundings does, however.

8. 89 percent; 7.2°C.

10. According to the second law of thermodynamics, heat cannot be transferred from a cooler to a hotter object by nonmechanical means. The temperature of the sun is about 6000°C, so you could not hope to obtain temperatures higher than that by use of the mirrors.

Index